Geobotany Studies

Basics, Methods and Case Studies

The series includes outstanding monographs and collections of papers on a range of topics in the following fields: Phytogeography, Phytosociology, Plant Community Ecology, Biocoenology, Vegetation Science, Eco-informatics, Landscape Ecology, Vegetation Mapping, Plant Conservation Biology, and Plant Diversity. Contributions should reflect the latest theoretical and methodological developments or present new applications on large spatial or temporal scales that will reinforce our understanding of ecological processes acting at the phytocoenosis and vegetation landscape level. Case studies based on large data sets are also considered, provided they support habitat classification refinement, plant diversity conservation or vegetation change prediction. ***Geobotany Studies: Basics, Methods and Case Studies*** is the successor to *Braun-Blanquetia*, a journal published by the University of Camerino from 1984 to 2011 in cooperation with the *Station Internationale de Phytosociologie* (Bailleul, France) and the *Dipartimento di Botanica ed Ecologia (*Università di Camerino, Italy) and under the aegis of the *Société Amicale Francophone de Phytosociologie*, the *Société Française de Phytosociologie*, the *Rheinold-Tüxen-Gesellschaft* and the *Eastern Alpine and Dinaric Society for Vegetation Ecology*. This series promotes the expansion, evolution, and application of the invaluable scientific legacy of the Braun-Blanquet school.

More information about this series at http://www.springer.com/series/10526

Salvatore Brullo • Cristian Brullo •
Salvatore Cambria • Gianpietro Giusso del Galdo

The Vegetation of the Maltese Islands

Springer

Salvatore Brullo
Department of Geological,
Biological and Environmental Sciences,
Section Plant Biology
University of Catania
Catania, Italy

Cristian Brullo
Department of Geological,
Biological and Environmental Sciences,
Section Plant Biology
University of Catania
Catania, Italy

Salvatore Cambria
Department of Geological,
Biological and Environmental Sciences,
Section Plant Biology
University of Catania
Catania, Italy

Gianpietro Giusso del Galdo
Department of Geological,
Biological and Environmental Sciences,
Section Plant Biology
University of Catania
Catania, Italy

ISSN 2198-2562 ISSN 2198-2570 (electronic)
Geobotany Studies
ISBN 978-3-030-34524-2 ISBN 978-3-030-34525-9 (eBook)
https://doi.org/10.1007/978-3-030-34525-9

This Springer imprint is published by the registered company Springer Nature Switzerland AG.
The registered company address is: Gewerbestrasse 11, 6330 Cham, Switzerland

Acknowledgements

We would like to express our thanks to our friends and colleagues, who over the years (1973–2017) accompanied one of the authors (S. Brullo) in the several field trips carried out in the Maltese Archipelago, in particular to Prof. Edwin Lanfranco (University of Malta), Prof. Giuseppe Angelo Ronsisvalle and Prof. Pietro Pavone (University of Catania) for their active collaboration. In addition, special thanks are also due to many other friends, colleagues and pupils who were of considerable help in the field surveys, such as Dr. Santa Pulvirenti and Dr. Giuseppina Bartolo (University of Catania), Dr. Sandro Lanfranco (University of Malta), Dr. Corrado Marcenò and Mr. Stephen Mifsud (Malta).

Contents

1	**Introduction**		1
2	**Physiography**		7
	2.1	Geographical Characteristics and History of Malta	7
	2.2	Geology	10
	2.3	Soils	14
	2.4	Climate	16
3	**History of the Flora and Vegetation**		21
	3.1	Flora	21
	3.2	Vegetation	29
4	**Syntaxonomical Arrangment of Vegetation**		33
5	**Woody Vegetation**		41
	5.1	*Pistacio lentisci-Quercetum ilicis* Brullo & Marcenò 1985	41
	5.2	*Asparago aphylli-Tetraclinidetum articulatae* ass. nov. hoc loco	44
	5.3	*Periploco angustifoliae-Euphorbietum dendroidis* Brullo, Di Martino & Marcenò 1977	46
	5.4	*Erico multiflorae-Antyllidetum melitensis* Brullo, Minissale & Spampinato 1997 corr.	46
	5.5	*Erico multiflorae-Coronilletum glaucae* ass. nov. hoc loco	50
	5.6	*Suaedo verae-Darnielletum melitensis* ass. nov. hoc loco	52
6	**Chasmophilous Vegetation**		57
	6.1	*Hyperico webbii-Chiliadenetum bocconei* Brullo & Marcenò 1979 corr.	57
	6.2	*Putorio calabricae-Micromerietum microphyllae* Brullo & Marcenò 1979	64
	6.3	*Eucladio verticillati-Adiantetum capilli-veneris* Br.-Bl. ex Horvatić 1934	66

7 Rocky Coast Vegetation 67
7.1 *Limonietum zeraphae* ass. nov. hoc loco 67
7.2 *Limonietum melitensis* ass. nov. hoc loco 71
7.3 *Crithmo maritimi-Limonietum virgati* Pirone 1995 71
7.4 *Anthyllido melitensis-Euphorbietum melitensis* ass. nov. hoc loco 74
7.5 *Crucianello rupestris-Helichrysetum melitensis* ass. nov. hoc loco 80

8 Psammophilous Vegetation 83
8.1 *Eryngio maritimi-Sporoboletum arenarii* (Arènes ex Géhu & Biondi 1994) Rivas Martinez & Cantò in Rivas Martinez et al. 2002 83
8.2 *Cypero capitati-Agropyretum juncei* (Kuhnholtz-Lordat 1923) Br.-Bl. 1933 86
8.3 *Medicagini marinae-Ammophiletum australis* Br.-BI. 1921 corr. Prieto & Diaz 1991 86
8.4 *Centaureo sphaerocephalae-Ononidetum ramosissimae* Br.-Bl. & M. Frei in M. Frei 1937 87
8.5 *Salsolo tragi-Cakiletum maritimae* Costa & Mansanet 1981, nom. corr. hoc loco 87

9 Salt Marshes Vegetation 91
9.1 *Arthrocnemo macrostachyi-Juncetum subulati* Brullo & Furnari 1976 91
9.2 *Agropyro scirpei-Inuletum longifoliae* Brullo in Brullo et al. 1988 nom. corr. hoc loco 92
9.3 *Halimiono portulacoidis-Suaedetum verae* (Molinier & Tallon 1970) Gèhu in Gèhu et al. 1984 93
9.4 *Inulo longifoliae-Juncetum maritimi* Brullo in Brullo et al. 1988 nom. corr. hoc loco 94
9.5 *Juncetum maritimo-acuti* Horvatic 1934 95
9.6 *Caricetum divisae* Br. -Bl. in Br.-Bl., Roussine & Nègre 1952 96
9.7 *Suaedo spicatae-Salicornietum patulae* Brullo & Furnari ex Géhu & Géhu-Franck 1984 corr. Alcaraz, Ríos, De la Torre, Delgado & Inocencio 1998 97
9.8 *Salsoletum sodae* Pignatti 1953 98
9.9 *Suaedetum spicatae* Pignatti 1953 corr. hoc loco 98
9.10 *Cressetum creticae* Brullo & Furnari 1976 100
9.11 *Lamprothamnietum papulosi* Corillion 1957 101
9.12 *Enteromorpho intestinalidis-Ruppietum maritimae* Westhoff ex R. Tx. & Böckelmann 1957 101
9.13 *Ruppietum drepanensis* Brullo & Furnari 1976 102

10 Hygrophilous Vegetation 103
10.1 *Phragmitetum australis* Savič 1926 103
10.2 *Typhetum domingensis* Brullo, Minissale & Spampinato 1994 103
10.3 *Polygono salicifolii-Phragmitetum australis* Barbagallo, Brullo & Furnari 1979 104
10.4 *Cyperetum distachyi* O. Bolós & R. Molinier 1984 106
10.5 *Bolboschoenetum compacti* Van Langend. 1931 corr. Bueno & F. Prieto in Bueno 1997 106
10.6 *Helosciadietum nodiflori* Maire 1924 106
10.7 *Apio nodiflori-Glycerietum plicatae* Brullo & Spampinato 1988 107
10.8 *Cyperetum longi* Micevski 1957 109
10.9 *Cypero longi-Caricetum cuprinae* R. Tx. ex Díaz Gonzalez & Fernández-Prieto 1994 109
10.10 *Caricetum hispidae* Brullo & Ronsisvalle 1975 112
10.11 *Festuco arundinaceae–Caricetum distantis* J. Duvign. 1967 . . . 113
10.12 *Phalarido coerulescentis-Schedonoretum arundinacei* ass. nov. hoc loco 113
10.13 *Schedonoro arundinacei-Caricetum divisae* ass. nov. hoc loco 114
10.14 *Potentillo reptantis-Panicetum repentis* ass. nov. hoc loco 115
10.15 *Ranunculetum trichophylli* Melendo, Cano & Valle 2003 119
10.16 *Lemnetum minoris* Oberd. ex Müller & Görs 1960 119

11 Seagrass Vegetation 121
11.1 *Posidonietum oceanicae* Molinier 1960 121
11.2 *Cymodoceetum nodosae* Giaccone & Pignatti 1967 122
11.3 *Halophiletum stipulaceae* Augier ex Brullo ass. nov. hoc loco 122

12 Rocky Pool Vegetation 125
12.1 *Charetum vulgaris* Corillion 1957 125
12.2 *Zanichellietum melitensis* ass. nov. hoc loco 126
12.3 *Damasonio bourgaei-Ranunculetum saniculifolii* ass. nov. hoc loco 126
12.4 *Crassulo vaillantii-Elatinetum gussonei* Bartolo, Brullo, Minissale & Spampinato 1990 128
12.5 *Lolio lepturoidis-Spergularietum marinae* ass. nov. hoc loco . . . 130
12.6 *Triglochino laxiflori-Romuleetum melitensis* ass. nov. hoc loco 131
12.7 *Damasonio bourgaei-Crypsietum aculeatae* Rivas-Martínez & Costa in Rivas-Martínez et al. 1980 corr. V. Silva & J.C. Costa in Costa et al. 2012 132

13 Woody Riparian Vegetation . . . 135
13.1 *Tamarici gallicae-Viticetum agni-casti* ass. nova hoc loco . . . 135
13.2 *Ulmo canescentis-Salicetum pedicellatae* Brullo & Spampinato, 1990 . . . 136

14 Annual Xerophilous Vegetation . . . 139
14.1 *Thero-Sedetum caerulei* Brullo 1975 . . . 139
14.2 *Vulpio ligusticae-Trisetarietum aureae* Brullo 1975 . . . 142
14.3 *Ononido sieberi-Stipelletum capensis* ass. nova hoc loco . . . 142
14.4 *Allietum lojaconoi* Brullo 1985 . . . 145
14.5 *Silenetum melitensis* ass. nova hoc loco . . . 145
14.6 *Phalaridetum paradoxae* ass. nov. hoc loco . . . 150
14.7 *Catapodio balearici-Linarietum pseudolaxiflorae* ass. nova hoc loco . . . 152
14.8 *Cutandio maritimae-Pseudorlayetum pumilae* ass. nov. hoc loco . . . 154

15 Xerophilous Grasslands . . . 157
15.1 *Leontodonto tuberosi-Lygeetum sparti* ass. nov. hoc loco . . . 157
15.2 *Convolvulo oleifolii-Lygeetum sparti* ass. nov. hoc loco . . . 160
15.3 *Hyparrhenietum hirto-sinaicae* A. & O. Bolos & Br.-Bl. in A. & O. Bolos 1950 . . . 163
15.4 *Chamaeleo gummiferis-Brachypodietum retusi* C. Brullo, Brullo, Giusso & Tomaselli 2006 . . . 163
15.5 *Diplotaxio tenuifoliae-Oryzopsietum miliaceae* Brullo 1984 . . . 168
15.6 *Feruletum melitensis* ass. nov. hoc loco . . . 171
15.7 *Romuleo melitensis-Ranunculetum bullati* ass. nov. hoc loco . . . 175

16 Chasmo-nitrophilous Vegetation . . . 179
16.1 *Oxalido-Parietarietum judaicae* (Br.-Bl. 1952) Segal 1969 . . . 179
16.2 *Parietario judaicae-Matthioletum incanae* Vigo & Terrada 1969 . . . 180
16.3 *Capparidetum rupestris* O. Bolòs & Molinier 1958 . . . 181
16.4 *Antirrhinetum siculi* Bartolo & Brullo 1986 . . . 183
16.5 *Hyoscyamo albi-Parietarietum judaicae* Segal 1969 . . . 184

17 Annual Halophilous Vegetation . . . 185
17.1 *Desmazerio pignattii-Senecionetum pygmaei* Brullo & Scelsi 1998 . . . 185
17.2 *Anthemido urvilleanae-Frankenietum pulverulentae* ass. nov. hoc loco . . . 188
17.3 *Parapholidetum filiformis* Brullo, Scelsi & Siracusa 1994 . . . 189

18 Ruderal Vegetation . . . 191
18.1 *Polycarpo tetraphylli-Spergularietum rubrae* Brullo & Marcenò 1976 . . . 191
18.2 *Trisetario aureae-Crepidetum bursifoliae* Brullo 1980 . . . 192

18.3 *Galio muralis-Trifolietum suffocati* ass. nov. hoc loco 193
18.4 *Spergulario rubrae-Filaginetum congestae* ass., nov. hoc loco 195
18.5 *Carthamo lanati-Onopordetum argolici* ass. nov. hoc loco 195
18.6 *Chrysanthemo coronarii-Malvetum parviflorae* Ferro 1980 . . . 197
18.7 *Carduetum marmorati* Brullo 1983 corr. hoc loco 197
18.8 *Chrysanthemo coronarii-Silybetum mariani* Brullo 1983 199
18.9 *Hordeo leporini-Sisymbrietum orientalis* Oberdorfer 1977 199
18.10 *Acantho mollis-Smyrnietum olusatri* Brullo & Marcenò 1985 201
18.11 *Lavatero creticae-Malvetum nicaeensis* ass. nov. hoc loco 203
18.12 *Spergulario marinae-Taraxacetum minimi* ass. nov. hoc loco . . . 207
18.13 *Lavateretum arboreae* Br.-Bl. & Molinier 1935 207
18.14 *Mesembryanthemetum crystallino-nodiflori* O. Bolos 1957 210
18.15 *Amarantho deflexi-Dysphanietum ambrosiodis* ass. nov. hoc loco 211
18.16 *Parietario lusitanicae-Veronicetum cymbalariae* Brullo & Marcenò 1985 213
18.17 *Galio aparine-Conietum maculati* Rivas-Martinez ex Lopez 1978 214
18.18 *Calystegio sylvaticae-Arundinetum donacis* Brullo, Scelsi & Spampinato 2001 214
18.19 *Nicotiano glaucae-Ricinetum communis* de Foucault 2013 215

19 Weedy Commensal Vegetation 219
19.1 *Rapistro rugosi-Meliloletum infestae* Bartolo, Brullo, Fagotto & Grillo 1983 219
19.2 *Calendulo tripterocarpae-Hypecoetum procumbentis* Bartolo, Brullo, Minissale & Spampinato 1990 222
19.3 *Fumario densiflorae-Veronietum hederifoliae* Brullo & Marcenò 1985a, b 222
19.4 *Fumarietum bicoloris* ass. nov. hoc loco 224
19.5 *Diplotaxio erucoidis-Brassicetum sylvestris* ass. nov. hoc loco 224
19.6 *Chrozophoro tinctoriae-Kickxietum integrifoliae* Brullo & Marcenò 1980 227
19.7 *Setario pumilae-Echinochloetum colonum* A. & O. Bolós ex O. Bolós 1956 230
19.8 *Setario ambigui-Cyperetum rotundi* Brullo, Scelsi & Spampinato 2001c 230
19.9 *Bromo diandri-Brassicetum sylvestris* Brullo & Marcenò 1985 232

20 Subnitrophilous Vegetation of Uncultivated Lands 235
20.1 *Chrysanthemo coronarii-Hippocrepidetum multisiliquosae* Brullo & Siracusa 1996 235
20.2 *Sullo coronariae-Ononidetum mitissimae* ass. nov. hoc loco . . . 236
20.3 *Meliloto messanensis-Hordeetum marini* Brullo 1983 238

21 Syndinamic Considerations 241

22 Maltese Habitat of Community Interest 251

References 255

Index 277

Chapter 1
Introduction

In this paper the results of a phytosociological investigations concerning the Maltese Archipelago, represented by some islands localized South of Sicily in the central Mediterranean, are provided. The vegetation of these islands is currently markedly altered by anthropic activities, because this territory was colonized since the Neolithic period (about 7000 years ago) by many peoples, who in the time have degraded its natural landscape. Despite this, these islands show still a relevant environmental interest, especially for the occurrence of a very interesting flora rich in endemisms and rare or relict species, that characterize many relevant plant communities, nowadays many of which are often reduced to small patches. The first our field research dates back to about 45 years ago, when mass tourism in these islands had not yet begun and continued until today. During this period, substantial changes in the landscape have been witnessed, with the increase in urban, industrial and tourist areas, as well as the modification of agricultural lands. From the biogeographical point of view, the Maltese Islands belonging to Holarctic Kingdom, Mediterranean Region, Western Mediterranean Subregion, Italo-Tyrrhenian Province, Sicilian Domain, Pelagian Sector, Maltese District (Brullo et al. 1995). As a whole, the territory has a total extension of 315.6 km^2, with 140 km coastline and is inhabited by a population with a density that exceeds 1500 people per km^2 on the larger island. Besides, most of its surface is used for agricultural activities (51%) or is urbanized (15%), while only 18% is currently covered by natural vegetation (Lanfranco et al. 2013). The topography of the islands is the result of geological, climatic, paleogeographic events that during the last 5–10 million years, since the archipelago emerged from the sea, have affected its surface (Fig. 1.1). The tectonic movements of the continental plates, as well as the repeated exposure of the Islands to erosion and meteoric infiltrations, have considerably altered the surface and the subsoil of these islands, with the formation of well-diversified environments, as incised river valleys, caves, plateaux, karstland, marshs, cliffs, budlands, rocky coasts, islets, dunes, temporary rocky pools, etc. (Figs. 1.2 and 1.3).

S. Brullo et al., *The Vegetation of the Maltese Islands*, Geobotany Studies,
https://doi.org/10.1007/978-3-030-34525-9_1

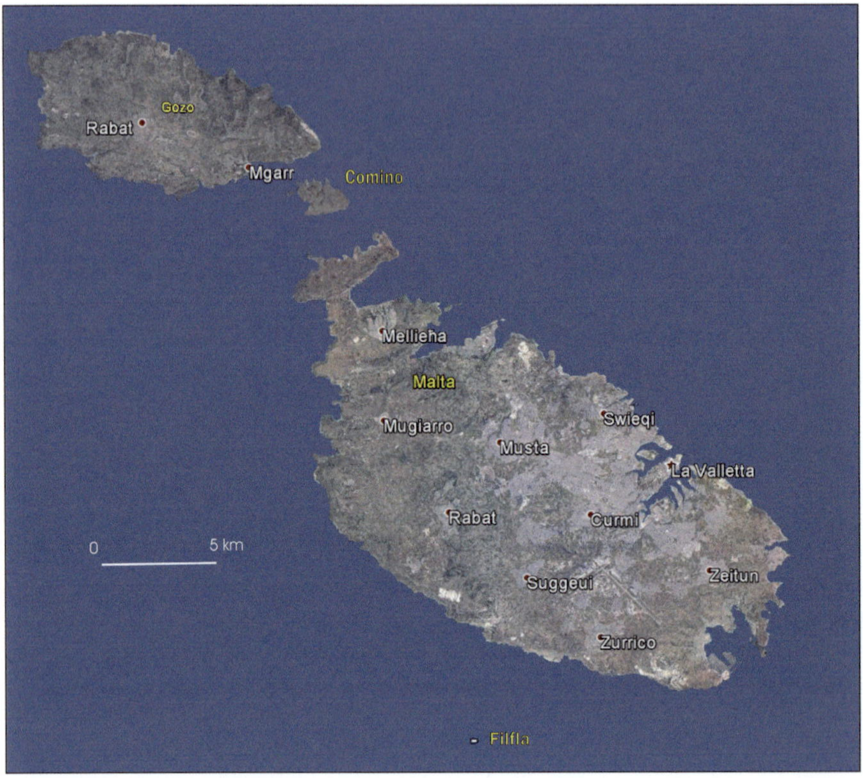

Fig. 1.1 Maltese Islands map from Google Earth (modified)

The Maltese Islands are mainly constitute of sedimentary rocks of marine origin, represented by some types of limestones, marls and clays. The soils are usually very superficial and young, since its evolution is inhibited by the harsh environmental conditions. The climate is Mediterranean dry with two main seasons, a hot dry summer and a mild wet winter.

Unfortunately, as in many other coastal insular territories of the Mediterranean area, also in Malta the strong human pressure plays a fundamental role in the alteration of the original natural landscape. Over the course of time, this has led to the loss of habitats with the rarefaction, sometimes until disappearance, of several wild plant species. It is to keep in mind, that currently there are many species of relevant taxonomic and phytogeographical interest become so rare that, due to threats of various kinds, run serious dangers of extinction. Phytogeographical considerations on the Maltese flora, especially concerning the endemic component, are reported by Lanfranco (1989), Junikka et al. (2006) and Lanfranco et al. (2013), who emphasized that are the most relevant species of this territory and the role by them played.

Fig. 1.2 Coastal habitats in the Maltese Islands: (**a**) Cliffs near Mtahleb, Malta, (**b**) Rocky coast near Dingli, Malta, (**c**) Azure Window at Dwejra Bay, Gozo, (**d**) Fungus Rock at Dwejra Bay, Gozo, (**e**) Blue Grotto, Malta, (**f**) Badlands of Gnejna Bay, Malta, (**g**) Dunes of Ramla Bay, Gozo, (**h**) Salt marsh of Ghadira, Malta

Fig. 1.3 Natural and synanthropic habitats in Malta: (**a**–**b**) Agricoltural landscape in Gozo, (**c**) Agricoltural landscape in Malta, (**d**) Oak woodland near Mellieha, Malta, (**e**) Karstland in Central Malta, (**f**) Rocky coast in Comino, (**g**) Filfla island, (**h**) Rocky pool in Malta

The authors in this book, apart from physiographic (geology, pedology, climate) and floristic information, examine all the plant communities surveyed during their phytosociological investigations on the Maltese Islands. For each association identified, the floristic, ecological, structural and dynamic characteristics, as well as a relevè table, are provided, further the nomenclatural and syntaxonomic aspects are taken into consideration too.

Chapter 2
Physiography

2.1 Geographical Characteristics and History of Malta

According to literature data (Haslam 1969; Schembri 1993) the Maltese Islands consist in a small archipelago aligned along an axis with NW-SE trend, situated in the Central Mediterranean about 96 km to the South of Sicily and about 284 km to Tunisia (Fig. 2.1). The main islands are Malta (245.7 km^2, coord. 35° 55′ 4″ N and 14° 24′ 35″ E), which is the largest and southern, Gozo (67.1 km^2, coord. 36° 2′ 39″ N and 14° 15′ 4″ E), which is the northernmost, and Comino (3.5 km^2, coord. 36° 00′ 45″ N and 14° 19′ 37″ E), which is located between these two islands. They are currently inhabited by about 445,000 people, most of which reside in Malta. Nearby these island there are numerous rocky islets, all uninhabited, among them the most extensive are Cominotto, Filfla, Islands of St. Paul, Fungus Rock, Gallis Roch, Delimara island, Xrobb I-Ghagin Rock, etc. Along the coast bays and inlets, often very deep, which are excellent harbors, are frequent some of which are characterized by sandy beaches. The eastern slopes of the islands are generally represented by low reefs and beaches, while the southern ones are mostly constituted by vertical cliffs, often exceeding 100 m from sea level. In the more depressed coastal stretches, separated from the sea by sandy or silty-clay deposits, are frequent salt marshes, periodically flooded by sea and meteoric waters. Along the low calcareous rocky shoreline often occur chequerboards of rock-cut saltpans, which are used by the local people for salt production as result of the seawater evaporation. It's believed that these saltpans existed since Roman times. The current landscape of the Maltese Islands is characterized by low hills, alternating with terraced plateaux, which are generally furrowed by more or less shallow valleys, locally called "Wied", which were formed prevalently by river erosion in periods characterized by very wet climatic regimes, as during the Pleistocene (Fig. 2.2). This is nothing but the result of tectonic movements and subsequent processes of surface erosion and karsism, which starts on the land immediately after its emersion from the sea. The islands do not reach high altitudes, the highest point in Malta is Ta′ Dmejrek near Dingli at

S. Brullo et al., *The Vegetation of the Maltese Islands*, Geobotany Studies,
https://doi.org/10.1007/978-3-030-34525-9_2

Fig. 2.1 Geographical map of the Maltese Islands

253 m, while in Gozo the highest point is Ta′ Dbiegi, reaching 195 m. As concerns the rivers, there are only few small watercourses that flow on the bottom of some valleys mainly during the wet season, which are dry for the rest of the year. In fact, usually no permanent rivers and lakes occur in the Maltese Islands, except of few streams fed by small springs of fresh water, such as those ones at Ras ir-Raħeb near Baħrija and at l-Imtaħleb. Most of the Maltese territory is today under cultivation with crops localized especially in land characterized by terracing with walls that surround them, preventing the soil erosion. Usually, the terrace walls follow the contours of the slopes radiating from the hill top and probably their realization dates back to ancient times and was continued until the nineteenth century. The terrace farming and in general the cultivated areas are more widespread on the freshest and humid slopes of hills and valleys or on flat surfaces where is possible the irrigation. The surfaces constituted by barren rocks very washed out and eroded, however without soil, are covered by a scattered and wild vegetation, represented by small shrubs and grasses. Only in few undisturbed and rocky stands it is still possible to find patches of natural woody vegetation, as maquis, garigues and shrublands.

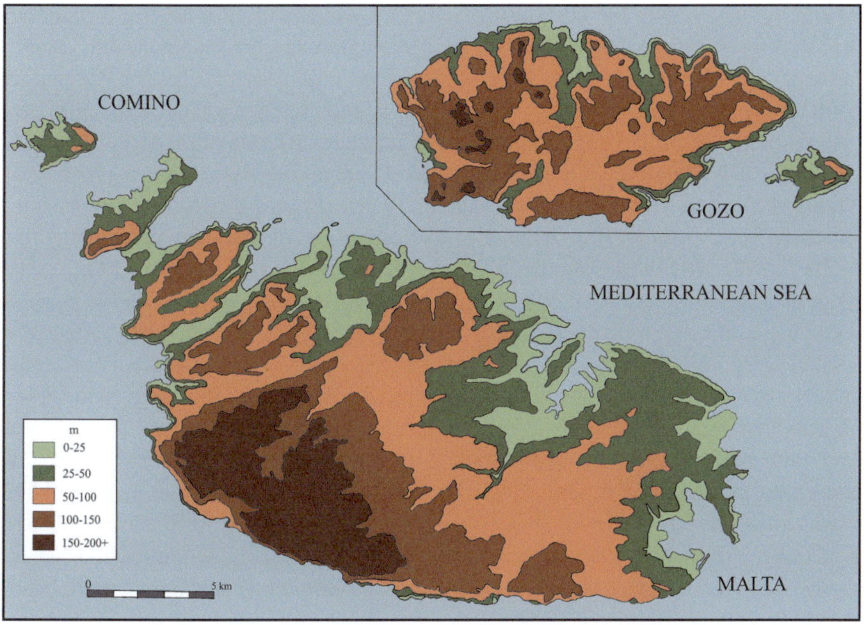

Fig. 2.2 Physical map of the Maltese Islands from ontheworldmap.com (modified)

Besides, examples of natural woodlands, such as those with *Quercus ilex*, today are very rare and can be considered as relicts of the forestal vegetation that originally covered part of these islands. It should be noted that the present landscape of the Maltese Islands is nothing but the result of the anthropic impact that has acted on this territory in the last 7000 years. According to literature (Renfrew 1972; Cutajar 1982), the first human colonization goes back to about 5200 BC and this prehistoric period ended in 750 BC. In particular, the earliest human civilization that reached these islands coming from Sicily dates back to Neolithic period (5200–4400 BC), with people having a hunter-gatherer culture, who probably did not have a strong impact on the natural environment. Towards the end of this period (4500–4100 BC) the population began to practice a primitive agriculture. During the Chalcolithic Age (4200–2500 BC), due to the arrival of other people from Sicily there was a marked improvement in agriculture that became more efficient and intensive, with a spread of pastoral activities. The Chalcolithic civilization of Malta, which is associated to the buildings of megalithic temples, due to intense sheep-grazing caused a depletion of natural resources with the degradation of vegetation cover and consequent denudation of the surfaces for soil loss. This was probably the main cause of the disappearance of this civilization towards 2500 BC, since it triggered off famines and diseases in the population of these islands, as hypothesized by the archaeologists. After 2500 BC, Malta remained unpopulated for several decades, until the arrival of a new migratory flow of people, called builders of Dolmen, who started the Maltese Bronze Age. This new civilization lasted about 1000 years, until 1450 BC, its end coincides with the eruption of Santorini which caused upheaval throughout the Mediterranean. After up to 750 BC, other peoples arrived in Malta, mainly farmers, who then became fully integrated with the culturally far superior Phoenician civilization, whose population colonized the islands later this date. After this prehistoric period, during which Malta suffered a first devastation of the original vegetal cover, the islands from 218 BC fell under Roman rule, becoming a prosperous colony (Ballou 1893; Cassar 2000). Malta remained part of the Roman Empire until the early sixth century AD. After a brief occupation of Vandals and Ostrogoths, Malta in 535 AD was conquered by the Byzantines, becoming a colony. The successive invasion was that of Muslims that began in 870 AD. The native population was massacred, remaining the island uninhabited for about 150 years. After, the islands were colonized by a Muslim community that, although short-lived, introduced, apart of the language, also many innovative crops and new irrigation techniques, that are still used by farmers. Around 1100, with the expulsion of the Muslims, the Maltese islands returned under Christian rule with the Normans, remaining part of the Kingdom of Sicily for about 440 years. From 1530 to 1798 it was ruled by the Knights of Malta who embellished and fortified the islands, protecting it from the Ottoman invasions. After a brief Napoleonic period, Malta since 1800 became a protectorate of the British Empire which lasted until 1965, when the state of Malta was created.

From this, it can easily deduce that all these historical vicissitudes, dating back to very old periods, have had as a more evident result a devastation of the natural heritage of the islands of the archipelago. In the last decades, due to the demographic and turistic infrastructures increase, these degradation processes of the landscape have had a sharp rise, reducing the areas characterized by natural vegetation, that today are limited to less accessible and unsuitable areas for human activities.

2.2 Geology

The Maltese Islands are constituted of marine sedimentary rocks, mainly represented by limestone, marls and clays, which were deposited during the Oligo-Miocene period (30–5 MA). Geologically, these substrata show close similarities with Tertiary limestones forming the Hyblaean plateau in South Sicily, Lampedusa in the Pelagian Islands and Cyrenaica. This suggests that all these territories were part of the same stratigraphic unit. Discontinuous and various Quaternary deposits occurs above the Tertiary strata, which are represented by infills of depressed surfaces, screes, sands and gravels. The archipelago is linked through the Sicilian Channel to the Hyblaean plateau, that rests on sedimentary platform, forming during the Triassic, of which however do not emerge outcrops. According to Shackleton et al. (1984), these Islands were connected with the Sicilian coast up to the maximum cold stage of the last glacial interval (18,000 BP) (Fig. 2.3). This platform is represented by submarine graben, which reaches a maximum depth of about 200 m and with an average depth of less than 90 m (Magri 2006). Besides, it should

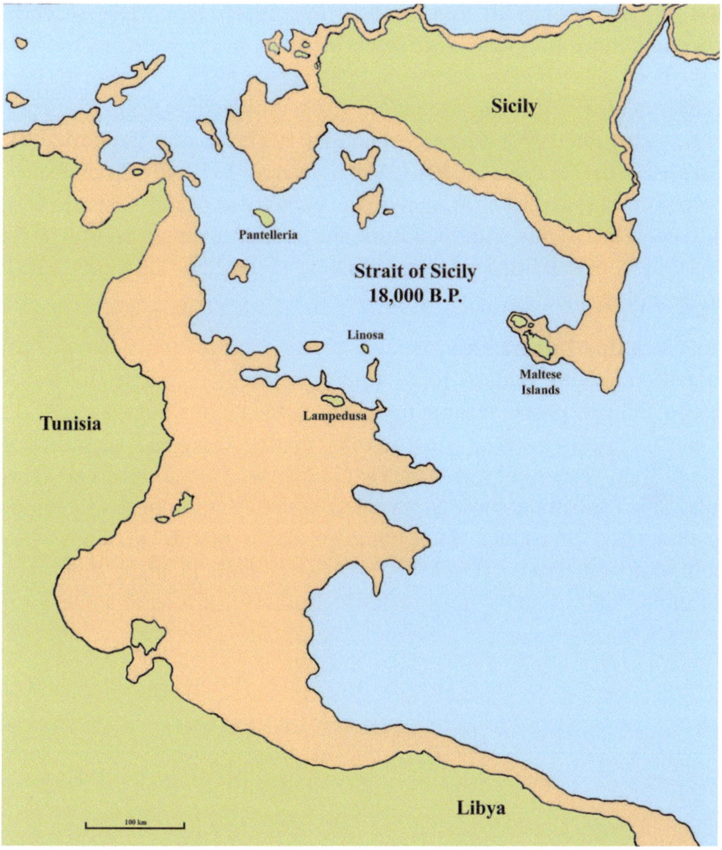

Fig. 2.3 Malta, Sicily and North Africa at lowest sea level during the last glacial maximum from Shackleton et al. (1984) (modified)

be noted that the Malta-Hyblaean platform belongs to Pelagian Block, which is part of African plate. The emersion of the Malta islands above sea level probably occurred between the upper Miocene and the lower Pliocene due to tilting of the NE block of the Malta graben. In fact, no rocks of this time have been observed, indicating therefore their definitive emersion.

As concerns the geology of Maltese islands, several authors have carried out detailed investigations on this topic, with the publication of important litho-stratigraphic, tectonic, morpho-structural and paleogeographic studies, as well as the drafting of geological maps. In particular Spratt (1843) was the first to publish a detailed geological description regarding the Maltese Archipelago, with a topographic map attached, correlated by faults and geological cross section. In a next edition of this work, Spratt (1852, 1854) published also a geological map of Malta and Gozo, made by Right Hon Earl od Ducie. The some revisioned map was later published by Adams (1870) and by Murray (1890) too. More recently, other geological and lithostratigraphic surveys, often with annexed modern geological maps, have been published by various authors (Hyde 1955; House et al. 1961; Felix 1973; Pedley et al. 1976, 1978; Alexander 1988; Antonelli et al. 1988; Pedley 1993; Hunt 1997; Schembri 1997; Zammit-Maempel 1997; Savona-Ventura 2001; Magri 2006; Cassar 2010; Devoto et al. 2012; Marriner et al. 2012; Eig 2015; Rotevan et al. 2016; Prampolinia et al. 2017). A map on the geology of Maltese Islands is shown in Fig. 2.4. According to this wide literature, the stratigraphic succession of the rocky sediments can be subdivided in five formation (Fig. 2.5). They are here listed in order of decreasing age:

a. Lower Coralline Limestones

This formation is the oldest rock outcrop currently visible in the Maltese Archipelogo and is exposed in strata up to 140 m thick, as in the sea cliffs. It belongs to Chattian (Oligocene) and is constitute by marine sediments deposited between 30 and 25 million years ago, which are rich in coralline algae and fossil animals. On the whole, these substrata show a stratified structure and consist of endurated to strongly endurated limestone. This outcrops occur mainly in the vertical cliffs frequent in south-western coast of Malta and Gozo, while in inland stands are mostly localized along valleys and gorges or faulted inliers, such as at some localities of Malta.

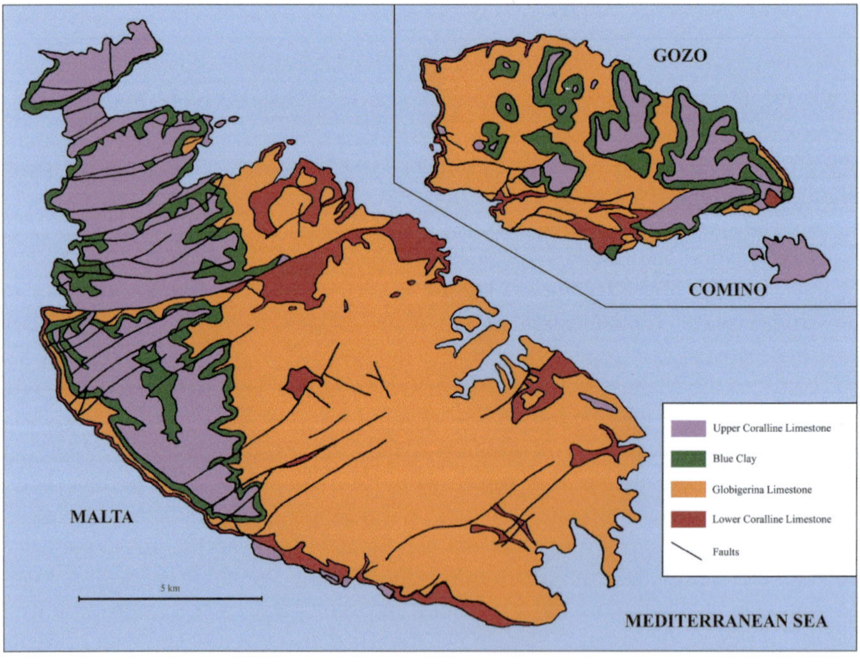

Fig. 2.4 Geological map of the Maltese Islands from House et al. (1961) (modified)

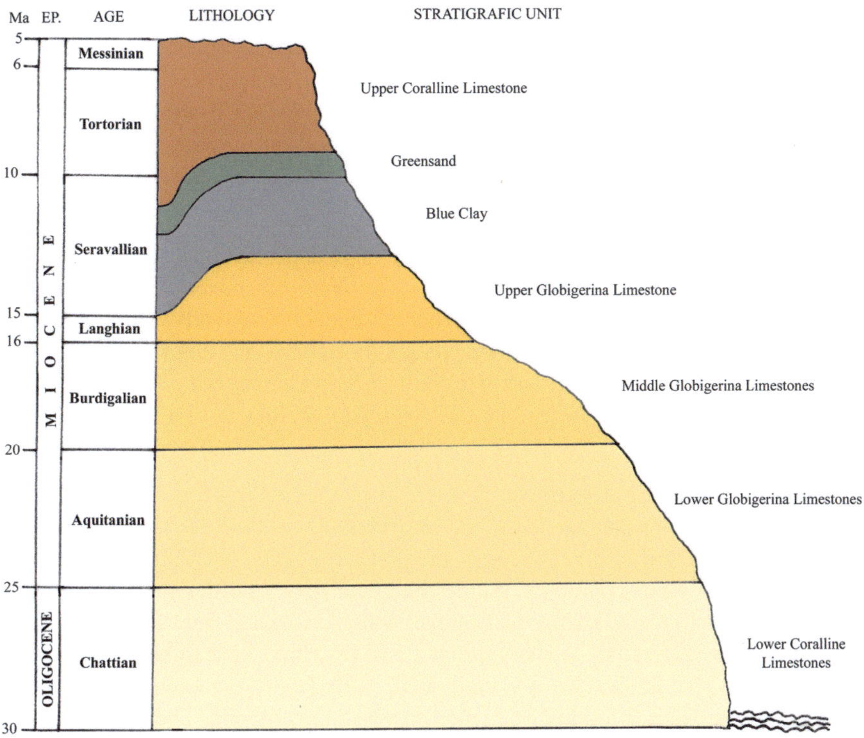

Fig. 2.5 Tectonic stratigraphic unit of the Maltese Islands according to Rotevan et al. (2016)

b. Globigerine Limestones

It is a stratified formation reaching sometime a thicknesser of 207 m, which spans the time between 25 and 13 Ma from Early Miocene to Middle Miocene. The rocks consist of more or less marly limestones, having a fine granulometry, with intercalation of chert, hard-grounds and pebble beds. These sedimentary rocks contain mainly microfossils, mainly Globigerina, and also numerous animal macrofossils, the same applies for the pebble strata. The Globigerine limestones forms usually undulating plains, since the very soft marl is easily eroded by atmospheric agents. These substrata, coinciding with the greater part of the cultivated lands, cover most of the surface of the islands. It is subdivided into three layers by two intervening pebble beds: (1) Lower—Consists of rocks belonging to Aquitanian (25–20); (2) Middle—The rocks rise to the Burdigalian (20–16 Ma); (3) Upper—Characterized by rocks between the Langhian and Seravallian (16–13 Ma).

c. Blue Clays

This formation, consisting in deposits of fine material, with an age that goes from the late Langhian to the early Serravallian (15–10 Ma), which is exposed in layers of thickness up to 65 m. It is constituted by dark marine clay rich in macrofossils, that, due to the strong surface erosion by the rainwater runoff, forming deep furrows on steep slopes, also for the scarce vegetation that covers these purely clayey surfaces. These badlands, which are often protected by a cap of Upper Coralline Limestone, are quite common in Malta and Gozo. In the central and eastern Malta the Blue clays substrata is absent or rare, since this formation can be removed by erosion or because Upper Coralline Limestone rests directly on Globigerina Limestone. This rock is the only impermeable layer of the Maltese islands.

d. Greensands
It consists of bioclastic limestones rich in glauconite deposited in a warm sea, which is represented by coarse, thickly bedded calcareous sands mixed to clay. The sea bottom in this period had risen and was relatively shallow and sandy. These substrata attaining usually a thickness not more than 1 m, with a maximum of 12 m. The Maltese Greensands deposited between the late Serravallian to the early Tortonian (12–8 Ma) and rest on the Blue Clay, although they are not always associated with the latter.

e. Upper Coralline Limestones
This formation, belonging to Tortonian and Messinian (11–5 Ma), represents the youngest Tertiary substrata occurring in the Maltese islands, reaching a thickness up to 162 m. It consists of a complex of limestones and shows a structure very similar to those one of the Lower Coralline Limestone, which originated from sea bottom with shallow waters. Infact, they are very rich in fossil Coralline algae mixed to numerous macrofossils. The limestones are endurated, with layers completely crystalline, showing at the basis a gradual transition towards Greensands. These rocks occur in Gozo and in western Malta, were they constitute the top of the isolated tabular hills or buttes, originated as a result of surface erosion.

f. Quaternary Rocks
During the Quaternary age, these rocks, that had already emerged from the sea, have undergone an intense alteration due to the action of meteoric phenomena, which have often led to their erosion and modification with terrestrial, aeolian and alluvial deposits. They are mainly represented by loams, breccias, coastal conglomerates, sands, paleosoils and lacustrine deposits. Currently, Pleistocene and Holocene deposits can be found in valleys, caves fissures and along the coast (Trechmann 1938; Shackleton et al. 1984; Hunt 1997; Gambin et al. 1998; Carroll et al. 2012).

2.3 Soils

Pedological investigations (Lang 1960, 1961; Sivarajasingham 1971; Vella 2001, 2005) on the Maltese Islands emphasized that the soils of this country are quite young or immature, due to purely calcareous composition of substrates and uniformly arid climate, their genesis is very slow. Besides, that determined a lack of significant development of a humus horizons. Moreover, for the millenary human activity, which has manipulated and substantially altered the morphology of these soils, they are quite artificial and difficult to catalog. The principal modes of soil disturbance are removing, plowing, manuring and terracing. For a correct classification of the Maltese Islands soils, the pedologists based their investigations on profiles not disturbed by human activities. In particular, the system proposed by Kubiena (1953) was followed, which allowed the identification of three main types of soil:

a. Xerorendzinas
The Maltese soils are formed from alterations of limestones, that due to high aridity of the climate, show a marked alkalinity, determining a weak evolution with a low humus content. These type of soils can be attributed to Xerorendzina whitish to red marly, with high calcium carbonate content (58–80%) derived from Globigerina limestone. It was considered by Vella (2001) an Entisol, that is a mineral soil with little or no evidence of pedogenetic horizons arising from a short pedogenesis period too.

b. Carbonate Raw Soils
It is also immature soil formed as a result of dry climatic conditions. It is represented by a grey soil with a very high calcium carbonate content (80–90%) and low in organic matter, derived from Greensand, Blue Clay, often mixed with Globigerina limestone and Upper Coralline limestones.

c. Terra Rossa Soils
It is a relict soils formed during the Pleistocene under woodlands and shrublands canopy. This red clayey soil originated when the wetter climate favored the leaching. This soil derived from both types of coralline limestone. Its reddish colour is due to the high iron oxide content. It is characterized by a low calcium carbonate content (2–15%), as well as a low humus content, which however is superior to that of the other soils.

More recently, the soils of the Maltese Islands have been classified by Sammut (2002–2004) according to the World Reference Base for Soil Resources (WRB Classification system). The seven identified soils are the following:

a. Arenosols (AR)
These soils were formed in recent sandy deposits occurring along the shoreline. Currently, they are represented at Ramla Beach (Gozo), as well as at Armier and Mellieha Bay (Malta), with examples also in Comino.

b. Calcisols (CL)
They are widespread in the all archipelago and are characterized by a high calcium carbonate concentration. These soils, very rich in lime, develop usually in countries affected by very dry climatic conditions. They for their dryness and stonyness do not lend much for agricultural use, but if proper cultural practices are made they can become highly productive. In this type of soil the Xerorendzinas can be included.

c. Cambisols (CM)
They are few developed soils showing only a subsoil showing a different colour (browner or redder) than the topsoil, but without other characteristics in the horizons. Cambisols, not frequent in the Maltese territory, they derive from medium and fine tessiture materials of various origins, represented by Quaternary deposits.

d. Leptosols (LP)
Leptosols originate from calcareous lithosols, such as those colonized by dwarf shrublands, in the Maltese islands occur mainly in the top of vertical cliffs and rocky outcrops. These soils are quite thin, since their natural evolution is usually limited by the erosion and generally they are not suitable for crops.

e. Luvisols (LV)
This type of reddish soils are developed in Malta during periods characterized by a wetter climate compared to the current one, like those coinciding with the Pleistocene glaciations. Therefore, they are relict soils having a low concentration of calcium carbonate, which clearly have no connection with the current pedoclimatic regime of these islands. The Luvisols, coinciding with the Terra Rossa Soils, are considered potentially fertile soils, but often require significant cultural actions. In the Maltese Archipelago they are frequent in the karst landscape.

f. Regosols (RG)
They gather soils, with a pedogenesis almost non-existent compared to parent material, which is reflected in completely indistinguishable horizons. In the Maltese Islands, the Spolic Regosols were recognized, which also known as urban soils or technosols, that occur usual in urban and industrial environments. Here, these soils are situated on made ground terraces overlying urban waste material.

g. Vertisols (VR)
This soils are characterized by high content of clay materials, which forms deep cracks in dry season. In Maltse territory they are localized on the Blue Clay outcrops, where the alternation during the year of the wet period with the dry one, causes a recurrent swelling and shrinkage of the silty-clay components, determining a self-mulching of the soil, which is constantly mixed. Therefore, the vertisols show an extremely deep horizon A, without horizon B.

2.4 Climate

The islands of the Maltese Archipelago, located in the central Mediterranean area, are characterized by a typically seasonal climate with very hot dry summers and wet cold winters. According to Rivas-Martinez et al. (2004, 2011), the Maltese territory falls in the Mediterranean pluviseasonal oceanic bioclimate ($Ic < =21$, $Io > 2.0$) of Thermomediterranean type ($Itc = 350$–450, $Tp(1) > 2150$) with dry ombrotype ($Io = 2.0$–3.6), which is the same of that one of Southern Sicily. As concerns the climate of Malta, it was analyzed in detail by several authors (Mitchell and Dewdney

Table 2.1 Pluvio-thermometric data of Maltese islands weather stations from: (A) Mitchell and Dewdney (1961), (B) Haslam (1969), (C) World Weather Information Service (2013), (D) Luqa Weather Average (2015), (E) Malta Weather (2013), (F) http://www.holiday-weather.com/gozo/averages/, (G) https://it.climate-data.org/location/123259/

	Average monthly temperatures for Malta (°C)														
	weather station	years	Jan	Feb	Mar	Apr	May	Jun	Jul	Aug	Sep	Oct	Nov	Dec	year
A	La Valletta	1841-1957	11.6	11.6	12.7	15	17.7	21.7	25	25	23.3	20.5	16.6	13.9	17.8
B	Malta s.l.	1931-1960	12.3	12.4	13.6	15.4	18.8	23.0	25.4	26.2	24.0	21.1	17.3	14.0	18.6
C	Luqa	1961-1990	12.2	12.4	13.4	15.5	19.1	23.0	25.9	26.3	24.1	20.7	17.0	13.9	18.6
D	Luqa	1981-2010	12.8	12.5	13.9	16.1	19.8	23.9	26.6	27.2	24.7	21.5	17.7	14.4	19.2
E	Balzan	1985-?	13.2	13.0	14.6	16.7	20.4	24.4	27.2	27.7	25.0	21.9	18.8	14.7	19.7
F	Gozo s.l.	?	12.0	12.5	13.5	15.5	19.0	23.0	26.0	26.5	24.0	20.5	17.0	14.0	18.6
G	Ghajnsielem	?	12.1	12.1	13.1	15.1	18.7	22.4	25.2	25.9	23.7	20.3	16.8	13.6	18.2

	Average monthly precipitation for Malta (mm)														
	weather station	years	Jan	Feb	Mar	Apr	May	Jun	Jul	Aug	Sep	Oct	Nov	Dec	year
A	La Valletta	1841-1957	88.6	65.5	64.1	23.6	10.6	2.3	1.0	4.3	28.5	79.0	90.4	97.3	521.7
B	Malta s.l.	1931-1960	90.6	53.7	41.6	21.7	?	1.8	0.5	7.0	31.1	102.6	67.7	77.9	498.6
C	Luqa	1961-1990	89.0	61.3	40.9	22.5	6.6	3.2	0.4	7.0	40.0	89.7	80.7	112..3	553.3
D	Luqa	1981-2011	98.8	60.1	44.2	20.7	16.0	4.6	0.3	12.8	58.6	82.9	92.3	109.2	595.8
E	Balzan	1985-?	94.7	63.4	37.0	26.3	9.2	5.4	0.2	6.0	67.4	77.2	108.6	107.7	603.1
F	Gozo s.l.	?	89.0	61.0	41.0	23.0	2.0	3.0	0.0	7.0	40.0	90.0	80.0	112.0	553.0
G	Ghajnsielem	?	78.0	50.0	38.0	23.0	8.0	3.0	0.0	5.0	35.0	96.0	80.0	88.0	504.0

1961; Haslam 1969; Chetcuti et al. 1992; Schembri 1997; Galdies 2011), who based on data from the nineteenth century up to a few years ago (1841–2011). As it is can be observed in the Table 2.1, where the pluvio-thermometric data of various weather stations of the Maltese islands (Malta and Gozo) dating to various periods are gathered, the climatic regime affecting this country is very similar to other Mediterranean territories with daily, seasonal and annual fluctuation, usually quite marked. In particular, the lowest temperatures are reached in January and February with monthly average values between 11.6 and 13.2 °C, slightly higher average values occur in March and December (12.7–14.7 °C). Higher values are reached in April–May and October–November with monthly averages of 15.0–21.9 °C, while the highest values are observed during the summertime (June–September) with monthly average of 21.7–27.7 °C, with higher peaks in August. The highest extreme temperature was reached in August 1999 with 43.8 °C, while the lowest ever recorded temperature was recorded in January 1981 with 1.4 °C. The annual precipitations are very variable from year to year, with yearly average for the periods listed in Table 2.1 between 498.6 and 603.1 mm. The high and low record of average annual rainfall is represented by 1031 mm (1859) and 140 mm (1894) respectively. The rainiest months coincide with the autumn ones (October–December) with average monthly values of 70.2–112.0 mm, decreasing slightly from January to February (98.8–50.0 mm). A faster rain decrease occurs from March to August, with minimum values close to zero in July, starting again with heavy rainstorms in September (28.5–67.4 mm). According to Mitchell and Dewdney (1961), despite the Maltese islands have a quite small surface without mountain ranges, the rains have a rather diversified distribution. In fact, during the period 1924–1939 the annual rainfall averages have a higher concentration in the central part of Malta (584–660 mm) with markedly lower values (584–508 mm) in the northern and southern parts, while in Gozo they are higher in the south-eastern part and lower in the north-western one. This is due to the fact that the rainfall is brought by north-westerly winds, whose conventional effect is strengthened by orographic uplift along the steep north-western coast of the islands. In this type of climate, clearly biseasonal, the rains are undoubtedly the most important element and their distribution pattern in space and time is of vital importance for natural vegetation, agriculture and people. In order to emphasize the biseasonally of this climate, the pluvio-thermic diagrams of some Maltese weather stations, regarding various periods, are drawn after Walter and Lieth (1967). Infact, in the all these diagrams the aridity period in these islands extends from last March to early September coinciding with a range of about 5 months (Fig. 2.6).

Besides, it should be noted that in addition to temperatures and rainfall, the climate of the Maltese islands is also influenced by other meteorological factors, such as winds, atmospheric humidity, evapo-transpiration, sea temperature and

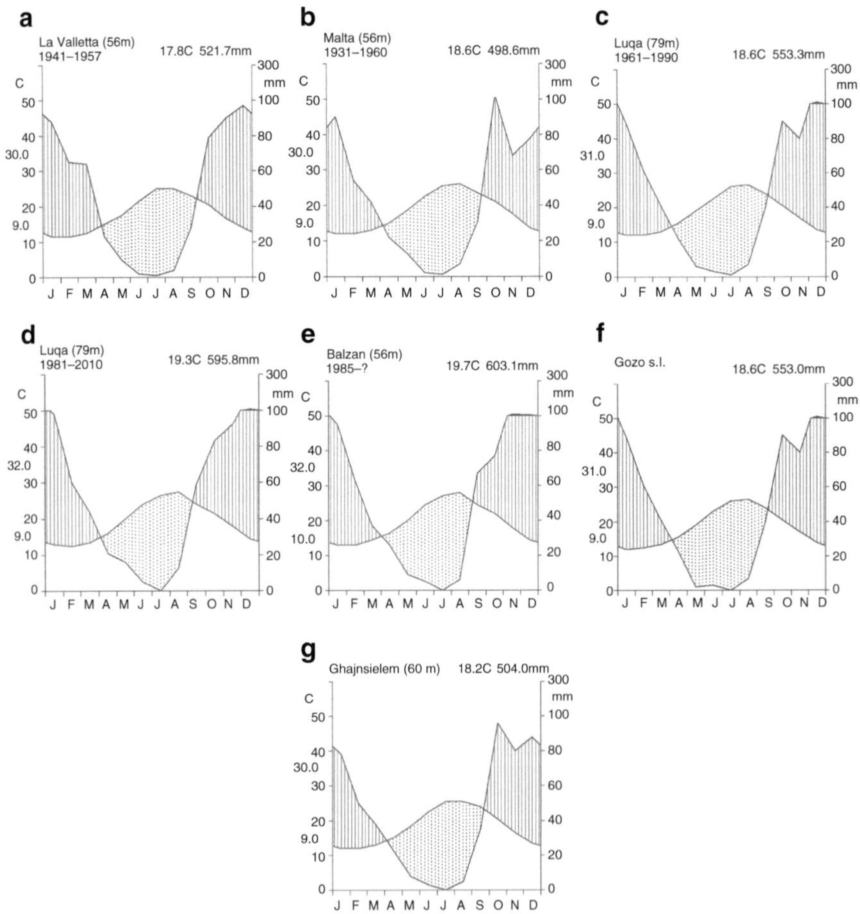

Fig. 2.6 Climograms of some Maltese weather stations, regarding different periods quoted in Table 2.1

duration of bright sunshine. These islands are quite windy due to the lower elevations and more or less flat topography, as well as for the proximity of the sea. Infact, only 7.7% of the days of an average year are calm, with a greater frequency and intensity in the cold season, while in the rest of the year they are less relevant. By far the most spread wind, frequent in all seasons, is the cool North-Westerly wind, better known as Mistral, blowing for an average year on 20–29% of the days, according to literature data (Mitchell and Dewdney 1961; Galdies 2011). Another important wind, although less frequent, is the wind blowing from the West (Ponente), which is quite mild and usually brings good weather. Others less significant winds are that one dry of North East (Gregale), that one warm of West Southwest (Libeccio), while more occasional ones are southern those, as Ostro and Scirocco, which are humid and hot winds, sometimes dust-ladened. Quite rare is the North wind (Tramontane), which is very cold.

As concerns the relative humidity of the atmosphere in the Maltese islands, it is generally quite high with monthly averages between 65–80% according to Mitchell and Dewdney (1961) or 61–87% according to Galdies (2011). The highest degree of humidity is reached in December–January, while the lowest level in July–August. Air humidity is an important factor indicating the percentage of water vapor contained in the atmosphere. In fact, a high degree of humidity favors the formation of clouds, fog and dew, while a low amount of humidity determines a dry air and clear weather.

A very important aspect for the life of plants is the water supply, that is closely linked to precipitation and temperatures as well as to the structure and nature of the soil. The water balance represents the relationships between the water requirement and the availability of water from rainfall that can be expressed as potential evapotranspiration (PE). According to the method of Thornthwaite and Mather (1955), Mitchell and Dewdney (1961) verified that in the Maltese territory the lowest values of PE occur from March to September, coinciding with over half of the year, months during which a high water deficit occurs. Instead during the autumn and winter period, higher PE values are reached, because the weather conditions are more humid and cold. Comparing the almost total absence of rainfall in summer and the surplus of water in winter, it emerges that for a good growth of herbaceous cover throughout the year it would be necessary in the Maltese Islands about twice the current rainfalls.

The average annual temperature of sea surrounding the Maltese islands (20 °C) it is one of the highest among those found in the European coasts of the Mediterranean. The monthly averages reach lower values (15–16 °C) from January to April, while the highest ones (24–26 °C) occur from July to September. In the other months the monthly averages are between 17 and 23 ° C.

Regards the duration of bright sunshine, the Maltese archipelago is considered a very sunny territory. Infact in summer the sunshine hours average are of 11.5–12.5 per day, while during the winter the average is over 5 h per day. Appreciable differences in the insolation were detected in relation to exposure, since in the same day it can observe gentle temperatures on the north-facing slopes compared to the south-facing ones, as well as different amounts of humidity on the soil surface.

On the whole, based on meteorological data, it is observed that compared to the past there has been a slight and progressive increase in average annual temperatures. Infact, according to (Galdies 2011, 2012; Government of Malta 2012; Galdies et al. 2016), in Malta the air temperature in the last hundred years it has undergone an increase of 0.7 °C, which is within the values of global average temperature increase that is around 0.8 °C above pre-industrial levels. It is significant to point out that the highest heating of about 1.5 °C occurred in the decades after 1970. The climate trends at regional scale, regarding in particular the Central Mediterranean, due to greenhouse effect of anthropogenic origin by the end of this century they foresee an overall temperature changes of over 4 °C (Hertig and Jacobeit 2008).

Chapter 3
History of the Flora and Vegetation

3.1 Flora

The flora of the Maltese islands was investigated by numerous authors, who since the seventeenth century have contributed to provide information on the indigenous and cultivated plants occurring in this Central Mediterranean territory. The earliest records date back to some prelinnaean authors (Abela 1647; Boccone 1674, 1697), who cited from Malta only few species or reported lists of species (Bonamicus 1670; Cavallini 1689). Later, some plants from Malta were mentioned by Linnaeus (1753), Forskåll (1775), Dumont D'Urville (1822), Giacinto (1811), Bertoloni (1829), Brenner (1838), etc., while the first checklists on Maltese flora were published by Zerapha (1827–1831), who recorded 644 indigenous and cultivated species, but with numerous incorrect identifications, Nyman (1844) listing 73 species, Grech Delicata (1853) published a list of 716 phanerogams, lastly Duthie (1872, 1874–1875) and Gulia (more works, see Sommier and Caruana Gatto 1915) increased the lists of previously published Maltese floras of several species not yet reported from the islands. During the twentieth century, several floras regarding the Maltese Islands were published, among these one of the most important is "Flora Melitensis nova" by Sommier & Caruana Gatto (1915), which is the most comprehensive flora of the period, where all the species occurring in the Maltese territory are listed with a detailed indication of the growth locality and of the authors who had quoted them, often with taxonomic comments. These authors report 902 species of phanerogams and 1085 species of cryptogams (fungi, algae, lichens, mosses, liverworts and ferns). Other Maltese floras were published by Borg (1927), in which almost all the species reported by Sommier and Caruana Gatto (1915) are examined, and successively by Lanfranco (1969a, b), Haslam et al. (1977), Weber and Kenzior (2006), Casha (2015), Lanfranco and Bonnett (2015) and by numerous other authors, who have published various floristic contributions, among the more recent can be quoted: Lanfranco (1970, 1971, 1972, 1973, 1975, 1989), Silverwood (1971), Kramer et al. (1972), Briffa (1986), Stevens and Lanfranco (1995), Bartolo et al. (2001),

S. Brullo et al., *The Vegetation of the Maltese Islands*, Geobotany Studies,
https://doi.org/10.1007/978-3-030-34525-9_3

Buttigieg and Lanfranco (2001), Mifsud (2007, 2008, 2009, 2010, 2011, 2012), Stevens (2003), Tabone (2007, 2008), Casha (2009), Sciberras and Sciberras (2010), Sciberras et al. (2012), Casha and Mifsud (2013), Mifsud et al. (2016a, b). Based on these floras, the autochthonous and allochthonous ferns and spermatophytes amount to about 1100 taxa (species and subspecies enclose), even if some of them have not been found anymore or are thought to have become extinct. Besides, there are numerous taxonomic contributions regarding the Maltese flora, where critical species and groups were investigated with the description of a lot of new taxa, many of which endemic, all based on specimens collected in these islands. Among these papers can be mentioned: (a) Bertoloni (1829) proposing for the *Centaurea spathulata* Zerapha nom. inval. the new name *C. crassifolia* (Fig. 3.1a), an endemic species currently included in the genus *Cheirolophus* by Susanna et al. (1999); (b) Sommier (1907), who described *Melitella pusilla* new genus and species considered initially endemic to Malta, but later found in various other countries of the Mediterranean and also in Australia, which was attributed by Merxmüller (1968) to the genus *Crepis* (Fig. 3.5a); (c) Béguinot (1907) described *Romulea melitensis* (Fig. 3.4b) as new species endemic of the Maltese Islands, providing successively a larger observations (Béguinot 1908), which more recently has been recorded from southern Sicily by Brullo et al. (2009a), while Mifsud (2015) believes that *R. melitensis* is an uncertain name replacing it with *R. variicolor*, which must be considered a superfluous name; (d) Sommier and Caruana Gatto (1915) described from Malta *Anacaptis urvilleana* (Fig. 3.1d), a neglected species of Orchidaceae, later revalued by Del Prete et al. (1984), proposing also as distinct species *Anthemis urvilleana* by Candolle de (1838) considered as var. *urvilleana* of *A. secundiramea* Biv. (Fig. 3.1b); (e) Ciferri and Giacomini (1950) considered a distinct species *Allium melitense* (Fig. 3.3b), taxon proposed by Borg (1927) as *A. ampeloprasum* L. var. *melitense* Sommier & Caruana nom. prov.; (f) Botschantzev (1976) basing on a herbarium specimen identified a new shrub species of Salsoleae from Malta, describing it as *Salsola melitensis*, by Brullo 1984a transferred to the genus *Darniella* Maire (Fig. 3.2c), which is noticeably distinct from *Salsola* L. for several significant morphologic features; (g) Brullo (1979) described *Chiliadenus bocconei* (Fig. 2.1b), endemic of North African origin first reported by Boccone (1674) and later recorded with different names, all illegitimate or invalid; (h) Brullo (1980b), proposed to consider the sea-lavender previously attributed to *Statice reticulata* auct. fl.melit. not L. (1753), by Del Guacchio et al. (2018) more recently proposed as a reject nomen, to a new species named *Limonium zeraphae* (Fig. 3.2d), which is endemic to the Maltese islands; (i) Brullo et al. (1982) described a new Maltese species of *Allium*, indicated as *A. lojaconoi* (Fig. 3.1e), which in the past was attributed to *A. parciflorum* Viv., which is exclusive of Sardinia and Corsica; (j) Brullo and Pavone (1985) attributed the populations of *Desmazeria sicula*, previously quoted from Maltese Islands and Southern Sicily, to a new species named *D. pignattii* (Fig. 3.4c), which must be considered a geographic vicariant of *D. sicula*; (k) Brullo et al. (1988a), in a contribution on Mediterranean *Limonium*, pointed out that the Maltese populations previously referred to *Statice cosyrensis* Guss. must be attributed to a new species named *Limonium melitense*,

Fig. 3.1 Endemic flora of the Maltese Islands: (**a**) *Cheirolopus crassifolius*, (**b**) *Anthemis urvilleana*, (**c**) *Cremnophyton lanfrancoi*, (**d**) *Anacamptis urvilleana*, (**e**) *Allium lojaconoi*, (**f**) *Hyoseris frutescens*

Fig. 3.2 Endemic flora of the Maltese Islands: (**a**) *Anthyllis hermanniae* subsp. *melitensis*, (**b**) *Chiliadenus bocconei*, (**c**) *Darniella melitensis*, (**d**) *Limonium zeraphae*, (**e**) *Ferula melitensis*, (**f**) *Helichrysum melitense*

morphologically well differentiated from *L. zeraphae*; (l) Brullo and Pavone (1987) attributed an archaic shrubby species of Chenopodiaceae, discovered in the Maltese island, to a new species *Cremophyton lanfrancoi* (Fig. 3.1c), including it within a monophytic genus, that, although morphologically and karyologically very differentiated from the other genera of the tribe Atripliceae, has been attributed by

Kadereit et al. (2010) to the genus *Atriplex*, but from the phylogenetic tree published by these authors *Cremonophyton lanfrancoi*, constitutes together with *Atriplex cana*, both representing an old lineage of *Atriplex*, a sister isolated clade well separated from all other investigated *Atriplex*, for which these two species must be clearly to attribute to two different genera distinct from *Atriplex*; (m) Brullo et al. (1988b) published a taxonomic note on the Maltese flora, where they examined some endemic or rare species from Malta, such as *Helichrysum melitense* (Pignatti) Brullo et al. (Fig. 3.2f), belonging to the group of *H. panormitanum* Tineo ex Guss., *Matthiola incana* (L.) M.T. Aiton subsp. *melitensis* Brullo et al. (Fig. 3.3d), *Elatine gussonei* (Sommier) Brullo et al. (Fig. 3.4a); (n) Brullo and Pavone (1988) in their review on the genus *Hyoseris* described from Gozo a new suffruticose species named *H. frutescens* (Fig. 3.1f), more recently recorded also from Malta (Sciberras et al. 2012); (o) Raffaelli and Ricceri (1988) recognized the specific rank of *Euphorbia melitensis* (Fig. 3.3a), critical species described by Parlatore (1869) from the Maltese islands, differentiating morphologically it from the related *E. bivonae* Steud. and *E. papillaris* (Boiss.) Raffaelli & Ricceri both occurring in Sicily; (p) Devillers and Devillers-Terschuren (1994) considered a distinct species (*Ophrys melitensis*) the orchid described from Malta by Salkowski (1992) as *O. sphegodes* Mill. subsp. *melitensis*; (q) Brullo et al. (2001a) referred the Maltese populations of *Zannichellia palustris* s.l. to a new endemic species named *Z. melitensis* (Fig. 3.3e), which represents a taxonomically very isolated species; (r) Brullo and Giusso (2006) attributed the Maltese population of *Anthyllis hermanniae* L. to a distinct endemic subspecies, named subsp. *melitensis* (Fig. 3.2a); (s) Peroni et al. (2013) referred the populations from Gozo of a rare *Polypodium* to *P. vulgare* subsp. *melitense*; (t) Mifsud et al. (2015) treated the *Sedum album* from Malta as an endemic subspecies, named subsp. *rupi-melitense*; (u) Brullo et al. (2017b) within taxonomic researches on the *Silene colorata* group, described from the Maltese Islands a new endemic species, represented by *Silene melitensis* (Fig. 3.3c), which is closely related to *Silene crassiuscula* from Sicily; (v) Brullo et al. (2018) emphasize that the Maltese populations of *Ferula communisis* must be treated as a morphologically well distinct species proposed as *F. melitensis* (Fig. 3.2e). Several others endemic species occurring apart from Malta also in some localities of Sicily and surrounding islands are: *Senecio pygmaeus* DC. (Maltese Islands, Lampedusa and southern Sicily) (Fig. 3.4d), *Linaria pseudolaxiflora* Lojac. (Maltese Islands and Linosa) (Fig. 3.3f), *Filago cossyrensis* Lojac. (Maltese Islands and Lampedusa), *Daucus rupestris* Guss. (Maltese Islands and Lampedusa) (Fig. 3.4f), *Daucus lopadusanus* Tineo (Maltese Islands and Lampedusa), *Iris sicula* Tod. (Maltese Islands and Sicily) (Fig. 3.4e), *Oncostema sicula* (Tineo ex Guss.) Speta (Maltese Islands and Sicily), *Ophrys caesiella* P. Delforge (Maltese Islands and southern Sicily), *Hymenolobus revelierei* (Jord.) Brullo subsp. *sommierianum* (Pamp.) Brullo (Maltese Islands, Lampedusa and Egadi Islands) (Fig. 3.5c), *Plantago afra* L. subsp. *zwierleinii* (Nicotra) Brullo (Maltese Islands and Sicily) and *Euphorbia sommeriana* C. Brullo & Brullo (Maltese Islands and Sicily) (Fig. 3.5b). Besides, some rare species having mainly a North African distribution are recorded from Malta, such as *Pteranthus dichotomus* Forssk.

Fig. 3.3 Endemic and subendemic flora of the Maltese Islands: (**a**) *Euphorbia melitensis*, (**b**) *Allium melitense*, (**c**) *Silene melitensis*, (**d**) *Matthiola incana* subsp. *melitensis*, (**e**) *Zanichellia melitensis*, (**f**) *Linaria pseudolaxiflora*

(Fig. 3.5h), *Plantago crypsoides* Boiss. (Figure 3.5f), *Fagonia cretica* L., *Enarthrocarpus pterocarpus* (Pers.) DC., *Crucianella rupestris* Guss. (Sicily, Lampedusa, Malta, Lybia and Egypt) (Fig. 3.4e), while other have a East Mediterranean distribution, such as *Hypericum aegypticum* L. subsp. *webbii* (Spach) N. Robson (Sicily, Lampedusa, Sardinia, Malta and Greece) (Fig. 3.5d) and *Convolvulus oleifolius* Desr. (Malta and East Mediterranean) (Fig. 3.5g).

Fig. 3.4 Subendemic flora of the Maltese Islands: (**a**) *Elatine gussonei*, (**b**) *Romulea melitensis*, (**c**) *Desmazeria pignattii*, (**d**) *Senecio pygmaeus*, (**e**) *Iris sicula*, (**f**) *Daucus rupestris*

Fig. 3.5 Rare and relevant species of the Maltese flora: (**a**) *Crepis pusilla*, (**b**) *Euphorbia sommeriana*, (**c**) *Hymenolobus revelieri* subsp. *sommieri*, (**d**) *Hypericum aegyptiacum* subsp. *webbii*, (**e**) *Crucianella rupestris*, (**f**) *Plantago crypsoides*, (**g**) *Convolvulus oleifolius*, (**h**) *Pteranthus dichotomus*

Some of these endemisms for their rarity and remarkable taxonomic role are included in the red lists of threatened species (IUCN 2017, Pasta et al. 2017). Among them in particular there are *Cheirolophus crassifolius*, *Cremnophyton lanfrancoi*, *Helichrysum melitense*, *Allium lojaconoi* and *Linaria pseudolaxiflora*, that give on the whole a remarkable environmental value to the Maltese islands, whose flora really deserves to be protected and preserved for future generations.

3.2 Vegetation

Currently, the Maltese Islands are characterized by a themophilous vegetation quite similar to that one of many other southern Mediterranean territories. The human impact, that started about 7000 years ago with the arrival of the first peoples on these islands, has progressively altered and often completely destroyed the original plant cover of the islands. Today, the landscape observable in these islands is strongly degraded by the anthropic activities, that over time have affected almost all the natural habitats that once characterized this territory. In particular, the plant communities that have undergone on a large scale the greatest devastation are clearly the woodlands and the maquis. According to palinological and archeobotanical investigations (Hunt 1997, 2000; Gambin et al. 1998; Haslam and Borg 1998; Hunt and Schembri 1999; Savona-Ventura 2001; Hunt and Vella 2004, 2005; Fenech 2009; Carroll et al. 2012; Marriner et al. 2012; Djamalìi et al. 2013), during the early Holocene the Maltese Islands were covered by a tree and shrubby vegetation dominated by evergreen and deciduous *Quercus*, pinewoods with *Pinus halepensis* and woody communities characterized by *Pistacia lentiscus*, *Olea europaea*, *Tetraclinis articulata*, alternate with open landscapes probably colonized by garrigues and grasslands. With the Neolithic, about 7000 years ago, following the arrival on the islands of the first peoples, the processes of deforestation began, which have become irreversible due to the intense summer heat and low precipitation as well as increasing anthropogenic activity. All this has favored the spread of the semi-natural vegetation types, as garrigue, steppe and maquis, which are characterized mainly by herbaceous and shrubby species. The only enviroment that have not been altered by the anthropic activities are the rocky ones, represented by sea cliffs, reefs, wied walls, which are notoriously the refuge habitats of most rare and endemic species occurring in the Maltese Archipelago. Based on the pollen diagrams, significant fluctuations of the vegetation assemblages were detected throughout the Holocene, which are linked primarily to climate changes, apart from the anthropogenic activities. In fact, on these islands in the past *Pinus halepensis* was widespread, dominating especially during the most arid and hot periods together with various sclerophyllous, such as *Pistacia lentiscus*, *Phyllirea*, *Juniperus/Tetraclinis*, etc. Conversely, in the wetter and colder periods the oak woodlands prevailed, with the occurrence of other mesophilous trees, as *Corylus*, *Fraxinus*, *Ostrya*, *Alnus*, *Cercis siliquastrum*, *Laurus nobilis*, etc., many of which are no longer part of the Maltese flora. The processes of anthropization, such as farm and pastoral activities, as well as urbanization, have instead led to an increase in herbaceous and nitrophilous species.

In the past, on the Malta vegetation were published several contributions, where the most relevant plant communities were examined from the physiognomic-structural, ecological and floristic point of view, while no or rarely phytosociological indications were given. In these papers (e.g., Sommier and Caruana Gatto 1915; Borg 1927; Haslam 1969; Haslam et al. 1977; Lanfranco 1984, 1995; Schembri 1994, 1997; Lanfranco and Schembri 1986; Anderson and Schembri 1989; Savona-

Ventura 2001; Casha 2015) the occurrence of some plant communitis are listed. Among them there are the following: (a) woodlands, represented by sclerophyllous forests, characterized by *Quercus ilex* or by *Pinus halepensis*, in particular the oak forests have a relict character with very old trees, while the second ones are today represented only by reforestation pinewood; (b) maquis with *Pistacia lentiscus*, *Olea europaea*, *Ceratonia siliqua*, *Euphorbia* dendroides, *Periploca angustifolia* and more rarely with *Tetraclinis articulata*, are recorded too; (c) garrigues, which are the result of woodlands and maquis degradation are especially widespread in the rocky calcareous places, which are characterized by several shrubs, such as *Thymbra capitata*, *Anthyllis hermanniae* subsp. *melitensis*, *Erica multiflora*, *Cistus* sp. pl., *Euphorbia melitensis*, *Convolvolus oleifolius*, etc.; (d) steppes linked to degradation of the shrubby communities are dominated by herbaceous perennial grasses, dominated mainly by *Hyparrhenia hirta*, *Lygeum spartum* or *Brachypodium retusum*, as well as by the purely subnitrophilous vegetation dominated by *Ferula melitensis*, *Asphodelus microcarpus* and *Charybdis pancration*; (e) ephemeral xerophilous communities are characterized by therophytes and geophytes, which are frequent in the more or less flat rocky surfaces, usually with very shallow soils and usually are differentiated by endemic or rare species (*Allium lojaconoi*, *Romulea melitensis*, *Desmazeria pignattii*, *Senecio pygmaeus*, *Anthemis urvilleana*, *Silene melitensis*, *Linaria pseudolaxiflora*, etc.); (f) coastal vegetation, which though quite localized and scattered shows a very relevant role, since it is linked to highly specialized habitats, such as dunes, salt marshes and rocky coasts, which are colonized respectively by psammophytes (*Elytrigia juncea*, *Sporobolus arenarius*, *Eryngium maritimum*, etc.), halophytes (*Suaeda vera*, *Arthrocnemum macrostachyum, Darniella melitensis*, *Juncus sp. pl.*, etc.) or small chasmophytes (*Limonium* sp. pl., *Crucianella rupestris*, *Cichorium spinosum* and *Crithmum maritimum*); (g) cliff vegetation, which is another very important and conservative habitat, widespread in the islands, where numerous endemic rupestrian species are localized, as *Cheirolopus crassifolius*, *Cremnophyton lanfrancoi*, *Darniella melitensis*, *Hyoseris frutescens*, *Chiladenus bocconii*, *Helichrysum melitense*, *Daucus rupestris*, etc.; (h) hygrophilous vegetation, which plays also an important role in the natural landscape of these islands, in particular it is localized in the temporary rocky pools frequent in limestone karstland or along the small watercourses, where in the first wet habitat occur *Elatine gussonei*, *Buillardia vaillantii*, *Damasonium bourgaei*, *Zanichellia melitensis*, *Ranunculus saniculifolius*, *Callitriche truncatula*, etc., while in the second stands it can be observed, *Typha dominigensis*, *Phragmites australis*, *Arundo donax*, *Cyperus* sp. pl., *Schedonorus arundinaceus*, *Carex* sp. pl., *Ranunculus* sp. pl., *Bolboschoenus compactus*, etc.; (i) synanthropic vegetation, currently widespread in all the islands, occurs especially in the crops localized in terraced slopes and on the bottom of the valleys, abandoned fields, ruderal stands, etc.; these habitats have allowed the establishment of a nitrophilous flora, usually very specialized, which is maintained over time with agricultural practices (plowing, fertilizing, irrigation, weeding) or grazing herds (especially sheep and goats), that have profoundly altered the natural vegetation, the result of which has been the degradation of the landscape with the impoverishment of native flora and fauna.

It must be emphatized that in recent decades various aspects of the Maltese vegetation have been examined from the phytosociological point of view. In particular, Brullo and Marcenò (1979) described the *Triadenio-Chiliadenetum bocconei*, rupestrian association occurring usually on the vertical cliffs near the sea and also the *Putorio calabricae-Micromerietum microphyllae* subass. *chiliadenetosum bocconei*, both belonging to *Dianthion rupicolae*. Later, Brullo (1985) reported for Malta the *Allietum lojaconoi*, ephemeral micro-association of the *Plantagini-Catapodion balearici*, while Brullo et al. (1997) within a study on the *Cisto-Micromerietea* detected a new shrubby association, proposed as *Erico multiflorae-Anthyllidetum melitensis* (sub *Erico multiflorae-Anthyllidetum hermanniae*) and Brullo et al. (2006) surveyed a perennial grassland, represented by *Chamaeleo gummiferis-Brachypodietum retusi*.

Some types of anthropogenic vegetation were described by Maugeri et al. (1979), who attributed the commensal vegetation of irrigated crops to *Setario ambigui-Cyperetum rotundi* and by Brullo and Marcenò (1980) including the summer-autumn vegetation of stubble and vineyards in a new association of *Diplotaxion erucoidis*, named *Chrozophoro tinctoriae-Kickxietum integrifoliae*, occurring also in Sicily. Other ruderal associations, as *Antirrhinetum siculi* and *Lavateretum arboreae* were reported from Malta by Brandes (1991) and Brandes and Brandes (1999). Finally, Devillers and Devillers-Terschuren (2001) presented a list of phytosociological syntaxa occurring in the Maltese Islands, where are quoted hygrophilous communities (*Cymodocetum nodosae*, *Posidonion oceanicae*, *Ruppietum drepanensis*, *Isoeto-Nanojuncetea*, *Potametea*, *Phragmition australis*, *Arundetum donacis*), psammophilous vegetation (*Salsolo-Cakiletum aegyptiacae*, *Echinophoro-Elymetum farcti*, *Echinophoro spinosae-Ammophiletum australis*), rocky coast vegetation (*Crithmo-Limonion*), woody vegetation (*Quercus ilex* woodland), maquis and garrigues, grasslands (*Brachypodietum retusi*, *Lygeum spartum* community, *Hyparrhenia hirta* community), ephemeral meadows (*Plantagini-Catapodion marini*), riparian vegetation (*Salix pedicellata* community, *Nerion oleandri*, *Tamaricion africanae*), chasmophilous vegetation (*Dianthion rupicolae*, *Putorio-Micromerietum microphyllae*, *Triadenio-Chiliadenetum boccone*), etc.

In order to improve the knowledge on the vegetation of the Maltese islands phytosociological researches were carried out according to the method of the Zurich-Montpellier Sigmatist School founded by J. Braun-Blanquet (1884–1980), which is widely illustrated by Braun-Blanquet (1928, 1964), with later modifications and integrations made to Géhu and Rivas-Martinez S (1981), Blasi et al. (2000), Rivas-Martinez (2005), Biondi (2011), Chytrý et al. (2011) and Pott (2011). As concerns this methodological approach, it is based on the concept of "association", term coined for the first time by Humboldt (see Humboldt and Bonpland 1807) as emphasized by Nicolson (2013), but whose definition was at first proposed by Flahaut and Schröter (1910), who stabilished that "An association is a vegetal grouping of determined floristic composition, presenting an uniform physiognomy that grows in uniform site conditions". Later, Braun-Blanquet (1928) provides a more detailed and clair definition: "The association defines a vegetal grouping more or less stable and in equilibrium with the environment, characterized by a particular floristic composition, in which some exclusive or almost exclusive elements (characteristic species) reveal with their presence a particular and autonomous ecology", which is unanimously adopted in phytosociology. In fact, the role of the diagnostic species is essential and fundamental in the identification of a association, allowing to differentiate it from the allied ones also from phytogeographical viewpoint, as well as ecologically, since the plants, both individually and together, give clear informations of autoecological and synecological kind. More recently, Biondi (2011) inserted in the concept of association also the quality of the dynamic and catenal relationships that each of them has with other communities. On the whole, the association is the fundamental unit of the synecology, based on a floristic, ecologic, structural and dynamic description, whose correct identification must be made on detailed field surveying. Besides, the phytosociological method is above all a system for classifying the plant communities, which are grouped hierarchically into alliances, orders and classes, with eventual subunits, which represent syntaxa of a higher rank than the association. The arrangement of the associations (subassociations included) in such a hierarchy makes a syntaxonomical system, which is the reference model for each pattern of vegetation. In particular, the phytosociological data provide a lot of information especially on the ecological, structural, physiognomical features of the investigated plant community, which are collected by the relevés (or plots). Each of them reports a list of species with cover value (sometimes also abundance value) and the measured environmental variables. A modern branch of the science of vegetation is the "Landscape Phytosociology", that analyzes the series of associations or, in general, complexes of plant communities (Rivas-Martinez 1976, 2005). Within a vegetational series linked to a progressive soil maturation on geomorphological base and a constant climate, the final stage coinciding with the highest form of expression, in a state of dynamic equilibrium with the environmental, represent the so-called climax. From the biogeographical viewpoint, an association statistically well characterized in its floristic set and ecology fits into a system of more or less allied synvicariant or geosynvicariant plant communities (Géhu and Rivas-Martinez 1981; Biondi 2011).

The results of about 45 years of phytosociological investigations carried out on the Maltese Archipelago, performed jointly at the same time with taxonomic study on the flora of these islands, have allowed to identify, basing on ca. 900 relevès, more than a hundred associations, many of them new to science, which provide a framework quite complete on the current plant cover of the territory. As concerns the questions related to the nomenclatural aspects, it was referred to the International Code of Phytosociological Nomenclature (Weber et al. 2000), while for the syntaxonomical arrangement of the vegetation it was based on the revisions published by Brullo et al. (2002b), Rivas-Martinez and coauthores (2011), Biondi et al. (2014) and Mucina et al. (2016).

Chapter 4
Syntaxonomical Arrangment of Vegetation

QUERCETEA ILICIS Br.-Bl. ex A. & O. Bolòs 1950
QUERCETALIA ILICIS Br.-Bl. ex Molinier 1934
QUERCION ILICIS Br.-Bl. ex Molinier 1934
Pistacio lentisci-Quercetum ilicis Brullo & Marcenò 1985
QUERCETALIA CALLIPRINI Zohary 1955
PERIPLOCION ANGUSTIFOLIAE Rivas-Martínez 1975
Asparago aphylli-Tetraclinidetum articulatae ass. nov.
Periploco angustifoliae-Euphorbietum dendroidis Brullo, Di Martino & Marcenò 1977

CISTO CRETICI-MICROMERIETEA JULIANAE Oberd. ex Horvatic 1958
CISTO CRETICI-ERICETALIA MANIPULIFLORAE Horvatic 1958
CISTO ERIOCEPHALI-ERICION MULTIFLORAE Biondi 2000
Erico multiflorae-Anthyllidetum melitensis Brullo, Minissale & Spampinato 1997 corr.
Erico multiflorae-Coronilletum glaucae ass.nov.

ASPLENIETEA TRICHOMANIS (Br.-Bl. in Meier & Br.-Bl. 1934) Oberd. 1977
ASPLENIETALIA GLANDULOSI Br.-Bl. in Meier & Br.-Bl. 1934
DIANTHION RUPICOLAE Brullo & Marcenò 1979
Hyperico webbii-Chiliadenetum bocconei Brullo & Marcenò 1979 corr.
Putorio calabricae-Micromerietum microphyllae Brullo & Marcenò 1979
subass. *chiliadenetosum bocconei* Brullo & Marcenò 1979

ADIANTETEA Bl.-Bl. in Br.-Bl. & Roussine & Nègre 1952
ADIANTETALIA CAPILLI-VENERIS Bl.-Bl. ex Horvatic 1934
ADIANTION CAPILLI-VENERIS Bl.-Bl. ex Horvatic 1934
Eucladio-Adiantetum capilli-veneris Br.-Bl. ex Horvatic 1934

CRITHMO MARITIMI-LIMONIETEA Br.-Bl. in Br- Bl., Roussine & Nègre 1952
CRITHMO MARITIMI-LIMONIETALIA Molinier 1934
CRITHMO MARITIMI-LIMONION Molinier 1934

S. Brullo et al., *The Vegetation of the Maltese Islands*, Geobotany Studies,
https://doi.org/10.1007/978-3-030-34525-9_4

Crithmo maritimi-Limonietum virgati Pirone 1995
CRUCIANELLION RUPESTRIS Brullo & Furnari 1990
Limonietum zeraphae ass. nov.
Limonietum melitensis ass. nov.
Anthyllido melitensis-Euphorbietum melitensis ass. nov.
Crucianello rupestris-Helichrysetum melitensis ass. nov.

EUPHORBIO PARALIAE-AMMOPHILETEA AUSTRALIS Géhu & Rivas Martinez in Rivas Martinez et al., 2011
AMMOPHILETALIA AUSTRALIS Br.-Bl. 1933
AMMOPHILION AUSTRALIS Br.-Bl.1921
SPOROBOLENION ARENARII (Géhu & Géhu-Frank ex Géhu & Biondi 1994) Rivas Martinez & Cantò in Rivas Martinez et al. 2002
Eryngio maritimi-Sporoboletum arenarii (Arènes ex Géhu & Biondi 1994) Rivas Martinez & Cantò in Rivas Martinez et al. 2002
ELYTRIGIENION JUNCEI Rivas-Martínez, Costa, Castroviejo & Valdes Bermajo 1980
Cypero capitati-Agropyretum juncei (Kühnholtz-Lordat 1923) Br.-Bl. 1933
AMMOPHILENION AUSTRALIS Rivas-Martínez & Géhu in Rivas-Martínez et al. 1990
Medicagini marinae-Ammophiletum australis Br.-Bl. 1921 corr. Prieto & Diaz 1991
CRUCIANELLETALIA MARITIMAE Sissing 1974
ONONIDION RAMOSISSIMAE Pignatti 1952
Centaureo sphaerocephalae-Ononidietum ramosissimae Br.-Bl. & Frei in Frei 1937

CAKILETEA MARITIMAE R.Tx. & Preising ex Br.-Bl. & R. Tx. 1952
THERO-ATRIPLICETALIA TRIANGULARIS Pignatti 1953
EUPHORBION PEPLIDIS R.Tx. ex Oberd. 1952
Salsolo tragi-Cakiletum maritimae Costa & Mansanet 1981 corr.

SARCOCORNIETEA FRUTICOSAE Br.-Bl. & R.Tx. ex A. & O. Bolòs 1950
SARCOCORNIETALIA FRUTICOSAE Br.-Bl.1933
ARTHROCNEMION MACROSTACHYI Rivas-Martínez & Costa 1984
Arthrocnemo macrostachyi-Juncetum subulati Brullo & Furnari 1976
subass. *juncetosum subulati* Brullo & Furnari 1976
subass. *arthrocnemetosum macrostachyi* Brullo & Furnari 1976
INULION LONGIFOLIAE Brullo & Furnari 1988 corr.
Agropyro scirpei-Inuletum longifoliae Brullo in Brullo et al. 1988 corr.
SUAEDION VERAE Brullo & Furnari 1988
Halimiono portulacoidis-Suaedetum verae (Molinier & Tallon 1970) Gèhu in Géhu et al. 1984

JUNCETEA MARITIMI Br.Bl. in Br.-Bl., Roussine & Nègre 1952
JUNCETALIA MARITIMI Br.Bl. ex Horvatic 1934
JUNCION MARITIMI Br.Bl. ex Horvatic 1934

Inulo longifoliae-Juncetum maritimi Brullo in Brullo et al. 1988 corr.
Juncetum maritimo-acuti Horvatic 1934
Caricetum divisae Br.Bl. in Br.-Bl., Roussine & Nègre 1952

THERO-SALICORNIETEA R. Tx. in R. Tx. & Oberd. 1958.
THERO-SALICORNIETALIA R. Tx. ex Géhu & Géhu-Franck 1984
SALICORNION PATULAE Géhu & Géhu-Franck ex Rivas-Martínez 1990
Suaedo spicatae-Salicornietum patulae Brullo & Furnari ex Géhu & Géhu-Franck 1984
THERO-SUAEDETALIA SPLENDENTIS Br.-Bl. & O. Bolòs 1958
THERO-SUAEDION SPLENDENTIS Br.-Bl. in Br.-Bl., Roussine & Nègre 1952
Suaedetum spicatae Pignatti 1953 corr.
Salsoletum sodae Pignatti 1953
Cressetum creticae Brullo & Furnari 1976

MAGNOCARICI ELATAE-PHRAGMITETEA AUSTRALIS Klika in Klika & Novák 1941
PHRAGMITETALIA AUSTRALIS Koch 1926
PHRAGMITION AUSTRALIS Koch 1926
Phragmitetum australis Savič 1926
Typhetum domininingensis Brullo, Minissale & Spampinato 1994
Polygono salicifolii-Phragmitetum australis Barbagallo, Brullo & Furnari 1979
SCIRPETALIA COMPACTI Hejny in Holub et al. 1967 corr. Rivas-Martinez et al. 1980
SCIRPION COMPACTI Dahl & Hadac 1941 corr. Rivas-Martínez et al. 1980
Cyperetum distachyi O. Bolós & R. Molinier 1984
Bolboschoenetum compacti Van Langendonck 1931 corr. Bueno & F. Prieto in Bueno 1997
NASTURTIO OFFICINALIS-GLYCERIETALIA FLUITANTIS Pignatti 1953
NASTURTION OFFICINALIS Géhu & Géhu-Franck 1987
Helosciadietum nodiflori Maire 1924
Apio nodiflori-Glycerietum plicatae Brullo & Spampinato 1990
MAGNOCARICETALIA ELATAE Pignatti 1953
MAGNOCARICION ELATAE Koch.1926
Cyperetum longi Micevski 1957
Cypero longi-Caricetum cuprinae R.Tx. ex Díaz Gonzalez & Fernández-Prieto 1994
Caricetum hispidae Brullo & Ronsisvalle 1975

MOLINIO-ARRHENATHERETEA R.Tx. 1937
POTENTILLO ANSERINAE-POLYGONETALIA AVICULARIS R.Tx. 1947
POTENTILLION ANSERINAE R.Tx. 1947
Festuco arundinaceae–Caricetum distantis J. Duvign. 1967
Phalarido coerulescentis-Schedonoretum arundinacei ass. nov.

Schedonoro arundinacei-Caricetum divisae ass. nov.
PASPALO DISTICHI-POLYGONION VIRIDIS Br.-Bl. in Br.-Bl., Roussine & Nègre 1952, nom. mut.
Potentillo repentis-Panicetum repentis ass. nov.

POTAMETEA Klika in Klika & Novak 1941
POTAMETALIA PECTINATI Koch 1926
RANUNCULION AQUATILIS Passarge 1964
Damasonio bourgaei -Ranunculetum saniculifolii ass. nov.
Zannichellietum melitensis ass. nov.
RANUNCULION FLUITANTIS Neuhäuls 1959
Ranunculetum trichophylli Melendo, Cano & Valle 2003

LEMNETEA MINORIS R.Tx. ex O. Bolòs & Masclans 1955
LEMNETALIA MINORIS R.Tx. ex O. Bolòs & Masclans 1955
LEMNION MINORIS R.Tx. ex O. Bolòs & Masclans 1955
Lemnetum minoris Oberd. ex Müller et Görs 1960

ZOSTERETEA Pignatti 1953
POSIDONIETALIA OCEANICAE Den Hartog ex Mucina in Mucina et al. 2016
POSIDONION OCEANICAE Br.-Bl. ex Molinier 1960
Posidonietum oceanicae Molinier 1960

HALODULO WRIGHTII-THALASSIETEA TESTUDINUM Den Hartog ex Rivas-Martines, Fernàndez-Gonzàlez & Loidi 1999
THALASSIO-SYRINGODETALIA FILIFORMIS Knapp ex Borhidi et al. 1979
CYMODOCEION NODOSAE Den Hartog ex Mucina in Mucina et al., 2016,
Cymodoceetum nodosae Giaccone & Pignatti 1967
Halophiletum stipulaceae Augier ex Brullo ass. nov.

CHARETEA INTERMEDIAE Fukarek 1961
CHARETALIA INTERMEDIAE Sauer 1937
CHARION VULGARIS (Krause ex Krause & Lang 1977) Krause 1981
Charetum vulgaris Corillon 1957
CHARION CANESCENTIS Krausch 1964
Lamprothamnetum papulosi Corillon 1957

RUPPIETEA MARITIMAE J.Tx. ex Den Hartog & Segal 1964
RUPPIETALIA MARITIMAE J.Tx. ex Den Hartog & Segal 1964
RUPPION MARITIMAE Br.-Bl. ex Westhoff in Bennema, Sissingh & Westhoff 1943
Enteromorpho intestinalidis-Ruppietum maritimae Westhoff ex R.Tx. & Böckelmann 1957
Ruppietum drepanensis Brullo & Furnari 1976

ISÖETO-NANOJUNCETEA Br.-Bl. & R.Tx. in Br.-Bl., Roussine & Nègre 1952
ISÖETETALIA Br.-Bl. 1935
ISÖETION Br.-Bl. 1935

Crassulo vaillantii-Elatinetum gussonei Bartolo, Brullo, Minissale & Spampinato 1990
Triglochino laxiflori-Romuleetum melitensis ass. nov.
Lolio lepturoidis-Spergularietum marinae ass. nov.
NANOCYPERETALIA Klika 1935
VERBENION SUPINAE Slavnic 1951
Damasonio bourgaei-Crypsietum aculeatae Rivas-Martínez & Costa in Rivas-Martínez et al. 1980 corr. Silva & Costa 2012

NERIO OLEANDRI-TAMARICETEA Br.-Bl. & O. Bolòs 1958
TAMARICETALIA AFRICANAE Br.-Bl. & O. Bolòs 1958
RUBO ULMIFOLII-NERION OLEANDRI O. Bolòs 1985
Tamarici gallicae-Viticetum agni-casti ass. nov.

ALNO GLUTINOSAE-POPULETEA ALBAE P.Fukarek & Fabijanic 1968
POPULETALIA ALBAE Br.-Bl. ex Tchou 1948
POPULION ALBAE Br.-Bl. ex Tchou 1948
Ulmo canescentis-Salicetum pedicellatae Brullo & Spampinato 1990

STIPO-TRACHYNIETEA DISTACHYAE Brullo in Brullo, Scelsi & Spampinato 2001
TRACHYNIETALIA DISTACHYAE Rivas-Martinez 1978
TRACHYNION DISTACHYAE Rivas-Martinez 1978
Thero-Sedetum caerulei Brullo 1975
subass. *sedetosum caespitosi* Brullo 1975
subass. *sedetosum litorei* Barbagallo, Brullo & Guglielmo 1980
Vulpio-Trisetarietum aureae Brullo 1975
Ononido sieberi-Stipelletum capensis ass. nov.
STIPO-BUPLEURETALIA SEMICOMPOSITI Brullo in Brullo, Scelsi & Spampinato 2001
PLANTAGINI-CATAPODION BALEARICI Brullo 1985 corr.
Allietum lojaconoi Brullo 1985
subass. *typicum*
subass. *anthemidietosum urvilleanae* subass. nov.
subass. *linetosum trigyni* subass. nov.
Silenetum melitensis ass. nov.
Phalaridetum paradoxae ass. nov.
Catapodio balearici-Linarietum pseudolaxiflorae ass. nov.

TUBERARIETEA GUTTATAE (Br.-Bl. in Br.-Bl. et al. 1952) Rivas Goday & Rivas-Martínez 1963
VULPIETALIA Pignatti 1953
ALKANNO-MARESION NANAE Rivas Goday ex Rivas Goday & Rivas-Martínez 1963 corr. Diaz-Garretas et al. 2001
Cutandio divaricatae-Pseudorlayetum pumilae ass. nov.

LYGEO SPARTI-STIPETEA TENACISSIMAE Rivas-Martinez 1978
LYGEO SPARTI -STIPETALIA TENACISSIMAE Br.-Bl. & O. Bolòs 1958
MORICANDIO-LYGEION SPARTI Brullo, De Marco & Signorello 1990
Leontodonto tuberosi-Lygeetum sparti ass. nov.
subass. *typicum*
subass. *fagonietosum creticae* subass. nov.
Convolvulo oleifolii-Lygeetum sparti ass. nov.
CYMBOPOGONO-BRACHYPODIETALIA RAMOSI Horvatic 1963
HYPARRHENION HIRTAE Br.-Bl., P. Silva & Rozeira 1956
Hyparrhenietum hirto-sinaicae A.& O. Bolòs & Br.-Bl. in A. & O. Bolòs 1950
THERO-BRACHYPODION RAMOSI Br.-Bl. 1925
Chamaeleo gummiferis-Brachypodietum retusi C. Brullo, Brullo, Giusso & Tomaselli 2006
BROMO-ORYZOPSION MILIACEAE O. Bolòs 1970
Diplotaxio tenuiflorae-Oryzopsietum miliaceae Brullo 1984
Feruletum melitensis ass. nov.
BELLIDO SYLVESTRIS-ARISARETALIA VULGARIS Biondi in Biondi et al. 2017
LEONTODONTO TUBEROSI-BELLIDION SYLVESTRIS Biondi et al. 2001
Romuleo melitensis-Ranunculetum bullati ass. nov.

PEGANO-SALSOLETEA Br.-Bl. & O. Bolòs 1958
SALSOLO VERMICULATAE-PEGANETALIA HARMALAE Br.-Bl. & O. Bolòs 1954
SALSOLO OPPOSITIFOLIAE-SUAEDION MOLLIS Rigual 1972
Suaedo verae-Darnielletum melitensis ass. nov.

PARIETARIETEA JUDAICAE Oberd. 1977
TORTULO-CYMBALARIETALIA Segal 1969
PARIETARION JUDAICAE SEGAL 1969
Oxalido corniculatae-Parietarietum judaicae (Br.-Bl. 1952) Segal 1969
Parietario judaicae-Matthioletum incanae Vigo & Terrada 1969
Capparietum rupestris O. Bolòs & Molinier 1958
Antirrhinetum siculi Bartolo & Brullo 1986
Hyosciamo albi-Parietarietum judaicae Segal 1969

SAGINETEA MARITIMAE Westhoff, Van Leeuwen & Adriani 1962
FRANKENIETALIA PULVERULENTAE Rivas-Martínez ex Castroviejo & Porta 1976
FRANKENION PULVERULENTAE Rivas-Martínez ex Castroviejo & Porta 1976
Desmazerio-Senecionetum pygmaei Brullo & Scelsi 1998
Anthemido urvilleanae-Frankenietum pulverulentae ass. nov.
Parapholidetum filiformis Brullo, Scelsi & Siracusa 1994

POLYGONO-POETEA ANNUAE Rivas-Martínez 1975
POLYGONO ARENASTRI-POETALIA ANNUAE R.Tx. in Gèhu et al.1972 corr. Rivas-Martinez et al. 1991
POLYCARPION TETRAPHYLLI Rivas-Martínez 1975
Polycarpo tetraphylli-Spergularietum rubrae Brullo & Marcenò 1976
Trisetario aurei-Crepidetum bursifoliae Brullo 1980
Galio muralis-Trifolietum suffocati ass. nov.
Spergulario rubrae-Filaginetum congestae ass. nov.

ARTEMISIETEA VULGARIS Lohmeyer et al. ex von Rochow 1951
CARTHAMETALIA LANATI Brullo in Brullo & Marcenò 1985
ONOPORDION ILLYRICI Oberd. 1954
Carthamo lanati-Onopordetum argolici ass. nov.

PAPAVERETEA RHOEADIS Brullo, Scelsi & Spampinato 2001
GLADIOLO ITALICI-RIDOLFIETALIA SEGETUM Mucina in Mucina et al. 2016
ROEMERION HYBRIDAE Br.-Bl. ex Rivas-Martínez et al. 1997
Rapistro rugosi-Melilotetum infestae Bartolo, Brullo, Fagotto & Grillo 1983
RIDOLFION SEGETUM Négre ex Rivas-Martínez et al. 1999
Calendulo tripterocarpae -Hypecoetum procumbentis Bartolo, Brullo, Minissale & Spampinato 1990

CHENOPODIETEA Br.-Bl. in Br.-Bl., Roussine & Nègre 1952
CHENOPODIETALIA Br.-Bl. in Br.-Bl. et al. 1936
FUMARION WIRTGENII-AGRARIAE Brullo in Brullo & Marcenò 1985
Fumario densiflorae-Veronietum hederifoliae Brullo & Marcenò 1985
Fumarietum bicoloris ass. nov.
Diplotaxio erucoidis-Brassicetum sylvestris ass. nov.
MALVION PARVIFLORAE (Rivas-Martínez 1978) Brullo in Brullo & Marcenò 1985
Lavatero creticae-Malvetum nicaensis ass. nov.
Spergulario marinae-Taraxacetum minimi ass. nov.
Lavateretum arboreae Br.-Bl. & Molinier 1935
MESEMBRYANTHEMION CRYSTALLINI Rivas-Martínez et al. 1993
Mesembryanthemum cristallino-nodiflori O. Bolós 1957
CHENOPODION MURALIS Br.-Bl. in Br.-Bl. et al. 1936 em. Brullo in Brullo & Marcenò 1985
Amarantho deflexi-Chenopodietum ambrosioidis ass.nov.
SOLANO-POLYGONETALIA CONVOLVULI (Sissingh in Westhoff et al. 1946) O. Bolós 1962
DIPLOTAXION ERUCOIDIS Br.-Bl. in Br.-Bl. et al.1936 em. Brullo & Marcenò 1980
Chrozophoro tinctoriae-Kickxietum integrifoliae Brullo & Marcenò 1980
subass. *euphorbietosum aleppicae* Brullo & Marcenò 1980
SPERGULO ARVENSIS-ERODION CICUTARII J.Tx. in Passarge 1964

Setario pumilae-Echinochloetum colonum A. & O. Bolòs ex O. Bolós 1956
Setario ambigui-Cyperetum rotundi Brullo, Scelsi & Spampinato 2001
BROMETALIA RUBENTI-TECTORUM (Rivas Goday & Rivas-Martinez 1973) Rivas-Martinez & Izco 1977
ECHIO PLANTAGINEI-GALACTITION TOMENTOSAE O. Bolòs & Molinier 1969
Chrysanthemo coronariae-Hyppocrepidietum multisiliquosae Brullo & Siracusa 1996
Sullo coronariae-Ononidetum mitissimae ass. nov.
Meliloto messanensis-Hordetum marini Brullo 1983
HORDEION LEPORINI Br.-Bl. in Br.-Bl. et al. 1936 corr. O.Bolòs 1962
Chrysanthemo coronarii-Malvetum parviflorae Ferro 1980
Carduetum marmorati Brullo 1983 corr.
Chrysanthemo coronarii-Silybetum mariani Brullo 1983
Hordeo leporini-Sisymbrietum orientale Oberd. 1954
URTICO-SCROPHULARIETALIA PEREGRINAE Brullo in Brullo & Marcenò 1985
VERONICO-URTICION URENTIS Brullo in Brullo & Marcenò 1985
Bromo diandri-Brassicetum sylvestris Brullo & Marcenò 1985
ALLION TRIQUETRI O. Bolòs 1967
Acantho mollis-Smyrnietum olusatri Brullo & Marcenò 1985
GERANIO PURPUREI-CARDAMINETALIA HIRSUTAE Brullo in Brullo & Marcenò 1985
VALANTIO MURALIS-GALION MURALIS Brullo in Brullo & Marcenò 1985
Parietario lusitanicae-Veronicetum cymbalariae Brullo & Marcenò 1985

EPILOBIETEA ANGUSTIFOLII R.Tx. & Preising ex von Rochow 1951
GALIO APARINES-ALLIARIETALIA PETIOLATAE Oberd. in Görs & Müller 1969
BALLOTO-CONION MACULATI Brullo in Brullo & Marcenò 1985
Galio aparinae-Conietum maculati Rivas-Martinez ex Lopez 1978
CONVOLVULETALIA SEPIUM R.Tx. ex Moor 1958
CYNANCHO ACUTI-CONVOLVULION SEPIUM Rivas Goday et Rivas-Martinez ex Rivas- Martinez 1977
Calystegio sylvaticae-Arundinetum donacis Brullo, Scelsi & Spampinato 2001

NICOTIANO GLAUCAE-RICINETEA COMMUNIS cl. nov.
NICOTIANO GLAUCAE-RICINETALIA COMMUNIS Rivas-Martínez, Fernández-González & Loidi 1999
NICOTIANO GLAUCAE-RICINION COMMUNIS Rivas-Martínez, Fernández-González & Loidi 1999
Nicotiano glaucae-Ricinetum communis de Foucault 2003

Chapter 5
Woody Vegetation

Currently, in the Maltese Islands the forest vegetation is very rare and localized, due to mainly for the millenarian anthropization of this territory. In fact, the woodland in the past were probably widespread especially in the inland and in the wadis, where they were mainly represented by *Quercus ilex* woods and perhaps also by pinewood of *Pinus halepensis*. As concern the first vegetation only few relict spots occur now in Malta, which are referable to *Pistacio lentisci-Quercetum ilicis*, while of the second one there is no trace of natural pine forests in this archipelago, since this gimnosperm is today exclusively cultivated in gardens or street edges, as well as it is used for reforestation. Conversely, the shrubby plant communities as maquis, garrigues and shrubbery are more frequent in these islands. In particular, it is possible to survey woody vegetation characterized by *Tetraclinis articulata*, belonging to *Asparago aphylli-Tetraclinetum articulatae,* thermophylous maquis with *Euphorbia dendroides*, represented by *Periploco angustifoliae-Euphorbietum dendroidis*, garrigues of *Cisto-Micromerietea* and halo-nitrophilous shrub vegetation of *Pegano harmalae-Salsoletea vermiculatae*.

5.1 *Pistacio lentisci-Quercetum ilicis* Brullo & Marcenò 1985 (Table 5.1)

The woodlands of *Quercus ilex* (Fig. 5.1a) are the only forestal communities currently present in the Maltese islands. They are now restricted to small surfaces, consisting mostly of old individuals, often senescent, of holm oak. These last wood relicts are confined to rocky places located on cooler slopes and well protected by winds or on the bottom of small valleys. From the phytosociological point of view, this vegetation can be attributed to *Pistacio-Quercetum ilicis*, thermophilous holm oak association described by Brullo and Marcenò (1985a, b) from Sicily, where grows usually on calcareous or limestone-marly subtrata and recorded also from

S. Brullo et al., *The Vegetation of the Maltese Islands*, Geobotany Studies,
https://doi.org/10.1007/978-3-030-34525-9_5

Table 5.1 *Pistacio lentisci-Quercetum ilicis*

Number of relevé	1	2	3	4	5	6
Plot size (m^2)	100	50	50	50	50	50
Plant cover (%)	100	100	100	100	100	100
Char. Association						
Pistacia lentiscus	1	+	1	2	2	3
Ceratonia siliqua	1	2	.	+	1	1
Prasium majus	1	1	.	+	1	1
Asparagus aphyllus	2	1	.	1	1	2
Teucrium fruticans	+	+	1	.	+	+
Olea europaea var. *sylvestris*	.	.	.	+	1	1
Rhamnus oleoides subsp. *oleoides*	+	.	.	.	.	.
Char. All. (*Quercion ilicis*) and Cl. (*Quercetea ilicis*)						
Quercus ilex	5	5	5	5	5	5
Rubia peregrina	2	2	2	2	2	1
Lonicera implexa	2	1	2	2	2	1
Asparagus acutifolius	1	+	1	+	1	1
Smilax aspera	2	3	2	1	+	1
Clematis cirrhosa	+	.	.	.	+	+
Rhamnus alaternus	.	.	.	+	1	1
Teucrium flavum subsp. *flavum*	+	.	2	.	.	.
Other species						
Brachypodium sylvaticum	.	.	1	2	2	1
Rubus ulmifolius	.	.	1	+	.	+
Acanthus mollis	+	1	.	.	.	.
Crataegus azalorus	.	.	+	.	.	+
Bituminaria bituminosa	.	.	.	.	+	+
Hyoseris radiata	.	.	.	.	1	+
Hedera helix	.	.	2	.	.	.
Capparis spinosa	+	.	.	.	.	.
Parietaria judaica	.	.	+	.	.	.
Euphorbia pinea	.	.	+	.	.	.
Ficus carica	.	.	.	1	.	.

Rel. 1.2: Malta, Mgubah, Selmun, 23.9.1985; Rel. 3: Malta, Ta Baldu, 24.9.1985; Rel. 4–6: Malta, Il Ballut, Wardija 26.9.1985

Sardinia by Arrigoni et al. (1990). and Biondi et al. (2001). This association is characterized by the dominance of *Quercus ilex*, which grows together with several sclerophylls, such as *Pistacia lentiscus*, *Ceratonia siliqua*, *Prasium majus*, *Asparagus aphyllus*, *Teucriun fruticans*, *Olea europaea* var. *sylvestris*, *Rhamnus oleoides*. These species falling in the order *Quercetalia calliprini* Zohary 1955, can be considered as differential species of this association (Brullo et al. 2009b). The characteristic species of *Quercion ilicis* are very frequent, among them there are *Rubia peregrina*, *Lonicera implexa*, *Asparagus acutifolius*, *Smilax aspera*, *Clematis*

Fig. 5.1 Woody vegetation in the Maltese Islands: (**a**) *Quercus ilex* woodland (*Pistacio-Quercetum ilicis*), (**b**) Maquis with *Euphorbia dendroides* (*Periploco angustifoliae-Euphorbietum dendroidis*), (**c**) Maquis with *Tetraclinis articulata* (*Asparago aphylli-Tetraclinidetum articulatae*), (**d**) *Darniella melitensis* shrubland (*Suaedo verae-Darnielletum melitensis*), (**e**) Garigue with *Erica multiflora* and *Anthyllis hermanniae* subsp. *melitensis* (*Erico multiflorae-Antyllidetum melitensis*), (**f**) Garigue with *Coronilla valentina* subsp. *glauca* (*Erico multiflorae-Coronilletum glaucae*)

cirrhosa, *Rhamnus alaternus* and *Teucrium flavum*. As concerns the bioclimate of this territory, the *Pistacio lentisci-Quercetum ilicis* represents an azonal woody vegetation of micro-climactic type. In fact, it for its mesophilous requirements was probably limited in the past to stands well-protected by the wind and the summer high temperatures, as the deeper and not much sunny valley. This association was replaced in the drier and open places, by more marked termophilous shrub communities, mainly those ones belonging to *Periplocion angusitfoliae*, such as *Periploco-Euphorbietum dendroidis* or *Asparago aphylli-Tetraclinetum articulatae*. Formerly as emphasized by several authors (e.g. Sommier and Caruana Gatto 1915; Borg 1927), woodlands dominated by *Pinus halepensis* were problably spread in the Maltese Island, but in remote times they were destroyed by human activities. To attest that there is in Malta a locality known as "Uied Znuber", toponym derived from the Maltese name of the Aleppo Pine.

5.2 *Asparago aphylli-Tetraclinidetum articulatae* ass. nov. hoc loco (Table 5.2)

Holotypus: rel. 7, hoc loco.

On the island of Malta it is possible to observe some relict and very fragmented examples of a termophilous maquis, characterized by *Tetraclinis articulata* (Fig. 5.1c). This species is widely distributed in North West Africa, while in Europe occurs only in a small place of the Iberian Peninsula and in Malta, where it is now represented by a limited number of individuals. On the few releves by us carried out in the stands colonized by this maquis, the vegetation can be included in the *Quercetalia calliprini* Zohary 1975, order belonging to *Quercetea ilicis*. From nomenclatural point of view, this syntaxon is a priority name of *Pistacio-Rhamnetalia alaterni* Rivas-Martinez 1975 nom. illeg., as emphasized by Brullo and Spampinato (2004). In particular, this vegetation, that for the occurence of *Periploca angustifolia* and *Tetraclinis articulata* must be attributed to *Periplocion angustifoliae* Rivas-Martinez 1975, shows some floristical and ecological relationships with the *Arisaro simorrhini-Tetraclinetum articulatae*, association described by Rivas-Martinez (1975) from Southern Spain. Nevertheless, these two communities differ significantly among them for their floristic set. In particular, the Maltese vegetation is differentiated by *Asparagus aphyllus*, as well as by a rich contingent of *Quercetalia calliprini* species, as *Prasium majus*, *Teucrium fruticans*, *Ceratonia siliqua*, *Rhamnus oleoides* subsp. *oleoides*, *Lonicera implexa*, etc. Therefore, it is proposed as *Asparagus aphylli-Tetraclinidetum articulatae* ass. nova. Other associations characterized by *Tetraclinis articulatae* were described from North Africa by Barbéro et al. (1981), Benabid (1984) and Fennane (1986). More recently, Rivas-Martínez and coauthores (2011) proposed to attribute the communities dominated by *Tetraclinis articulata* to a new alliance named *Periploco angustifoliae-Tetraclinidion articulatae*, but this syntaxon must be considered a synonym of the *Periplocion angustifolae*, since it is floristically and ecologically very similar, moreover its distribution area falls within the latter alliance. As concerns the distribution, this conifer, declared the Maltese National tree in 1992, in the past was common in Malta, as attested by Abela (1647), while currently occurs only in

Table 5.2 *Asparago aphylli-Tetraclinetum articulatae*

Number of relevé	1	2	3	4	5	6	7	8	9	10
Plot size (m^2)	50	40	20	40	25	25	30	20	50	50
Plant cover (%)	70	50	90	90	100	100	100	90	100	100
Char. Association										
Asparagus aphyllus	+	1	1	1	2	1	2	2	2	2
Char. All. (*Periplocion angustifoliae*)										
Tetraclinis articulata	3	2	3	3	3	2	4	3	3	3
Periploca angustifolia	+	1	1	3	4	3	2	1	1	3
Char. Ord. (*Quercetalia calliprini*) and Cl. (*Quercetea ilicis*)										
Arisarum vulgare	1	1	2	1	2	2	2	.	.	.
Prasium majus	+	2	1	2	1	2	1	2	2	1
Teucrium fruticans	2	1	2	2	3	2	2	1	2	2
Olea europaea var. *sylvestris*	.	1	1	2	1	1	1	1	1	.
Ceratonia siliqua	1	.	2	2	3	4	2	1	2	3
Rubia peregrina	.	+	+	1	1	1	1	1	2	6
Rhamnus oleoides subsp. *oleoides*	.	+	3	2	2	1	2	3	4	2
Lonicera implexa	.	.	2	1	2	2	1	2	1	1
Pistacia lentiscus	.	2	.	1	1	.	.	2	2	2
Rhamnus alaternus	1	1	.	.	.	.	.	.	.	.
Smilax aspera	.	+	.	.	.	.	.	1	2	.
Teucrium flavum subsp. *flavum*	.	.	.	.	.	.	.	.	6	1
Other species										
Hyparrhenia hirta	+	1	1	1	1	.	1	.	.	.
Erica multiflora	.	.	1	1	1	+	+	2	1	2
Capparis spinosa	1	.	+	.	+	1	.	.	.	.
Sedum sediforme	+	+	.	.	+	.	.	.	.	.
Chiliadenus bocconei	1	.	.	+	+	.	.	.	.	.
Allium subhirsutum	.	+	.	.	+	+	.	.	.	.
Asphodelus ramosus	.	.	+	+	.	+	.	.	.	.
Charybdis pancration	.	.	.	1	+	+	.	.	.	.
Euphorbia melitensis	.	.	.	1	1	+	.	.	6	1
Convolvulus oleifolius	.	.	1	.	+	.	.	.	.	.
Phagnalon graecum	+	.	.	.	+	.	.	.	.	.
Ficus carica	+	.	.	.	.	.	.	.	1	.
Arum italicum	.	.	.	.	.	.	.	.	1	.
Anthyllis hermanniae subsp. *melitensis*	.	.	.	.	.	.	.	.	.	1
Antirrhinum siculum	1	.	.	.	.	.	.	.	.	.

Rel. 1: Malta, il-Maqluba, 27.6.1973; Rel. 2: Malta, Wied il-Ghasel, 27.6.1973; Rel. 3–7: Malta, Ta Pernellu (Mellieha), 31.12.2997; Ril. 8–10: Malta, Mellieha, 22.5.2017

few location of the island. It is probably that this association replaced in the stands with more mesic conditions, especially in the shaded slopes or with a northern exposure, the *Periploco angustifoliae-Euphorbietum dendroidis*, vegetation with more markedly xerophilous requirements.

5.3 *Periploco angustifoliae-Euphorbietum dendroidis* Brullo, Di Martino & Marcenò 1977 (Table 5.3)

The rupestrian or quite rocky places, frequent along the coast and in the inland, usually no affected by human activities, are often colonized by sclerophyllous shrubs, dominated by *Euphorbia dendroides* (Fig. 5.1b), associated with *Periploca angustifolia, Prasium majus, Teucrium fruticans, Pistacia lentiscus, Lonicera implexa, Rhamnus oleoides* subsp. *oleoides, Teucrium flavum*, etc. For its floristic set and ecology, this vegetation must be attributed to *Periploco angustifoliae-Euphorbietum dendroidis*, markedly xerophilous association distributed in the Sicilian Channel Islands (Brullo et al. 1977, 2009b; Brullo and Marcenò 1983, 1985a; Brullo and Piccione 1980; Bartolo et al. 1990; Brullo and Siracusa 1996). From the syntaxonomical point of view, it falls within the *Periplocion angustifoliae*, one of the most xeric alliances of *Quercetalia calliprini*. It is an association with very peculiar edaphic and climatic requirements, since linked to extremely arid environmental conditions of insular type, within of the infra-Mediterranean belt. It was surveyed both on calcareous and volcanic islands of Sicily Channel, on places characterized by shallow or strictly rocky soils. In fact, in the inland areas potentially covered by woodlands, the *Periploco-Euphorbietum dendroidis* shows a clear edaphophilous behavior, localizing in the semi-rocky habitats often in contact with associations of *Asplenietea trichomanis*. In the latter case, this association shows some ecological relations with *Euphorbietum dendroidis* Guinochet in Guinochet and Drounieau 1944, strictly edaphophilous vegetation characterized also by the dominance of *Euphorbia dendroides*, which is widely distributed in the Circum-Tyrrhenian and Anfi-Adriatic territories (Trinajstić 1974, 1975; Trinajstić and Sugar 1977; Brullo and Marcenò 1985a; Bianco et al. 1984; Biondi and Gèhu 1987; Brullo et al. 2009b). In the Maltese Islands due to the marked anthropization of the whole territory, the *Periploco angustifoliae-Euphorbietum dendroidis* is not very frequent. Usually, it is very degraded and floristically impoverished and the best examples were found only in the impervious and just accessible rocky places.

5.4 *Erico multiflorae-Antyllidetum melitensis* Brullo, Minissale & Spampinato 1997 corr. (Table 5.4)

Syn: *Erico multiflorae-Anthyllidetum hermanniae* Brullo, Minissale & Spampinato 1997.

In the Maltese Archipelago the shrubby vegetation characterized by microphyllous xerophytes is currently infrequent and limited to rocky or particularly abrupt places. These garigues are characterized by *Erica multiflora, Thymbra capitata, Rosmarinus officinalis, Fumaria thymifolia, Fumaria arabica, Cistus creticus, Cistus monspeliensis*, etc. This vegetation falls in the *Cisto eriocephali-Ericion multiflorae* Biondi 2000, alliance according Mucina et al. (2016) belonging to *Cisto cretici-Micromerietea julianae* Oberd. ex Horvatić 1958. It is well

Table 5.3 *Periploco angustifoliae-Euphorbietum dendroidis*

Number of relevé	1	2	3	4	5	6	7	8	9	10	11	12	13
Plot size (m^2)	50	50	40	100	100	30	30	50	30	50	50	50	50
Plant cover (%)	70	80	90	80	70	70	70	80	70	90	80	90	90
Char. Association													
Euphorbia dendroides	2	1	1	2	1	3	3	1	3	4	3	3	2
Char. All. (*Periplocion angustifoliae*)													
Periploca angustifolia	1	+	1	3	3	1	1	2	2	2	+	+	1
Diff. subassociation													
Phlomis fruticosa	.	.	.	.	.	.	.	.	.		3	3	4
Char. Ord. (*Quercetalia calliprini*)													
Teucrium fruticans	+	2	+	1	2	1	2	2	2	1	1	2	2
Asparagus aphyllus	1	2	2	1	1	1	1	1	1	2	1	+	1
Prasium majus	1	2	1	2	2	1	2	.	1	2	2	1	2
Rhamnus oleoides subsp. *oleoides*	2	+	1	1	1	.	+	.	.	1	.	+	.
Ceratonia siliqua	2	+	1	1	1	.	+	.	.	.	1	.	.
Olea europaea var. *sylvestris*	.	.	1	2	1	.	.	.	.		.	1	.
Char. Cl. (*Querceta ilicis*)													
Pistacia lentiscus	.	.	2	2	1	.	2	3	1	1	2	2	1
Teucrium flavum subsp. *flavum*	1	.	1	1	.	2	.	2	+	2	.	+	1
Smilax aspera.	2	+	1	.	+	.	+	.	.	1	+	+	.
Lonicera implexa	+	2	3	.	1	1	.	.	.	.	+	1	.
Arisarum vulgare	+	.	+	1	.	+	.	.	.	.	1	1	+
Rubia peregrina	1	.	1	.	+	.	+	.	.	2	+	.	.
Dioscorea communis	+	.	+	.	.	.	.	1	.	.	.	.	.
Melica arrecta	.	.	.	+	+	+	.	.	.	.	.	.	.
Other species													
Thymbra capitata	2	1	+	2	1	.	1	2	2	1	1	2	2
Asphodelus ramosus	1	1	1	.	+	+	+	.	1	.	2	2	1
Erica multiflora	2	2	1	2	1	1	.	.	.	1	+	1	.
Anthyllis hermanniae subsp. *melitensis*	1	1	1	1	+	.	.	2	2	1	.	.	.
Jacobaea maritima subsp. *bicolor*	1	1	1	.	.	2	.	.	.	1	+	.	+
Bituminaria bituminosa	.	+	+	+	.	.	+	.	+	+	+	.	.
Phagnalon graecum	1	.	.	1	.	.	1	.	1	.	.	.	.
Sedum sediforme	+	+	.	.	.	.	.	1	.	.	+	.	.
Hyparrhenia hirta	.	.	+	1	.	.	.	.	2	1	.	.	.
Charybdis pancration	.	.	1	.	.	.	.	.	1	.	1	.	.
Capparis spinosa	+	+	.	.	.	.	.	.	.	1	.	.	.
Acanthus mollis	.	+	.	+	.	.	+	.	.	.	.	.	.
Rubus ulmifolius	.	.	1	+	.	.	.	.	.	.	.	.	.
Carex cuspidata	+	.	+	.	.	.	.	.	.	.	.	.	.
Andropogon distachyos	.	.	+	1	.	.	.	.	.	.	.	.	.
Lotus cytisoides	.	.	.	+	.	.	+	.	.	.	.	.	.
Carlina corymbosa	+	.	.	.	.	.	.	.	+	.	.	.	.

(continued)

Table 5.3 (continued)

Number of relevé	1	2	3	4	5	6	7	8	9	10	11	12	13
Plot size (m^2)	50	50	40	100	100	30	30	50	30	50	50	50	50
Plant cover (%)	70	80	90	80	70	70	70	80	70	90	80	90	90
Pallenis spinosa	.	.	.	.	.	.	+	.	.	.	1	.	.
Foeniculum pipieritum	.	.	.	.	.	.	.	.	.	1	.	.	.
Rosmarinus officinalis	.	.	.	+	.	.	.	.	.	.	.	.	.
Oncostema sicula	.	.	.	+	.	.	.	.	.	.	.	.	.
Chiliadenus bocconei	1	.	.	.	.	.	.	.	.	.	.	.	.
Allium subhirsutum	+	.	.	.	.	.	.	.	.	.	.	.	.
Euphorbia melitensis	.	.	.	.	.	.	.	1	.	.	.	.	.
Leontodon tuberosus	.	.	.	.	.	.	+	.	.	.	.	.	.
Asperula aristata	.	.	.	.	.	.	+	.	.	.	.	.	.
Daucus carota	.	.	.	.	.	.	.	.	+	.	.	.	.
Reichardia picroides var. *picroides*	.	.	.	.	.	.	.	.	+	.	.	.	.
Chamaeleon gummifer	.	.	.	.	.	.	.	.	+	.	.	.	.
Fumana arabica	.	.	.	.	.	.	.	.	+	.	.	.	.
Allium melitense	.	.	.	.	.	.	.	.	+	.	.	.	.

Rel. 1: Malta, Mistra Valley, 9.4.1984; Rel. 2: Malta, Salina Bay, 26.6.1973; Rel. 4–5: Malta, Ghar Lapsi, 10.4.1984; Rel. 6: Malta, Naxxar, 12.4.1987; Rel. 7: Gozo, Xlendi Valley, 13.4.1987; Rel. 8: Comino, 27.9.1985; Rel. 9: Malta, Mnajdra near Qrendi, 1.5.2007; Rel. 10: Gozo, Triq Kalati Punic, 19.5.2017; Rel. 11: Malta, Victoria Lines, 12.4.1987; Rel. 12–13: Malta, Ghar Lapsi, 10.4.1984

differentiated from the allied association occurring in Central Mediterranean for the presence of *Anthyllis hermanniae* subsp. *melitensis*, taxon endemic to Maltese Archipelago. At first, this vegetation was described by Brullo et al. (1997) as *Erico multiflorae-Antyllidetum hermanniae*, but this name according to art. 43 of ICPN (Weber et al. 2000) must be corrected in *Erico multiflorae-Antyllidetum melitensis*, since *Anthyllis hermanniae* in the Maltese Islands is represented by subsp. *melitensis* (see Brullo and Giusso del Galdo 2006). This association can be considered a geographic vicariant of other syntaxa described within the above mentioned alliance in the other Mediterranean countries (Horvatić 1958; Brullo et al. 1977, 1997; Brullo and Marcenò 1983; Barbagallo 1983; Marcenò and Colombo 1982). Actually, this association is very rare in this archipelago, where it was surveyed only in few places of the inland (Fig. 5.1e). From the syndinamic point of view it is mostly considered as a secondary community replacing the holm oak, pine forests and sandarac maquis, as a result of substitution processes. A semirupestrian variant is differentiated by occurrence of *Chiliadenus bocconei* and seems connected to the degradation of *Asparago aphylli-Tetraclinetum articulatae*. As regards the pine forests, this vegetation dominated by *Pinus halepensis*, though now represented only by reforestation, was in a very ancient past widespread in these islands, as also hypothesized by Sommier and Caruana Gatto (1915). As confirmation of this hypothesis, natural and old *Pinus halepensis* woods are recorded in some localities of Southern Sicily by

Table 5.4 *Erico multiflorae-Anthyllidietum melitensis*

Number of relevé	1	2	3	4	5	6	7	8	9	10	11	12	13	14
Plot size (m^2)	50	50	20	15	50	100	20	50	50	50	50	50	10	10
Plant cover (%)	80	90	90	80	70	80	80	90	90	70	80	70	80	70
Char. Association														
Anthyllis hermanniae subsp. *melitensis*	1	2	1	+	1	2	2	2	2	3	4	2	3	3
Char. All. (*Cisto cretici-Ericion manipuliflorae*) and Cl. (*Cisto cretici-Micromerietea julianae*)														
Erica multiflora	3	4	1	3	+	+	3	4	4	3	3	3	4	3
Thymbra capitata	2	2	2	3	3	2	2	2	1	3	3	3	3	3
Fumana thymifolia	1	2	1	1	1	2	1	+	+	1	2	1	1	1
Fumana arabica	1	1	.	.	.	1	.	1	2	.	.	1	1	1
Cistus creticus subsp. *creticus*	.	.	2	2	+	.	3	.	.	.	.	.	1	1
Asperula aristata	.	.	1	1	1	.	1	.	.	+	+	.	.	.
Rosmarinus officinalis	3	2	4	2	+	.	.	.	.	.	.	.	.	.
Phagnalon graecum	.	.	.	.	.	.	.	.	.	2	1	1	2	1
Micromeria graeca	1	1	.	.	.	.	.	..	.	.	.	.	.	.
Cistus creticus subsp. *eriocephalus*	.	.	.	.	.	.	.	1	1	.	.	.	.	.
Cistus monspeliensis	.	.	.	.	3	4	.	.	.	.	.	.	.	.
Convolvulus oleifolius	.	.	.	.	.	.	.	.	.	.	.	.	1	2
Phagnalon rupestre	.	.	.	.	.	1	.	.	.	.	.	.	.	.
Other species														
Asparagus aphyllus	1	1	.	.	1	1	1	1	+	1	+	.	2	2
Teucrium flavum subsp. *flavum*	1	+	1	1	1	.	.	+	.	+	.	+	2	2
Teucrium fruticans	1	+	.	.	1	2	1	.	+	1	1	+	.	2
Prasium majus	1	+	.	.	.	.	.	1	1	1	1	+	1	1
Hyparrhenia hirta	+	+	.	.	.	.	.	1	+	1	1	+	1	2
Asphodelus ramosus	2	1	.	.	1	2	.	.	1	2	2	1	.	.
Lonicera implexa	2	1	1	2	.	.	2	2	1	.	.	.	.	.
Pistacia lentiscus	2	2	.	+	.	.	1	.	.	+	1	.	.	.
Carlina corymbosa	.	.	.	.	1	1	+	.	.	1	1	1	.	.
Anthyllis vulneraria subsp. *maura*	+	+	.	.	.	.	.	.	.	+	1	+	.	.
Rhamnus oleoides subsp. *oleoides*	.	.	.	.	+	1	1	.	.	.	.	.	1	1
Rechardia picroides	.	.	.	.	.	+	.	.	.	+	+	+	.	.
Brachypodium retusum	.	.	+	1	3	.	2	.	.	.	.	.	.	.
Dactylis hispanica	.	.	.	.	2	.	.	.	.	1	1	+	.	.
Convolvulus elegantissimus	.	.	.	.	.	+	.	.	.	1	+	+	.	.
Chamaeleon gummifer	.	.	.	.	+	+	.	.	.	+	1	.	.	.
Carex cuspidata	+	+	.	.	+	.	.	.	.	.	.	.	.	.
Cuscuta epithymum	.	.	.	.	.	.	.	.	.	+	+	1	.	.
Euphorbia pinea	.	.	.	.	.	.	.	.	.	+	1	+	.	.
Jacobaea maritima subsp. *bicolor*	.	+	.	.	.	.	.	.	.	+	+	.	.	.
Bituminaria bituminosa	1	+	.	.	.	.	.	.	.	+	.	.	.	.
Pallenis spinosa	.	.	.	.	.	.	.	.	.	+	+	+	.	.

(continued)

Table 5.4 (continued)

Number of relevé	1	2	3	4	5	6	7	8	9	10	11	12	13	14
Plot size (m^2)	50	50	20	15	50	100	20	50	50	50	50	50	10	10
Plant cover (%)	80	90	90	80	70	80	80	90	90	70	80	70	80	70
Foeniculum piperitum	.	.	.	1	.	.	.	.	.	+	+	.	.	.
Micromeria microphylla	.	.	.	.	.	.	.	.	.	.	.	+	+	1
Lotus cytisoides	.	.	.	.	.	.	.	.	.	.	.	.	1	1
Charybdis pancration	.	.	.	.	.	.	.	.	.	.	.	.	1	2
Chiliadenus bocconei	.	.	.	.	.	.	.	.	.	.	.	.	+	1
Capparis spinosa	.	.	.	.	.	.	.	.	.	.	.	.	1	1
Hyoseris radiata	+	.	.	.	.	.	.	+	.	.	.	.	.	.
Allium melitense	.	.	.	.	+	.	.	.	.	.	+	.	.	.
Charybdis pancration	.	+	.	.	+	.	.	.	.	.	.	.	.	.
Smilax aspera	.	.	.	.	.	.	1	1	.	.	.	.	.	.
Rubia peregrina	.	.	.	.	.	.	+	+	.	.	.	.	.	.
Convolvulus cantabrica	+	.	.	.	.	.	.	.	.	.	.	.	.	.
Euphorbia dendroides	.	.	.	.	.	.	.	.	.	.	.	+	.	.
Chiladenus bocconei	.	.	.	.	.	.	.	.	.	.	.	+	.	.
Loncomelos narbonensis	.	.	.	.	+	.	.	.	.	.	.	.	.	.
Leontodon tuberosus	+	.	.	.	.	.	.	.	.	.	.	.	.	.
Arisarum vulgare	.	+	.	.	.	.	.	.	.	.	.	.	.	.
Periploca angustifolia	.	.	.	.	.	.	.	.	.	.	.	+	.	.
Pinus halepensis	.	.	.	.	.	.	1	.	.	.	.	.	.	.

Rel. 1–2: Malta, Bingemma, 12.4.1984; Rel. 3–4: Malta, Ta Baldu (Dingli), 24.9.1985; Rel. 5: Malta, Ta Wied Rini (Bahrija), 26.9.1985; Rel. 6: Gozo, Nadur, 27.11.1985; Rel. 7: Malta, Ras-il-Griebeg (Mellieha), 28.9.1985; Rel. 8: Malta, Salina Bay, 9.4.1984; Rel. 9: Malta, St. Paul's Bay, 11.4.1987; Rel. 10–11: Malta, Mtahleb, 26.9.1985; ril. 12: Malta, Mnajdra (presso Qrendi), 01.05.2007; ril. 13–14: Gozo, Triq Kalati Punic. 19/05/2017

Bartolo et al. (1978, 1985) and Bazan et al. (2010), which colonized the sunny and exposed stands, mainly on substrata constitute by marly limestones or marls. On the basis of geomorphological and climatic affinity between the Maltese Islands and Hyblean territory (S Sicily), it was probably that a very similar Hyblean pine forest, described as *Thymo-Pinetum halepensis* De Marco & Caneva 1985, previously occurred also in Malta.

5.5 *Erico multiflorae-Coronilletum glaucae* ass. nov. hoc loco (Table 5.5)

Holotypus: rel. 1, hoc loco

On the rocky ridges that rise above the high sea cliffs, it is quite frequent a garigue characterized by the occurence of *Coronilla valentina* subsp. *glauca* (Fig. 5.1f). This small shrub is usually a member of a vegetation rich in *Erica multiflora*,

Table 5.5 *Erico multiflorae-Coronilletum glaucae*

Number of relevé	1	2	3	4	5	6
Plot size (m^2)	100	50	50	50	50	20
Plant cover (%)	80	90	100	90	70	70
Fall	10	30	45	40	20	70
Exposure	S	SO	SO	SE	S	S
Char. Association						
Coronilla valentina subsp. *glauca*	4	4	5	4	3	3
Daucus rupestris	2	1	+	1	2	2
Convolvulus oleifolius	3	3	2	1	1	.
Char. All. (*Cisto cretici-Ericion manipuliflorae*) and Cl. (*Cisto cretici-Micromerietea julianae*)						
Erica multiflora	3	1	1	3	3	2
Thymbra capitata	3	2	2	2	1	2
Phagnalon graecum	2	2	2	1	1	1
Asperula aristata	.	.	.	.	+	.
Other species						
Rechardia picroides	1	1	+	+	+	+
Lonicera implexa	1	+	+	2	2	2
Teucrium flavum subsp. *flavum*	2	+	1	1	+	2
Teucrium fruticans	1	1	1	+	1	2
Asphodelus ramosus	1	1	2	1	1	1
Pistacia lentiscus	+	1	+	+	+	+
Hyparrhenia hirta	2	1	1	2	1	2
Charybdis pancration	1	+	1	+	+	1
Carlina sicula	2	2	+	1	+	+
Asparagus aphyllus	1	1	1	1	+	+
Chamaeleon gummifer	2	1	1	1	1	1
Periploca angustifolia	+	1	1	2	.	1
Prasium majus	1	1	+	+	.	1
Sedum sediforme	+	1	1	+	+	.
Cuscuta epithymum	+	+	.	+	+	.
Limonium melitense	.	+	+	+	.	.
Melica arrecta	+	+	.	+	.	.
Allium melitense	+	.	+	+	.	.
Allium subhirsutum	.	.	+	+	.	.
Hypericum aegyptiacum subsp. *webbii*	.	.	.	.	1	1
Anacamptis pyramidalis	+	.	.	+	.	.
Micromeria microphylla	+	.	.	.	.	.
Centaurium erythraea	+	.	.	.	.	.
Sixalix maritima	.	.	.	.	+	.
Jacobaea maritima subsp. *bicolor*	.	.	.	.	.	+
Cheirolophus crassifolius	+	.	.	.	.	.
Smilax aspera	.	.	.	+	.	.

Rel. 1–6: Malta, Wadi Babu, near Blue Grotto (Zurrieq), 01.05.2007

Coridothymus capitatus, *Phagnalon graecum*, which is well differentiated from *Erico multiflorae-Antyllidetum melitensis*, for the absence of *Anthyllis hermanniae* subsp. *melitensis*, *Rosmarinus officinalis*, *Fumana* sp. pl., *Cistus* sp. pl., etc., while are well represented two species quite peculiar of Maltese flora. They are *Convolvulus oleifolius*, species with East Mediterranean distribution, and *Daucus rupestris*, endemic to Malta and Lampedusa. These species together with *Coronilla valentina* subsp. *glauca* are proposed as characteristics of a new association, named *Erico multiflorae-Coronilletum glaucae*. Besides, it differs from the previous association also from the ecological point of view, since it is linked to subhalophilous conditions. In fact, this vegetation grows exclusively in stands near the sea, directly affected by sea aereosol, as it is attested by the occurence of some halophytes, such as *Limonium melitense*, *Daucus rupestris*, *Hypericum aegyptiacum* subsp. *webbii*, *Allium melitense*, etc. As concerns its sindynamic role, the association occupies an intermediate position between the salt-tolerant communities of the *Crithmo maritimi-Limonietea*, located in the places closest to the sea and the maquis of *Periploco angustifoliae-Euphorbietum dendroidis* inward.

5.6 *Suaedo verae-Darnielletum melitensis* ass. nov. hoc loco (Table 5.6)

Holotypus: rel. 17, hoc loco

Along the coastal belt, in the clayey-marly outcrops, and halo-nitrophilous shrub vegetation is frequent, which is physiognomically characterized by *Darniella melitensis*, *Suaeda vera* and *Atriplex halimus* (Fig. 5.1d). This plant community, which constitutes a high and dense woody vegetation, can be attributed for its floristic and ecological peculiarity to *Pegano harmalae-Salsoletea vermiculata* Br.-Bl. & O. Bolós 1958, class distributed in the arid territories of the southern Mediterranean (Brullo et al. 2012). Associations belonging to this class were previously recorded from Spain (Braun-Blanquet and de Bolós 1958; de Bolós et al. 1970; Peinado et al. 1989; Rivas-Martínez et al. 2001, 2002), Italian Peninsula (Biondi 1988; Biondi et al. 1992, 1994a; Pirone 1995; Brullo et al. 2001c, 2012), Sardinia (Biondi et al. 1988, 1994b; Biondi and Mossa 1992), Corsica (Géhu and Biondi 1986; Géhu et al. 1986b, 1988), Sicily (Bartolo et al. 1982, 1990; Brullo et al. 1985), Libya (Brullo and Furnari 1996) and Canary Islands (Rivas-Martínez et al. 1993).

As regards the Maltese Archipelago, the vegetation attribuible to *Pegano harmalae-Salsoletea vermiculatae* differs mainly from the other associations known in literature for the occurence of *Darniella melitensis*, endemic of this islands, which is proposed as a characteristic of a new association proposed as *Suaedo verae-Darnielletum melitensis*. In particular, *Darniella melitensis* is localized also in the calcareous coastal cliffs within the rupestrian association *Triadenio-Chiliaderetum bocconei*, which represents its primary habitat. In fact, in these stands

Table 5.6 *Suaedo verae-Darnielletum melitensis*

Number of relevé	1	2	3	4	5	6	7	8	9	10	11	12	13	14	15	16	17	18	19	20	21
Plot size (m^2)	30	50	50	20	50	20	20	15	100	100	50	50	50	50	20	20	50	50	20	20	20
Plant cover (%)	90	100	90	70	70	70	70	70	90	90	90	90	70	100	100	100	100	90	50	50	70
Char. Association																					
Darniella melitensis	2	2	3	3	3	3	3	3	4	4	5	5	4	5	2	3	5	3	3	3	4
Char. All. (*Salsolo vermiculatae-Peganion harmalae*) and Cl. (*Pegano-Salsoletea*)																					
Atriplex halimus	1	4	2	2	.	.	1	+	.	.	2	2	1	2	2	4	2	2	2	2	2
Suaeda vera	3	2	3	2	3	3	2	2	3	2	.	.	.	.	5	3	1	4	.	.	.
Other species																					
Limbarda crithmoides subsp. *longifolia*	3	2	1	+	1	2	3	1	1	1	.	1	+	1	2	1	2	3	.	2	1
Limonium virgatum	+	1	+	+	.	+	1	+	1	+	+	.	1	.	.	1	1	.	.	.	.
Asparagus aphyllus	.	+	+	.	.	.	.	.	.	+	1	1	+	1	+	.	+	1	.	.	.
Beta maritima	.	.	.	.	.	.	.	.	.	.	1	.	+	2	.	+	+	1	1	+	1
Mesembryanthemum nodiflorum	1	+	1	1	.	+	+	.	.	.	.	.	.	+	.	.	.	.	.	.	.
Moraea sisyrinchium	1	1	+	1	1	.	+	.	.	.	.	.	.	.	.	.	.	.	.	.	.
Dactylis hispanica	1	+	.	.	+	.	.	1	.	.	.	.	.	.	.	.	+	1	.	.	.
Capparis spinosa	.	.	.	.	.	.	.	.	.	.	1	+	.	+	.	+	1	1	.	.	.
Catapodium balearicum	+	.	+	+	+	.	.	+	.	.	.	.	.	.	.	.	.	.	.	.	.
Oxalis pes-caprae	.	+	.	1	1	.	.	.	+	1	.	.	.	.	.	.	.	.	.	.	.
Plantago coronopus	.	+	+	.	+	+	.	+	.	.	.	.	.	.	.	.	.	.	.	.	.
Anthemis urvilleana	+	.	+	+	+	+	.	.	.	.	.	.	.	.	.	.	.	.	.	.	.
Sonchus oleraceus	+	.	+	+	.	.	.	.	+	+	.	.	.	.	.	.	.	.	.	.	.
Daucus gingidium	+	.	.	+	+	.	.	+	.	.	.	.	.	.	.	.	.	.	.	.	.
Parapholis incurva	+	.	+	+	.	.	+	.	.	.	.	.	.	.	.	.	.	.	.	.	.

(continued)

Table 5.6 (continued)

Number of relevé	1	2	3	4	5	6	7	8	9	10	11	12	13	14	15	16	17	18	19	20	21
Plot size (m^2)	30	50	50	20	50	20	20	15	100	100	50	50	50	50	20	20	50	50	20	20	20
Plant cover (%)	90	100	90	70	70	70	70	70	90	90	90	90	70	100	100	100	100	90	50	50	70
Charybdis pancration	1	+	.	.	+	.	.	+	.	.	.	.	.	.	.	.	.	.	.	.	.
Allium commutatum	.	.	.	.	.	.	.	.	.	.	.	.	.	+	+	.	+	+	.	.	.
Phagnalon graecum	.	.	.	.	.	.	.	.	.	.	+	1	+	.	.	.	.	.	.	.	.
Matthiola tricuspidata	+	.	.	+	+	.	.	.	.	.	.	.	.	.	.	.	.	.	.	.	.
Frankenia hirsuta	+	+	.	.	.	.	+	.	.	.	.	.	.	.	.	.	.	.	.	.	.
Lavatera arborea	.	.	.	.	.	.	.	.	1	+	.	.	.	+	.	.	.	.	.	.	.
Daucus rupestris	.	.	.	.	.	.	.	.	.	.	.	.	.	.	.	.	.	.	1	1	1
Tamarix africana	.	.	.	.	.	.	.	.	2	+	.	.	.	.	.	.	.	.	.	.	.
Glebionis coronaria	.	.	.	.	.	.	.	.	+	+	.	.	.	.	.	.	.	.	.	.	.
Avena barbata	.	.	.	.	.	.	.	.	+	+	.	.	.	.	.	.	.	.	.	.	.
Crucianella rupestris	.	.	.	.	.	.	1	1	.	.	.	.	.	.	.	.	.	.	.	.	.
Jacobaea maritima subsp. *bicolor*	.	.	.	.	.	.	+	2	.	.	.	.	.	.	.	.	.	.	.	.	.
Galactites elegans	.	.	.	.	.	.	.	.	+	.	.	.	.	.	.	.	.	.	.	.	.
Reichardia picroides var. *picroides*	.	.	.	.	.	.	.	.	.	.	.	.	.	.	.	.	2	1	.	.	.
Limonium melitense	.	.	.	.	.	.	.	.	.	.	.	.	.	.	.	.	.	.	.	1	1
Asphodelus ramosus	.	.	1	.	.	.	.	.	.	.	.	.	.	.	.	.	.	.	.	.	.
Foeniculum piperitum	.	.	.	.	.	.	.	.	.	.	.	.	.	1	.	.	.	.	.	.	.
Oryzopsis miliacea	.	.	.	.	.	.	.	.	.	.	.	.	.	1	.	.	.	.	.	.	.
Plantago serraria	.	.	.	.	.	.	+	.	.	.	.	.	.	.	.	.	.	.	.	.	.

Rel. 1–5: Malta, Delimara, 10.4.1984; Rel. 6–8: Malta, St. Thomas Bay, 26.9.1985; Rel. 9–10: Malta, Marsaxlokk (St. George Bay), 1.5.2007; Rel. 11–13: Malta, St. Thomas, Marsalforn, 18.5.2017; Rel. 14: Malta, Krobb il-Ghagin, Marsaxlokk 18.5.2017; Rel. 15–18: Malta, Delimara, 18.5.2017; Rel. 19–21: Malta, Migra I. Ferha, Rabat, 20.5.2017

this species grows together with numerous relict chasmophytes belonging to the ancient Tertiary flora. Therefore, in the marly-clayy outcrops this endemism seems to have a secondary role, since these coastal habitats are clearly affected by antropic disturbance, supporting the settlement of shrubby Chenopodiaceae, which are suitable to colonize halo-nitrophilous stands. As concerns the alliance, the Maltese vegetation belongs to the *Salsolo oppositifoliae-Suaedion mollis* Rigual 1972 as well as the other Central Mediterranean associations examined by Brullo et al. (2012).

Chapter 6
Chasmophilous Vegetation

In the Maltese Islands the rupestrian communities linked to the crevices of the calcareous rocky walls, represented by the very high cliffs facing the sea and the wadis, are very widespread. In these stands reside several relict endemic species, often having an old Tertiary origin, which are member of very peculiar chasmophilous vegetation of the *Asplenietea trichomanis*. Another very specialized rock habitat is that interested by percolating or splashed waters, as shady and damp walls, colonized by bryo-pteridophytic communities of *Adiantetea*.

6.1 *Hyperico webbii-Chiliadenetum bocconei* Brullo & Marcenò 1979 corr. (Table 6.1)

Syn.: *Triadenio-Chiliadenetum bocconei* Brullo & Marcenò 1979.

Floristically, one of the most interesting plant community of the Maltese Islands is represented by the casmophilous vegetation colonizing the limestone cliffs that characterize the coastline of this archipelago (Fig. 6.1). In these rupicolous stands, usually constituting by spectacular cliffs (often over 100 m high), takes refuge a rich and rare flora, mainly endemic, which is represented by very specialized species, many of them belonging to the old Mediterranean Tertiary flora. According to Bocquet et al. (1978), the rupestrian stands and in particular the limestone cliffs of the Mediterranean coasts can be considered as refuge habitat, which have allowed a lot of species belonging to the old Messinian flora to survive the Pliocene transgression, which caused the flooding and the subsequent filling of the Mediterranean Basin. This caused the destruction of most of the flora, that in the late Miocene as a consequence of the Messinian crisis was widespread in almost all the Mediterranean. Only the populations localized in the upper part of the mountain ranges or cliffs not died out, even if they underwent a geographic isolation (Scossiroli and Ronsisvalle 1991). Later even the Quaternary glaciations contributed to a further depauperation

S. Brullo et al., *The Vegetation of the Maltese Islands*, Geobotany Studies,
https://doi.org/10.1007/978-3-030-34525-9_6

Table 6.1 *Triadenio webbii-Chiliadenetum bocconei*

Number of relevé	1	2	3	4	5	6	7	8	9	10	11	12	13	14	15	16	17	18	19
Plot size (m^2)	40	50	50	40	50	100	50	50	50	40	50	40	80	50	50	50	30	100	50
Plant cover (%)	40	15	20	40	40	30	30	40	50	40	30	20	40	40	20	30	50	30	40
Fall	90	90	90	90	90	90	90	90	90	90	90	90	90	90	90	90	90	90	90
Exposure	N	SW	S	N	NE	SE	S	SW	W	SW	S	S	N	NW	SW	SW	SW	SW	SW
Char. Association																			
Hypericum aegyptiacum subsp. *webbii*	2	1	2	1	2	2	1	2	1	1	2	2	1	2	2	1	2	1	1
Darniella melitensis	.	.	+	.	.	+	+	+	1	.	+	+	.	+	1	+	.	+	+
Daucus rupestris	.	+	.	1	+	2	1	2	+	+	1	1	.	2	1	1	2	+	+
Chiliadenus bocconei	+	2	1	2	2	+	+	.	1	2	+	1	+	+	+	+	1	1	+
Cheirolophus crassifolius	.	.	.	.	.	.	.	.	3	1	2	2	3	2	1	2	3	2	2
Hyoseris frutescens	.	.	.	1	+	+	1	1	.	.	.	.	.	.	.	.	.	.	.
Cremnophyton lanfrancoi	.	.	.	.	.	.	.	.	.	.	.	.	.	.	.	.	.	.	.
Helichrysum melitense	.	.	.	.	.	2	2	3	.	.	.	.	.	.	.	.	.	.	.
Matthiola incana subsp. *melitensis*	.	.	.	.	.	1	+	.	.	.	.	.	.	.	.	.	.	.	.
Char. All. (*Dianthion rupicolae*) and Cl. (*Asplenietea trichomanis*)																			
Melica minuta	+	.	+	+	1	.	.	.	.	+	+	.	+	.	.	.	.	+	.
Antirrhinum siculum	1	+	.	+	.	.	.	.	+	.	.	.	.	.	.	.	1	1	+
Teucrium flavum subsp. *flavum*	1	.	.	+	+	.	.	.	.	.	.	.	.	.	.	.	.	.	+
Umbilicus rupestris	.	.	.	+	+	.	.	+	.	.	.	.	.	.	.	.	.	.	.
Sedum dasyphyllum	.	.	.	.	.	.	.	.	.	.	.	.	.	.	.	.	+	.	.
Ceterach officinarum	.	.	.	.	+	.	+	.	.	.	.	.	.	+	.	.	.	.	.
Silene fruticosa subsp. *fruticosa*	.	.	.	1	.	.	.	.	.	.	.	.	.	.	.	.	.	.	.
Other species																			
Capparis spinosa	1	+	.	+	+	+	.	+	1	.	+	.	.	.	.	.	.	.	+
Limbarda crithmoides subsp. *longifolia*	+	+	.	.	.	1	+	1	.	.	.	.	.	+	.	.	+	.	.
Crucianella rupestris	.	1	+	1	.	+	+	+	.	.	.	.	.	1	+	.	+	.	.

Sedum sediforme	+	.	.	.	.	.	.	.	+	+	.	+	+	.	+	+	1	+	+
Lotus cytisoides	+	.	.	1	+	+	+	+	.	.	.	.	.	.	.	.	+	.	.
Erica multiflora	+	.	.	1	2	.	.	.	+	+	.	1	1	+	.	.	1	+	+
Ficus carica	1	+	.	+	+	.	.	.	1	.	.	.	.	.	.	.	.	.	+
Teucrium fruticans	.	.	.	+	.	.	.	.	.	.	.	.	+	.	.	.	.	+	+
Limonium zeraphae	.	.	+	.	.	+	+	.	.	.	.	.	.	+	+	.	.	.	.
Jacobaea maritima subsp. *bicolor*	.	+	.	+	1	.	.	.	1	1	.	.	1	.	+	.	+	.	+
Crithmum maritimum	.	.	.	.	.	1	+	+	.	.	2		.	.	.	.	.	.	.
Thymbra capitata	.	.	.	.	.	.	.	.	.	1	.	.	.	1	1	+	+	.	.
Roccella cf. tinctoria	1	.	.	3	1	.	.	.	.	.	.	.	.	.	.	.	.	.	.
Bituminaria bituminosa	.	.	.	1	+	.	.	.	.	1	.	.	.	.	.	.	.	+	.
Rhamnus oleoides subsp. *oleoides*	+	+	1	+	.	.	.	.	+	+	.	.	.	.	.	.	.	.	.
Reichardia picroides var. *picroides*	.	+	+	.	.	.	.	.	.	.	+	.	+	.	.	.	.	.	.
Hyparrhenia hirta	.	.	+	.	.	.	+	+	+	+	.	.	.	.	.	.	.	.	+
Phagnalon rupestre	+	.	.	.	.	.	.	.	.	.	.	.	.	.	.	.	.	+	+
Limonium melitense	.	.	.	.	.	.	.	.	.	.	.	.	.	.	.	.	.	.	.
Euphorbia dendroides	.	.	+	.	.	.	+	+	.	.	.	.	.	.	.	.	.	.	.
Convolvulus oleifolius	+	.	.	.	.	.	.	.	+	.	.	.	.	.	.	.	.	.	.
Asphodelus ramosus	.	.	.	2	+	.	+	.	+	.	.	.	.	.	.	.	.	.	+
Asperula aristata	+	.	.	.	.	.	.	.	.	.	.	.	.	.	.	.	.	.	.
Euphorbia melitensis	.	.	.	.	.	.	.	.	.	.	+	.	.	.	+	+	+	.	.
Prasium majus	.	+	+	.	.	.	.	.	.	.	.	.	+	.	.	.	.	.	.
Hyoseris radiata	+	+	+	.	.	.	.	.	.	.	.	.	.	.	.	.	.	.	.
Parietaria lusitanica	.	.	.	.	.	.	.	.	.	.	.	.	.	.	.	.	.	+	.
Arthrocnemum macrostachyum	.	.	.	.	.	.	.	.	.	.	.	.	.	.	.	.	.	.	.
Antirrhinum tortuosum	.	.	.	.	.	.	.	.	.	.	.	.	.	.	+	.	.	.	.
Coronilla valentina subsp. *glauca*	.	.	.	.	.	.	.	.	.	.	.	.	.	.	.	.	.	.	.

(continued)

Table 6.1 (continued)

Number of relevé	20	21	22	23	24	25	26	27	28	29	30	31	32	33	34	35	36	37	38	39
Plot size (m^2)	100	50	50	50	50	50	50	100	100	100	20	20	10	50	50	50	50	50	50	20
Plant cover (%)	40	25	30	20	15	30	35	30	30	40	70	80	80	80	70	80	90	80	40	50
Fall	90	90	90	90	90	90	90	90	90	70	90	90	90	90	90	90	70	90	90	90
Exposure	SW	SW	SW	SW	SW	S	S	SW	S	S	N	NW	SE	W	S	S	W	W	W	W
Char. Association																				
Hypericum aegyptiacum subsp. *webbii*	2	+	1	2	1	2	2	1	1	1	.	2	2	2	2	1	.	2	1	1
Darniella melitensis	1	+	+	+	+	+	1	3	2	3	2	2	3	3	3	4	4	2	3	3
Daucus rupestris	+	.	1	.	.	1	+	1	1	+	1	2	2	1	.	.	1	2	2	2
Chiliadenus bocconei	1	+	1	+	+	.	.	+	1	.	.	.	.	1	+	.	.	+	2	2
Cheirolophus crassifolius	3	2	3	2	2	2	1	+	.	+	.	.	1	2	+	.	.	.	.	
Hyoseris frutescens	.	.	.	.	+	+	+	.	+	.	.	.	.	.	1	.	+	1	.	
Cremnophyton lanfrancoi	.	.	.	.	.	+	1	.	+	2	3	2	3	3	1	2	2	.	.	
Helichrysum melitense	.	.	.	.	.	.	.	.	.	.	.	.	.	.	.	.	1	2	.	
Matthiola incana subsp. *melitensis*	.	.	.	.	.	1	1	.	.	.	.	.	.	.	.	.	2	1	.	.
Char. All. (*Dianthion rupicolae*) and Cl. (*Asplenietea trichomanis*)																				
Melica minuta	.	+	.	.	.	+	+	.	.	.	.	.	.	.	.	.	.	.	.	
Antirrhinum siculum	.	+	.	+	.	.	.	.	.	.	.	.	.	.	.	.	.	.	.	
Teucrium flavum subsp. *flavum*	.	.	.	.	.	+	+	.	.	.	.	.	.	.	.	.	.	.	+	+
Umbilicus rupestris	.	.	.	.	.	.	.	.	+	.	.	.	.	.	.	.	.	.	.	
Sedum dasyphyllum	1	.	+	.	.	.	.	.	.	.	.	.	.	.	.	.	.	.	.	
Ceterach officinarum	.	.	.	.	.	.	.	.	.	.	.	.	.	.	.	.	.	.	.	
Silene fruticosa subsp. *fruticosa*	.	.	.	.	.	.	.	.	.	.	.	.	.	.	.	.	.	.	.	
Other species																				
Capparis spinosa	1	+	.	.	+	+	+	.	+	.	1	2	.	+	1	1	+	1	1	1

Limbarda crithmoides subsp. *longifolia*	.	.	.	.	+	+	+	.	+	.	2	1	2	2	+	.	3	1	.	1
Crucianella rupestris	.	.	.	.	+	+	+	.	+	.	.	+	1	.	1	.	.	2	.	
Sedum sediforme	1	+	.	.	.	+	+	1	+	.	.	.	.	.	.	.	.	.	.	+
Lotus cytisoides	.	.	.	.	+	+	+	.	1	.	+	+	+	1	.	.	3	.	.	
Erica multiflora	.	.	.	.	.	+	+	.	.	.	.	.	.	.	.	.	.	.	+	
Ficus carica	.	+	+	+	.	.	+	.	.	.	1	2	.	.	.	.	.	.	.	
Teucrium fruticans	+	.	.	+	.	+	+	.	+	.	.	.	.	.	.	.	.	3	.	
Limonium zeraphae	+	.	.	.	+	.	.	+	.	.	.	.	.	.	.	.	.	.	1	
Jacobaea maritima subsp. *bicolor*	.	.	.	.	.	.	.	.	.	.	.	.	.	.	.	.	.	.	.	
Crithmum maritimum	.	.	.	.	.	.	.	.	.	.	1	1	2	.	.	.	.	+	.	
Thymbra capitata	.	.	.	.	.	+	1	.	+	.	.	.	.	.	.	.	.	.	.	
Roccella cf. tinctoria	+	.	+	+	.	.	+	.	.	.	.	.	.	.	.	.	.	.	1	1
Bituminaria bituminosa	+	.	.	.	.	+	+	.	.	.	.	.	.	.	.	.	.	.	.	
Rhamnus oleoides subsp. *oleoides*	.	.	.	.	.	.	.	.	.	.	.	.	.	.	.	.	.	.	.	
Reichardia picroides var. *picroides*	.	.	.	.	.	.	.	.	.	.	.	.	.	+	.	.	+	.	.	
Hyparrhenia hirta	.	.	.	.	.	.	.	.	.	.	.	.	.	.	.	.	.	.	.	
Phagnalon rupestre	.	.	+	.	.	+	.	.	.	.	.	.	.	.	.	.	.	.	.	
Limonium melitense	.	.	.	.	.	.	.	.	.	.	1	2	2	.	1	2	.	.	.	1
Euphorbia dendroides	.	+	.	.	.	.	.	+	.	.	.	.	.	.	.	.	.	.	.	
Convolvulus oleifolius	.	.	.	.	.	1	.	+	+	.	.	.	.	.	.	.	.	.	.	
Asphodelus ramosus	.	.	.	.	.	.	.	.	.	.	.	.	.	.	.	.	.	.	.	
Asperula aristata	.	.	.	.	.	+	+	.	+	.	.	.	.	.	.	.	.	+	.	
Euphorbia melitensis	.	.	.	.	.	.	.	.	.	.	.	.	.	.	.	.	.	.	.	
Prasium majus	.	.	.	.	.	.	.	.	.	.	.	.	.	.	.	.	.	.	.	
Hyoseris radiata	.	.	.	.	.	.	.	.	.	.	.	.	.	.	.	.	.	.	.	
Parietaria lusitanica	.	.	.	+	.	.	.	.	.	.	.	.	.	.	.	.	.	.	.	
Arthrocnemum macrostachyum	.	.	.	.	.	.	.	.	.	.	.	+	1	.	.	.	.	.	.	

(continued)

Table 6.1 (continued)

Number of relevé	20	21	22	23	24	25	26	27	28	29	30	31	32	33	34	35	36	37	38	39
Plot size (m^2)	100	50	50	50	50	50	50	100	100	100	20	20	10	50	50	50	50	50	50	20
Plant cover (%)	40	25	30	20	15	30	35	30	30	40	70	80	80	80	70	80	90	80	40	50
Fall	90	90	90	90	90	90	90	90	90	70	90	90	90	90	90	90	70	90	90	90
Exposure	SW	SW	SW	SW	SW	S	S	SW	S	S	N	NW	SE	W	S	S	W	W	W	W
Antirrhinum tortuosum	.	.	.	.	.	+	.	.	.	.	.	.	.	.	.	.	.	.	.	
Coronilla valentina subsp. *glauca*	.	.	.	.	.	.	.	.	.	.	.	.	.	.	1	.	.	.	.	

Rel. 1: Malta, Bahrija Valley, 25.6.1973; Rel. 2–3: Malta, Ghar Lapsis, 27.6.1973; Rel. 4: Gozo, Mgarr ix-Xini Valley, 29.6.1973; Rel. 5–8: Gozo, Xlendi Valley, 29.6.1973; Rel. 9–14: Malta, Wied Babu, 27.6.1973; Rel. 15–17: Malta, Wied ix-Xaqqa, 27.6.1973; Rel. 18–19: Malta, Il Fawwara, 27.6.1973; Rel. 20: Malta, Rdum Dikkiea, 28.6.1973; Rel. 21–23: Malta, Il Fawwara, 28.6.1973; Rel. 24: Gozo, Bardan, 29.6.1973; Rel. 27–28: Gozo, Ta Cenc, 29.6.1973; Rel. 29: Malta, Ghar Hasan Cave, 24.9.1985; Rel. 30–32: Malta, Mtahleb, 24.9.1985; Rel. 33: Malta, Dingli, 21.5.2017; Rel. 34–35: Gozo, Ta Cenc, 25.9.1985; Rel. 36–37: Gozo, Dwerja, 11.4.1984; Rel. 38: Malta, Dingli, 1.5.2007; Rel. 39: Comino, 27.9.1985

Fig. 6.1 Examples of chasmophilous vegetation belonging to *Hyperico webbii-Chiliadenetum bocconei*: (**a**) Cliffs near Mtahleb, (**b**) Cliffs near Dingli, (**c**) Cliffs of Blue Grotto, (**d**) Dwejra Bay

of this ancient flora, mainly due to their thermophilous character. In particular, their isolation favored the speciation processes that led to the differentiation of new taxa well distinct from the ancestral populations. As concerns the Maltese cliffs there is a significant concentration of these species, clearly having a Tertiary origin. Some of them can be considered as very rare relicts of the Mediterranean paleoflora, often taxonomically quite isolated, while other ones are microendemics: they are: **a)** *Cheirolophus crassifolius*, archaic species belonging to a genus allied to *Centaurea*, with several species occurring only in the Canary Islands (Susanna et al. 1999), while previously Dostal (1976) included it in the monospecific genus *Paleocyanus*; **b)** *Cremnophyton lanfrancoi*, taxonomically very isolated species, belonging to a monospecific endemic genus, having some correlations with *Atriplex* (Brullo and Pavone 1987); **c)** *Chiliadenus bocconei*, chasmophyte belonging to a very peculiar genus of the tribe Inulaee, currently represented by several rupicolous species, distributed in the well circumscribed areas of the southern and western Mediterranean, as well as in some Saharan mountains (Brullo 1979); **d)** *Darniella melitensis*, it is a succulent Chenopodiaceae belonging to a genus having, like the previous species, a typically Southern-Mediterranean and Saharan distribution (Brullo 1984a); **e)** *Hyoserys frutescens*, chasmophyte widespread on the island of Gozo, while it is rare in Malta, which for its suffruticous habit can be considered as the most primitive species of the genus *Hyoseris* (Brullo and Pavone 1988); **f)** *Helicrysum* *melitense*, species belonging to the *H. panormitanum* cycle, but probably due to the geographical isolation of its populations, it is morphologically well distinct from the other members of this group occurring in Sicily (Brullo et al. 1988b); **g)** *Matthiola incana* subsp. *melitensis*, the natural populations of *Matthiola incana* growing in the cliffs of the Maltese archipelago, are represented by an endemic subspecies, also it originated as a result of a geographical isolation (Brullo et al. 1988b); **h)** *Daucus rupestris* it is also an endemism, but not exclusive of Malta, since is grows also in the neighboring calcareous islands of Lampedusa and Lampione; **i)** *Hypericum aegyptiacum* subsp. *webbii*, represents a tertiary relic with an areal extended to Lampedusa, Sardinia, Ionian islands of Greece and Crete (Robson 1993), which previously by Brullo and Marcenò (1979) was identified as *Triadenia aegyptiaca*. All these species in the Maltese archipelago are spread mainly in the coastal cliffs, and occur in many places of the islands of Malta and Gozo, as well as on the rocky walls of the wadis, especially near its outlet to the sea. These species were proposed by Brullo and Marcenò (1979) as characteristics of a special chasmophilous association endemic of Malta, named *Triadenio-Chiliadenetum bocconei*, but it must be corrected in *Hyperico webbii-Chiliadenetum bocconei*. From the syntaxonomical view-point, it falls within the *Asplenietea trichomanis* (Br.-Bl. in Meier & Br.-Bl. 1934) Oberd. 1977 and particularly in the alliance *Dianthion rupicolae* Brullo Marcenò 1979, which has a South Tyrrhenian distribution (Brullo et al. 2004b). Among the species of this alliance, *Antirrhinum siculum* and *Silene fruticosa* occur in Malta. This association was quoted from Malta also by Devillers and Devillers-Terschuren (2001).

6.2 *Putorio calabricae-Micromerietum microphyllae* Brullo & Marcenò 1979 (Table 6.2)

This chasmophilous association replaces the previous one in the rocky walls of some wadis in the island of Malta. On the whole, it shows a clearly more mesophilic requirements, localizing mainly in the slope with northern exposure or quite fresh and shaded. This association is characterized by *Putoria calabrica*, *Micromeria microphylla* and *Asperula aristata*, which differentiate it very well from the *Triadenio aegyptiae-Chiliadenetum bocconei*. According to Brullo and Marcenò (1979) and Brullo et al. (2004a, b), the *Putorio calabricae-Micromerietum microphyllae* is distributed, apart from Malta, mainly in the Hyblaean territory (South Sicily), an area showing geomorphological similarities with the valleys occurring in the Maltese islands. Floristically, the Maltese vegetation is much more impoverished than that one of Sicily, mainly for the lack of several species of the *Dianthion rupicolae* alliance and higher rank. In fact, as concern the differentation at syntaxonomical level, the Hyblaean vegetation is referred to the subass. *dianthetosum rupicolae* Brullo & Marcenò 1979, while the Maltese vegetation is attributed to subass. *chiliadenetosum bocconei* Brullo & Marcenò 1979,

Table 6.2 *Putorio calabricae-Micromerietum microphyllae*

Number of relevé	1	2	3	4	5	6	7	8	9
Plot size (m^2)	100	100	100	100	100	100	100	100	100
Plant cover (%)	20	40	40	30	30	40	40	40	40
Fall	90	90	90	90	90	90	90	90	90
Exposure	N	N	N	N	N	NE	NW	NW	N
Char. Association									
Putoria calabrica	2	2	3	1	+	+	2	1	2
Micromeria microphylla	+	1	+	+	1	+	1	+	+
Asperula aristata	1	+	1	+	1	+	+	+	+
Diff. subassociation									
Chiliadenus bocconei	2	3	2	2	2	2	2	1	2
Char. All. (*Dianthion rupicolae*) and Cl. (*Asplenietea trichomanis*)									
Melica minuta	+	+	1	1	1	+	+	+	+
Antirrhinum siculum	+	+	+	.	+	+	+	.	+
Umbilicus rupestris	..	.	.	.	.	+	.	+	.
Ceterach officinarum	.	.	.	+	.	.	.	.	.
Other species									
Teucrium fruticans	1	+	1	1	+	2	1	1	1
Ficus carica	+	+	+	+	+	+	+	+	+
Capparis spinosa	1	+	.	1	+	+	1	+	1
Euphorbia dendroides	+	+	+	.	+	+	+	1	+
Erica multiflora	+	1	1	1	+	1	.	1	.
Hyoseris radiata	.	.	1	+	.	1	+	1	+
Parietaria judaica	+	+	+	+	+	.	.	.	.
Sedum sediforme	.	.	.	+	+	+	1	+	.
Roccella cf. *tinctoria*	.	.	.	.	.	+	+	1	1
Hyparrhenia hirta	+	.	+	+	+	.	.	.	.
Bituminaria bituminosa	+	+	.	+	.	.	.	+	.
Adianthum capillus-veneris	1	1	+	+	.	.	.	.	.
Ceratonia siliqua	.	.	+	+	.	+	.	+	.
Sonchus tenerrimus	.	.	.	.	+	+	+	+	.
Phagnalon rupestre	.	.	.	.	.	+	+	+	+
Rhamnus oleoides subsp. *oleoides*	+	+	+	.	.	.	.	.	.
Clinopodium nepeta	+	.	1	+	.	.	.	.	.
Sixalix maritima	+	.	+	.	+	.	.	.	.
Trachynia distachya	.	+	+	.	+	.	.	.	.
Jacobaea maritima subsp. *bicolor*	.	.	+	.	.	+	.	.	+
Asparagus acutifolius	.	.	+	.	.	+	.	+	.
Reichardia picroides var. *picroides*	.	.	.	.	+	.	+	+	.
Brachypodium retusum	.	.	.	.	.	+	+	+	.
Antirrhinum tortuosum	.	.	.	.	.	.	.	+	.

Rel. 1–9: Malta, Misran Ghonoq, 28.6.1973

having *Chiliadenus bocconei* as differential species, which is absent in Sicily (Brullo and Marcenò 1979; Devillers and Devillers-Terschuren 2001).

6.3 *Eucladio verticillati-Adiantetum capilli-veneris* Br.-Bl. ex Horvatić 1934 (Table 6.3)

The vertical limestone surfaces affected by water-splashed during the whole year are colonized by a hygrophilous vegetation dominated by moss carpets, ferns and various chasmo-chomophytes. Physiognomically, among the chasmophytes a significant role is played by *Adiantum capillus-veneris*, which grows usually with *Samolus valerandi* and some bryophytes, such as *Eucladium verticillatum* and *Cephalozia bicuspidata*. This plant community, very rare in the Maltese Islands, belongs to *Adiantetea capilli-veneris* Br.-Bl. in Br.-Bl. et al. 1952 and must be referred to *Eucladio verticillati-Adiantetum capilli-veneris*, association widespread in the all Mediterranean area (Horvatić 1934; Braun-Blanquet et al. 1952; Brullo et al. 1989; Deil 1996, 1998; Dakskobler et al. 2014). According to Lalov et al. (2008) and Mifsud et al. (2016b), in some usually inaccessible cavities and fissures of cliff screes of the Maltese archipelago it was observed *Asplenium sagittatum* (=*Phyllitis sagittata*) very rare fern growing on damp, shaded, sheltered rocky stands often together with *Adianthum capillus-veneris*. Very shady bryo-pteridophytic communities of *Adiantetea* characterized by *Phyllitis scolopendrium* and *P. sagittata* were described from Sicily and South Italy by Brullo et al. (1993) and referred to *Thamnobryo alopecuri-Phyllitidetum scolopendrii*.

Table 6.3 *Eucladio verticillati-Adiantetum capilli-veneris*

Number of relevé	1	2	3	4	5	6	7
Plot size (m^2)	1	2	1.5	2	1	1	2
Plant cover (%)	100	100	50	90	100	90	100
Char. Association							
Eucladium verticillatum	3	2	3	2	2	3	5
Char. All. (*Adiantion*) e Cl. (*Adiantetea*)							
Adiantum capillus-veneris	4	4	2	5	4	3	2
Samolus valerandi	1	2	1	1	1	2	.
Cephalozia bicuspidata	.	2	1	1	2	2	.
Other species							
Limbarda crithmoides subsp. *longifolia*	.	+	+	1	2	1	.
Hyoseris frutescens	.	1	+	.	.	+	.
Symphyotrichum squamatum	.	.	+	+	.	.	.
Hypericum aegyptiacum subsp. *webbii*	.	.	.	+	.	.	.

Rel. 1: Malta, Bahrija Valley, 25.6.1973; Rel. 2–6: Gozo, Dwejra, 13.4.1987; Rel. 7: Malta, Girgenti Valley, 25.6.1973

Chapter 7
Rocky Coast Vegetation

The rocky shorelines directly affected by sea agents are colonized by perennial halophylous communities dominated by various species of *Limonium*, usually associated to *Crithmum maritimum*, *Cichorium spinosum* and *Crucianella rupestris*. This vegetation, clearly belonging to *Crithmo maritimi-Limonietea*, is widespread along the coast of all the islands, where it forms a more or less developed band in the stands closest to the seashore (see Brullo et al. 2017a). In the coastal places more distant from the sea, these communities are replaced by a pulvinate vegetation dominated by small shrubs showing a denser coverage. The latter, always included into *Crithmo maritimi-Limonietea*, can be considered a subhalophilous communities affected mainly by sea aerosol.

7.1 *Limonietum zeraphae* ass. nov. hoc loco (Table 7.1)

Holotypus: ril. 3, hoc loco.

Along the rocky coasts of the Maltese Islands it is very frequent a halophilous vegetation quite specialized, characterized by the dominance of small chamaephytes, often having a pulvinate habit. It is an aerohaline community, differentiated by the occurrence of *Limonium* species, genus in the Maltese islands represented by *L. zeraphae*, *L. melitense* and *L. virgatum*, which are usually associated with numerous other halophytes, always linked to rocky substrata, such as *Crithmum maritimum*, *Lotus cytisoides*, *Silene sedoides*, *Plantago macrorhiza*, *Daucus gingidium*, *Frankenia hirsuta*, *Senecio bicolor*, etc. These species allow to include this community in the *Crithmo-Limonietea*, syntaxon widespread in the Mediterranean-Atlantic territories. As concerns the alliance, for the constant occurrence of some peculiar chamaephytes, among them *Crucianella rupestris*, *Cichorium spinosum* and *Hypericum aegyptiacum* subsp. *webbii*, this vegetation falls within the *Crucianellion rupestris* Brullo & Furnari 1990. This syntaxon, as already underlined by Brullo and Furnari (1981, 1990) and Bartolo and Brullo

S. Brullo et al., *The Vegetation of the Maltese Islands*, Geobotany Studies,
https://doi.org/10.1007/978-3-030-34525-9_7

Table 7.1 *Limonietum zeraphae*

Number of relevé	1	2	3	4	5	6	7	8	9	10	11	12	13	14	15	16	17	18	19	20	21	22	23	24	25
Plot size (m²)	20	10	50	30	20	50	50	20	10	10	50	10	10	10	20	25	20	10	10	20	10	15	50	50	50
Plant cover (%)	40	50	70	40	40	40	50	50	80	70	50	20	20	25	30	30	30	20	30	30	50	30	70	35	40
Char. Association																									
Limonium zeraphae	2	3	2	2	1	1	1	1	2	3	2	1	1	2	2	2	1	2	1	2	3	2	2	2	3
Char. All. (*Crucianellion rupestris*)																									
Crucianella rupestris	1	.	2	1	+	2	2	2	+	+	1	.	+	1	1	1	1	1	2	1	1	+	2	1	2
Cichorium spinosum	1	1	2	1	2	2	1	.	.	+	1	1	+	+	1	+	2	1	2	.	.	.	.	1	2
Hypericum aegyptiacum subsp. *webbii*	2	.	.	2	.	.	.	.	.	.	.	+	+	.	+	.	1	1	+	.	.	.	2	.	.
Daucus rupestris	.	.	.	1	.	.	+	+	.	.	.	.	.	.	.	.	.	.	.	.	.	.	3	.	.
Char. Cl. (*Crithmo maritimi-Limonietea*)																									
Crithmum maritimum	2	2	3	.	1	+	2	1	1	1	2	1	2	1	.	2	2	2	1	1	.	+	2	.	.
Silene sedoides	+	+	+	1	.	.	+	+	1	.	+	+	+	+	1	.	.	+	+	+	+	+	+	1	1
Lotus cytisoides	+	.	2	.	+	1	.	+	+	+	.	.	1	+	+	.	1	1	1	1	1	2	2	1	1
Reichardia picroides var. *maritima*	.	+	1	+	+	1	+	+	+	+	1	.	+	+	+	.	+	+	+	.	+	.	.	+	1
Plantago macrorhiza	.	.	1	.	+	.	+	+	.	+	1	+	+	+	1	.	.	.	.	.	+	.	+	+	1
Frankenia hirsuta	.	.	.	.	.	.	+	+	1	+	.	.	.	+	.	1	.	+	.	.	.	+	+	2	1
Limonium virgatum	.	.	.	.	2	+	1	1	.	+	1	.	.	.	.	+	+	+	.	+	.	.	.	.	.
Daucus gingidium	.	.	+	.	+	1	.	+	+	.	.	.	.	.	.	.	1	1	+	+	.	.	.	.	+
Jacobaea maritima subsp. *bicolor*	.	.	.	.	.	+	+	+	.	+	+	.	.	.	.	.	1	1	+	+	.	.	.	.	.
Limonium melitense	+	.	.	+	.	.	.	.	.	.	.	.	+	1	.	.	.	.	.	.	.	.	.	.	.
Sonchus glaucescens	.	.	.	.	.	+	+	.	.	.	.	.	+	.	.	.	.	.	.	.	.	.	.	.	.
Allium commutatum	.	.	.	.	.	.	.	.	.	.	.	.	.	.	.	.	.	.	.	.	.	.	+	.	.

Other species																									
Limbarda crithmoides subsp. *longifolia*	1	2	2	2	1	2	2	3	4	3	2	2	2	+	2	2	+	+	1	1	1	1	2	2	2
Parapholis incurva	.	.	+	.	+	+	+	.	+	+	+	.	.	+	+	.	+	+	+	+	.	.	.	.	.
Anthemis urvilleana	.	.	.	.	.	+	+	1	+	.	.	+	.	+	.	.	.	.	.	.	1	+	.	+	+
Catapodium balearicum	.	.	+	.	.	+	+	.	.	.	.	.	.	+	+	.	+	+	+	.	.	.	.	.	.
Plantago coronopus	.	.	2	+	.	.	+	.	.	.	.	.	.	.	.	.	.	+	+	.	2	2	.	.	.
Arthrocnemum macrostachyum	.	.	.	.	.	+	.	.	.	.	.	+	.	2	.	.	.	.	.	+	2	+	.	.	.
Capparis spinosa	.	.	.	.	.	.	+	+	.	.	.	.	.	.	.	.	.	+	+	.	.	.	.	+	+
Euphorbia pinea	.	.	.	.	.	.	+	+	.	.	.	.	.	.	.	.	.	+	+	.	.	.	.	+	+
Asparagus acutifolius	.	.	.	.	.	.	.	+	.	.	.	.	+	+	+	.	.	.	.	.	.	.	.	.	.
Dactylis hispanica	.	.	.	.	.	.	+	+	.	.	.	.	.	.	.	.	.	.	.	.	.	.	.	.	.
Halimione portulacoides	.	.	.	.	.	.	.	.	.	.	.	.	.	.	.	.	.	.	.	.	+	+	.	.	.
Sarcocornia fruticosa	.	.	.	.	.	.	.	.	.	.	.	.	.	.	.	.	.	.	.	.	.	.	+	.	.

Rel. 1: Malta, Ghar Lapsi, 24.9.1985; Rel. 2: Malta, Zonqor Point (Marsascala), 26.9.1985; Rel. 3: Malta, Marfa Point, 27.9.1985; Rel. 4–5: Malta, St. Andrews, 24.6.1973; Rel. 6–7: Malta, Dragunara, 24.6.1973; Rel. 8: Malta, Ras il-Qammieh, 25.6.1973; Rel. 9–11: Malta, Il Blata l Bajdai (Floriana), 25.6.1973; Rel. 12–14: Malta, Qrejten Point, 25.6.1973; Rel. 15: Malta, Ramla Tat-Torri (Marfa), 26.6.1973; Rel. 16–19: Malta, Wied Babu, 27.6.1973; Rel. 20: Malta, Ghar Lapsi, 27.6.1973; Rel. 21–22: Malta, St. Dimitri Point, 14.4.1987; Rel. 23: Gozo, Dwejra Point, 30.4.2007; Rel. 24–25: Malta, Ramla Talt-Torri (Marfa), 29.04.2007

Fig. 7.1 Rocky coast with *Crucianellion rupestris* communities: (**a**) *Limonietum zeraphae*, (**b**) *Anthyllido melitensis-Euphorbietum melitensis*, (**c**) *Crucianello rupestris-Helichrysetum melitensis*

(1993), shows a central-eastern Mediterranean distribution, where it is well represented along the coasts of Libya, Malta, Lampedusa and western Sicily. In the Maltese islands within the *Crucianellion rupestris* it can be distinguished, especially in relation to the nature of the substrate, some associations characterized by the dominance of distinct species of *Limonium*. The compact calcareous rocks, mainly represented by coralline limestones, are colonized by a more or less scattered vegetation, where a dominant role is played by *Limonium zeraphae* (Fig. 7.1a). This species, belonging to the *L. bocconei* cycle (see Brullo 1980b, 1988a), allows to differenciate a new endemic association, proposed as *Limonietum zeraphae*. It is very spread along the eastern and southern coast of Malta and Gozo, while it is rare to Comino. The characteristics of alliance and class are frequent and often abundant, among them there are *Crucianella rupestris*, *Cichorium spinosum*, *Crithmum maritimum*, *Lotus cytisoides*, *Hypericum aegyptiacum* subsp. *webbii*, *Limonium virgatum*, *Plantago macrorhiza*, etc. In particular, the *Limonietum zeraphae* is linked to more or les flat coastal surfaces, exposed to marine agents, showing a marked edaphic xericity. Therefore, this association can be considered as the more halo-xerophilous one among the communities characterized by *Crithmum* and *Limonium* sp. pl. occurring in the Maltese archipelago.

7.2 *Limonietum melitensis* ass. nov. hoc loco (Table 7.2)

Holotypus: : ril. 10, hoc loco.

The previous association is vicariate along the rocky coast characterized by softer and friable substrates represented mainly by globigerine limestones, like marls or marly limestones, more rarely limestones, from another community differentiated by *Limonium melitensis*. It colonizes also the more or less compact limestones, but limitedly to shady or well protected and not very sunny stands. In particular, *L. melitensis*, species endemic from Malta too, but showing more mesic ecological requirements than *L. zeraphae*, is proposed as characteristic of a new association, named *Limonietum melitensis*. For the occurrence of *Crucianella rupestris*, *Cichorium spinosum*, *Daucus rupestris* and *Hypericum aegyptiacum* subsp. *webbii*, it falls also in the *Crucianellion rupestris*. This association is physiognomically well diversified from the *Limonietum zeraphae*, since *L. melitensis* shows an erect habit with dense and a little divergent branches, while *L. zeraphae* show a prostrate habit with very curved branches (Brullo 1988a). Besides, *L. melitensis* within the association tends to have a greater coverage, emphasizing a less edaphic xericity of the stands. The *Limonietum melitensis* is mainly distributed on the northern and northwest coasts of Malta, while in Gozo and Comino it is spread anywhere. As concerns its syndinamic role, this association, as well as the previous one, is usually in contact inward with the *Antllyllido melitensis-Euphorbietum melitensis*.

7.3 *Crithmo maritimi-Limonietum virgati* Pirone 1995 (Table 7.3)

On the clays and the clay-marls, well represented along the south-eastern coast of Malta, the halophilous communities of *Crithmo-Limonietea* are physiognomically characterized by *Limonium virgatum*, while *L. melitense* and *L. zeraphae* are totally absent, as well as the species of *Crucianellion rupestris*. Actually, this is not a impoverished aspect of the two previous associations, but it is a distinct community especially in ecological terms. From the floristic viewpoint, it is characterized by *Limonium virgatum*, apomictic species widespread in the all Mediterranean area, while phytogeograpically other rilevant species are missing. For its floristic set and well defined edaphic requirements, this vegetation can be referred to *Crithmo maritimi-Limonietum virgati*, association described by Pirone (1995) from Aprutium coasts in Central Italy. Syntaxonomically, this association must be included in the *Crithmo-Limonion* and not in the *Crucianellion rupestris* alliance as the other two Maltese associations. The *Crithmo maritimi-Limonietum virgati* shows some relations with the *Reichardio-Crithmetum maritimi*, association described by de Bolós (1962) for Catalonia (South Spain), from which it differs for the ecology and floristic set. As regards its syndinamic position, this association shows catenal contact with subhalophilous community of *Pegano-Salsoletea*, among them the *Suaedo-*

Table 7.2 *Limonietum melitensis*

Number of relevé	1	2	3	4	5	6	7	8	9	10	11	12	13	14	15	16	17	18	19	20	21	22	23	24	25	26	27
Plot size (m^2)	50	20	30	30	40	50	20	70	20	50	50	20	30	30	30	30	20	20	20	50	10	10	20	15	10	50	10
Plant cover (%)	70	50	70	50	50	70	70	50	50	50	50	40	40	50	70	50	50	40	40	30	70	70	70	70	90	40	50
Char. Association																											
Limonium melitense	3	3	2	3	2	2	3	3	2	3	2	3	2	2	3	2	2	2	2	2	3	3	2	3	2	2	2
Char. All. (*Crucianellion rupestris*)																											
Crucianella rupestris	2	2	2	1	2	2	2	1	3	2	+	2	2	1	2	.	3	2	2	1	1	2	1	2	+	+	2
Cichorium spinosum	1	+	.	1	+	2	.	.	+	1	2	1	.	+	1	+	1	+	.	1	.	+	.	.	.	+	+
Hypericum aegyptiacum subsp. *webbii*	.	.	.	.	.	.	+	.	+	.	1	+	+	+	.	+	1	+	.	+	.	.	.	.	.	+	2
Daucus rupestris	2	1	2	+	1	.	.	.	+	2	2	+	1	.	.	.	.	.	.	.	.	.	.	.	.	.	.
Char. Cl. (*Crithmo maritimi-Limonietea*)																											
Limbarda crithmoides subsp. *longifolia*	3	2	2	2	2	3	2	2	2	2	2	2	1	3	2	3	2	1	2	.	2	2	1	3	1	2	2
Lotus cytisoides	+	+	1	+	.	1	1	1	+	+	+	+	.	1	1	1	1	2	+	1	2	1	1	1	1	1	2
Crithmum maritimum	2	2	3	1	2	2	2	+	+	1	1	.	2	2	3	1	2	2	2	1	.	.	.	.	.	.	.
Silene sedoides	.	.	.	.	.	+	+	.	1	1	+	1	+	+	.	.	1	+	1	1	.	.	.	.	.	1	1
Plantago macrorhiza	.	.	.	.	+	.	+	+	.	1	+	.	+	.	.	.	+	1	+	+	2	+	.	.	1	.	.
Daucus gingidium	.	.	.	+	.	.	.	1	.	.	.	+	+	+	+	.	+	1	+	1	.	+	.	1	1	.	.
Reichardia picroides var. *maritima*	.	.	.	.	.	.	.	1	+	+	+	+	.	+	+	.	+	+	+	+	.	.	.	.	.	.	.
Jacobaea maritima subsp. *bicolor*	1	+	1	+	+	1	.	1	.	+	2	.	1	.	.	.	.	.	.	.	.	.	.	.	.	.	.
Frankenia hirsuta	.	.	.	.	.	+	.	.	+	+	.	.	.	.	.	1	.	.	.	.	+	1	.	+	+	+	+
Limonium zeraphae	.	.	.	.	.	.	.	.	1	+	1	+	+	1	.	.	.	.	.	.	.	.	.	.	.	.	.
Limonium virgatum	.	.	.	.	.	1	.	.	.	.	.	.	.	.	.	.	1	+	+	.	.	.	.	.	.	.	.
Other species																											
Anthemis urvilleana	.	.	.	.	.	.	.	.	.	.	.	.	.	.	.	+	.	+	+	+	1	+	1	1	1	.	.
Desmazeria pignattii	+	.	+	+	.	.	.	.	+	.	+	+	.	.	.	+	.	.	.	.	.	.	.	+	.	.	.
Capparis spinosa	+	.	+	+	+	.	.	.	.	.	.	.	.	.	.	.	1	+	+	.	.	.	.	.	.	.	.

Halimione portulacoides	.	.	.	.	.	.	.	2	.	.	.	.	.	.	.	.	.	.	.	.	3	3	4	2	5	.	.
Euphorbia pinea	.	.	.	.	.	.	.	.	.	.	.	.	.	.	.	.	+	.	+	.	.	+	.	1	.	.	.
Senecio leucanthemifolius	.	.	.	.	.	.	.	.	.	.	.	.	.	.	.	.	.	.	.	+	.	+	.	1	+	.	.
Parapholis incurva	.	.	.	.	.	.	.	.	.	.	.	.	.	.	.	1	+	.	+	+	.	.	.	.	.	.	.
Catapodium balearicum	.	.	.	.	.	.	.	.	.	.	.	.	.	.	.	+	+	.	+	+	.	.	.	.	.	.	.
Hyoseris frutescens	.	.	.	.	.	.	.	.	+	+	.	.	.	.	.	.	.	.	.	+	.	.	.	+	.	.	.
Arthrocnemum macrostachyum	.	.	+	2	3	.	.	.	.	.	.	.	.	.	.	.	.	.	.	.	.	.	.	.	.	.	.
Matthiola incana subsp. *melitensis*	.	.	.	.	.	.	.	.	.	+	1	.	.	.	.	.	.	.	.	1	.	.	.	.	.	.	.
Mesembryanthemum nodiflorum	.	.	.	.	.	.	.	.	.	.	.	.	.	.	.	+	.	+	+	.	.	.	.	.	.	.	.
Allium melitense	.	.	.	.	.	.	+	.	+	+	.	.	.	.	.	.	.	.	.	.	.	.	.	.	.	.	.
Lygeum spartum	.	.	.	.	.	.	.	.	.	.	.	.	.	.	.	.	.	.	.	.	.	.	.	.	.	1	+

Rel. 1–5: Malta, Paradise Bay (Mellieha), 25.9.1985; Rel. 6: Malta, Madelein Tower, 23.9.1985; Rel. 7: Malta, Ghar Lapsi, 24.9.1985; Rel. 8: Gozo, Marsalforn Bay, 25.9.1985; Rel. 9–10: Gozo, Dwejra, 25.9.1985; Rel. 11–14: Comino, 27-9-1985; Rel. 15: Malta, Golden Bay, 24.9.1985; Rel. 16: Malta, Gnejna Bay, 26.6-1973; Rel. 17–19: Gozo, Dwejra, 29.6.1973; Rel. 20: Malta, Marfa Ramla Bay, 25.9.1985; Rel. 21–25: Gozo, San Dimitri Point, 14.4.1987; Rel. 26–27: Malta, Ghar Lapsi (Siggewi), 1.5.2007

Table 7.3 *Crithmo maritimi-Limonietum virgati*

Number of relevé	1	2	3	4	5
Plot size (m^2)	10	5	10	10	6
Plant cover (%)	40	40	50	30	60
Char. Association					
Limonium virgatum	2	2	3	2	3
Char. All. (*Crithmo maritimi-Limonion*) and Cl. (*Crithmo maritimi-Limonietea*)					
Limbarda crithmoides subsp. l*ongifolia*	1	1	2	1	1
Lotus cytisoides	1	+	1	+	2
Reichardia picroides var. *maritima*	1	1	2	2	1
Crithmum maritimum	+	+	2	+	.
Daucus gingidium	.	+	1	+	.
Jacobaea maritima subsp. *bicolor*	.	1	1	.	.
Silene sedoides	.	.	+	.	+
Other species					
Plantago coronopus	1	1	+	.	2
Catapodium balearicum	.	+	+	.	+
Dactylis hispanica	.	1	.	1	.
Euphorbia pinea	+	.	.	.	.
Capparis spinosa	+	.	.	.	.

Rel. 1–2: Malta, St. Thomas Bay, 12.4.1987; Rel. 3–5: Malta, Sliema, 26.9.1985

Darnielletum melitensis, which generally replaces it in the inner part of the clayey coast.

7.4 *Anthyllido melitensis-Euphorbietum melitensis* ass. nov. hoc loco (Table 7.4)

Holotypus: ril 3, hoc loco.

In the inland belt of the low rocky coast, the community dominated by *Crithmum mariitmum* and *Limonium* sp. pl. are replaced by scrublands dominated by nanophanerophytes and chamaephytes, usually with pulvinate habit. This community shows its optimum on a little acclive slopes, mainly on calcareous and marly substrata. Currently, it is possible to survey this vegetation on all islands, limitedly to not or a little disturbed places, overall not affected by intense human activities. In this vegetation some significant species of Maltese flora are localized or otherwise find here their optimum. Among them there are some endemisms, as *Euphorbia melitensis*, *Anthyllis hermanniae* subsp. *melitensis* and *Allium melitense*, as well as two species with a typical Eastern Mediterranean distribution, as *Phagnalon graecum* and *Convolvulus oleifolius*. In particular, *Euphorbia melitensis,* belonging to the cycle of *E. bivonae* group (Raffaelli and Ricceri, 1988), is widespread in all islands and occurs exclusively in this vegetation (Fig. 7.1b). As concerns *Anthyllis*

Table 7.4 *Anthyllido melitensis-Euphorbietum melitensis*

Number of relevé	1	2	3	4	5	6	7	8	9	10	11	12	13	14	15	16	17	18	19	20	21	22	23	24	25	26	27	28	29	30	31
Plot size (m^2)	20	10	10	10	15	20	10	10	50	30	50	50	20	40	20	25	50	50	20	10	20	20	20	50	50	100	50	20	40	20	10
Plant cover (%)	60	80	80	70	70	70	70	70	50	50	60	70	90	50	90	70	80	70	50	50	40	50	50	90	90	90	80	90	90	60	90
Fall	-	-	-	-	-	-	-	-	-	-	-	-	-	-	-	-	-	-	-	-	-	-	-	-	-	-	-	35	70	-	-
Exposure	-	-	-	-	-	-	-	-	-	-	-	-	-	-	-	-	-	-	-	-	-	-	-	-	-	-	-	E	SE	-	-
Char. Association																															
Phagnalon graecum	1	+	1	2	1	1	1	1	1	.	+	1	2	1	1	2	2	1	+	+	+	+	+	2	2	+	2	1	1	.	.
Euphorbia melitensis	2	2	3	.	1	1	3	2	+	+	+	.	+	.	+	2	1	+	.	+	.	1	3	2	2	.	2	3	2	3	4
Anthyllis hermanniae subsp. *melitensis*	3	2	3	3	2	3	2	3	.	.	.	3	4	2	+	.	3	4	2	3	3	+	+	3	4	3	3	.	.	.	.
Convolvulus oleifolius	2	1	2	1	1	+	1	1	.	1	.	.	.	.	+	+	.	+	1	1	1	1	.	1	2	2	+	+	+	.	1
Allium melitense	.	.	.	.	.	.	.	.	.	.	+	.	.	.	.	+	.	+	+	+	.	+	+	+	+	1	+	.	.	.	.
Char. All. (*Crucianellion rupestris*)																															
Crucianella rupestris	2	1	1	2	3	1	+	3	3	2	2	.	+	1	+	.	.	.	2	2	1	+	+	+	1	2	+	1	2	2	2
Cichorium spinosum	1	3	2	1	2	+	1	1	2	1	1	.	.	.	.	.	.	+	+	1	1	.	.	+	+	1	.	1	1	.	+
Hypericum aegyptiacum subsp. *webbii*	1	2	3	.	+	+	.	+	.	.	+	+	.	.	.	+	1	.	.	+	.	+	.	.	.	.	.	2	2	3	3
Daucus rupestris	.	.	.	.	1	.	1	+	.	.	.	.	.	.	.	.	.	.	.	.	.	.	.	.	.	.	.	2	4	.	.
Char. Cl. (*Crithmo maritimi-Limonietea*)																															
Daucus gingidium	1	.	+	+	.	+	1	1	1	1	2	1	1	+	1	1	+	+	1	+	+	1	1	+	1	2	+	.	.	.	.
Reichardia picroides var. *maritima*	+	+	+	.	+	+	+	+	1	1	+	.	+	.	.	+	+	+	+	+	+	.	.	+	1	1	+	1	1	.	.
Jacobaea maritima subsp. *bicolor*	1	2	1	1	1	+	2	2	1	1	.	.	.	.	.	.	+	.	.	+	+	1	+	1	.	+	.	2	2	2	2
Lotus cytisoides	.	+	.	.	+	+	.	+	.	1	.	+	.	.	1	.	.	.	1	+	+	+	+	1	+	2	.	2	1	1	1
Plantago macrorhiza	.	.	.	.	.	.	.	.	+	.	+	.	.	+	.	1	+	+	+	1	+	.	.	.	.	.	.	.	.	.	.
Limonium zeraphae	.	.	.	.	.	.	.	.	+	.	.	.	.	.	.	.	.	.	+	.	+	+	.	.	.	.	.	.	.	.	.
Frankenia hirsuta	.	.	.	.	.	.	.	.	+	.	.	.	.	.	.	.	.	.	.	+	.	.	.	.	.	.	.	.	.	+	.

(continued)

Table 7.4 (continued)

Number of relevé	1	2	3	4	5	6	7	8	9	10	11	12	13	14	15	16	17	18	19	20	21	22	23	24	25	26	27	28	29	30	31
Plot size (m^2)	20	10	10	10	15	20	10	10	50	30	50	50	20	40	20	25	50	50	20	10	20	20	20	50	50	100	50	20	40	20	10
Plant cover (%)	60	80	80	70	70	70	70	70	50	50	60	70	90	50	90	70	80	70	50	50	40	50	50	90	90	90	80	90	90	60	90
Fall	-	-	-	-	-	-	-	-	-	-	-	-	-	-	-	-	-	-	-	-	-	-	-	-	-	-	-	35	70	-	-
Exposure	-	-	-	-	-	-	-	-	-	-	-	-	-	-	-	-	-	-	-	-	-	-	-	-	-	-	-	E	SE	-	-
Sonchus glaucescens	.	.	.	.	.	.	.	.	+	+	.	.	.	.	.	.	.	.	.	.	.	.	.	.	.	.	.	.	.	.	.
Limonium melitense	.	.	.	.	.	.	.	.	.	.	.	.	.	.	.	.	.	.	.	.	.	.	.	.	.	.	.	.	.	+	.
Crithmum maritimum	.	.	.	.	.	.	.	.	+	.	.	.	.	.	.	.	.	.	.	.	.	.	.	.	.	.	.	.	.	.	.
Silene sedoides	.	.	.	.	.	.	.	.	.	.	.	.	.	.	.	.	.	.	.	.	.	.	.	.	.	.	.	.	.	+	.
Other species																															
Lygeum spartum																														1	+
Thymbra capitata	2	3	2	3	2	4	1	2	2	3	3	2	2	2	4	4	4	3	2	3	3	2	2	3	3	3	3	1	+	.	2
Dactylis hispanica	+	1	.	.	1	1	+	2	2	2	2	2	1	2	1	1	1	+	+	.	.	.	1	1	1	+	2	+	+	.	1
Carlina sicula	.	+	.	1	+	+	.	1	1	+	+	+	+	+	.	1	+	+	+	+	+	.	.	+	+	+	+	1	+	.	+
Asparagus aphyllus	.	1	+	1	+	+	1	1	.	+	+	.	.	.	+	.	.	+	+	.	.	1	+	2	1	+	+	1	+	.	.
Charybdis pancration	+	1	1	1	.	+	+	.	.	.	.	.	.	.	.	+	.	+	+	.	+	+	+	1	1	+	1	1	2	.	+
Euphorbia pinea	1	.	+	+	+	.	+	+	2	2	1	.	.	.	.	.	1	1	+	1	+	.	.	+	+	1	+	.	.	.	.
Asphodelus ramosus	.	+	+	1	1	1	1	+	+	+	+	+	.	+	.	.	.	+	1	+	+	+	.	.	.	.	.	.	.	.	.
Limbarda crithmoides subsp. *longifolia*	1	1	.	1	+	+	2	2	.	.	+	+	.	.	+	1	.	.	.	+	.	.	.	.	.	.	.	1	2	1	.
Teucrium fruticans	1	+	1	.	.	.	.	.	.	.	.	.	.	.	.	1	.	+	.	.	1	1	1	1	1	.	1	1	1	.	.
Brachypodium retusum	.	.	.	3	1	1	+	+	.	1	+	+	.	+	1	.	.	.	.	.	+	+	.	.	.	.	.	.	.	.	.
Anthyllis vulneraria subsp. *maura*	.	.	.	.	.	.	.	.	.	.	.	+	.	.	1	+	+	+	.	.	.	+	+	+	+	2	.	.	.	.	.
Chamaeleon gummifer	.	.	.	+	.	+	+	.	.	.	.	1	.	+	2	+	+	+	.	.	.	.	.	.	.	.	.	.	.	.	.
Moraea sisyrinchium	.	.	+	.	+	+	.	.	+	+	1	.	.	.	+	+	+	.	.	.	.	.	.	.	.	.	.	.	.	.	.
Hyparrhenia hirta	.	+	.	.	.	.	.	.	.	.	.	.	.	.	.	.	1	+	.	.	.	2	1	2	2	+	2	.	.	.	.

Pallenis spinosa	.	.	+	.	.	.	.	.	.	.	.	+	+	+	+	.	.	.	.	.	+	.	.	+	+	.	+	.	.	.	.
Romulea melitensis	.	+	.	+	+	.	.	.	+	.	+	+	.	.	.	.	.	+	+	.	.	.	.	.	.	.	.	.	.	.	.
Asperula aristata	.	.	.	.	.	.	.	.	.	.	.	.	+	+	.	.	.	.	.	.	.	+	+	1	+	+	.	.	.	.	.
Allium lojaconoi	.	.	.	.	.	.	.	.	+	.	+	.	.	.	.	+	.	.	.	+	+	.	.	.	.	.	.	.	.	.	.
Anthemis urvilleana	.	.	.	.	.	+	.	.	+	+	+	.	.	.	.	.	.	.	.	.	+	.	.	.	.	.	.	.	.	.	.
Catapodium balearicum	.	.	.	+	.	.	.	.	.	.	.	.	.	+	.	.	.	.	+	+	+	.	.	.	.	.	.	.	.	.	.
Sulla capitata	.	.	2	.	.	.	.	.	.	.	.	.	1	1	.	1	.	+	.	.	.	.	.	.	.	.	.	.	.	.	.
Linum strictum	.	.	.	.	.	.	+	.	.	.	.	+	+	.	.	.	+	+	.	.	.	.	.	.	.	.	.	.	.	.	.
Prasium majus	.	.	.	.	.	.	.	.	.	.	.	.	.	.	.	.	.	.	.	.	.	.	.	1	+	+	1	+	.	.	.
Erica multiflora	.	.	.	.	.	.	.	.	.	.	.	.	.	.	.	.	.	.	1	+	.	2	1	.	.	.	.	.	.	.	.
Parapholis incurva	.	.	.	.	.	.	.	.	.	.	.	.	.	+	.	.	.	.	+	+	+	.	.	.	.	.	.	.	.	.	.
Anacamptis pyramidalis	.	.	.	.	.	.	.	.	.	.	.	.	.	.	.	.	.	.	.	.	.	.	.	+	+	.	+	+	.	.	.
Cuscuta epithymum	.	.	.	.	.	.	.	.	.	.	.	.	.	.	.	.	.	.	.	.	.	.	.	+	+	+	+	.	.	.	.
Chamaeleon gummifer	.	.	.	.	.	.	.	.	.	.	.	.	.	.	.	.	.	.	.	.	.	.	.	+	+	1	+	.	.	.	.
Capparis spinosa	.	.	.	.	.	.	.	.	+	+	+	.	.	.	.	.	.	.	.	.	.	.	.	.	.	.	.	.	.	.	.
Lagurus ovatus subsp. *ovatus*	.	.	.	.	.	.	.	.	.	.	.	.	+	+	+	.	.	.	.	.	.	.	.	.	.	.	.	.	.	.	.
Valantia muralis	.	.	.	.	.	.	.	.	+	+	.	.	.	+	.	.	.	.	.	.	.	.	.	.	.	.	.	.	.	.	.
Sixalix maritima	.	.	.	.	.	.	.	.	.	.	.	.	.	.	.	.	.	.	.	.	.	.	.	+	+	+	.	.	.	.	.
Bituminaria bituminosa	.	.	.	.	.	.	.	.	.	.	.	.	.	.	.	.	.	.	.	.	.	.	.	+	+	+	.	.	.	.	.
Melica arrecta	.	.	.	.	.	.	.	.	.	.	.	.	.	.	.	.	.	.	.	.	.	.	.	2	+	.	.	+	.	.	.
Fumana arabica	.	.	.	.	.	.	.	.	.	.	.	.	.	.	.	.	.	.	.	.	.	.	.	2	2	.	.	.	.	.	.
Fumana thymifolia	.	.	.	.	.	.	.	.	.	.	.	.	.	.	.	.	.	.	.	.	.	.	.	2	2	.	.	.	.	.	.
Teucrium flavum subsp. *flavum*	.	.	.	.	.	.	.	.	.	.	.	.	.	.	.	.	.	.	.	.	.	.	.	2	2	.	.	.	.	.	.
Sanguisorba minor	.	.	.	.	.	.	.	.	.	.	.	.	.	.	.	.	.	.	.	.	.	.	.	+	.	.	.	+	.	.	.
Darniella melitensis	.	.	.	.	.	.	.	.	.	.	.	.	.	.	.	.	.	.	.	.	.	.	.	.	.	.	.	3	3	.	.
Lonicera implexa	.	.	.	.	.	.	.	.	.	.	.	.	.	.	.	.	.	.	.	.	.	.	.	.	.	.	.	+	+	.	.
Pistacia lentiscus	.	.	.	.	.	.	.	.	.	.	.	.	.	.	.	.	.	.	.	.	.	.	.	+	.	.	.	.	.	.	.

(continued)

Table 7.4 (continued)

Number of relevé	1	2	3	4	5	6	7	8	9	10	11	12	13	14	15	16	17	18	19	20	21	22	23	24	25	26	27	28	29	30	31
Plot size (m^2)	20	10	10	10	15	20	10	10	50	30	50	50	20	40	20	25	50	50	20	10	20	20	20	50	50	100	50	20	40	20	10
Plant cover (%)	60	80	80	70	70	70	70	70	50	50	60	70	90	50	90	70	80	70	50	50	40	50	50	90	90	90	80	90	90	60	90
Fall	-	-	-	-	-	-	-	-	-	-	-	-	-	-	-	-	-	-	-	-	-	-	-	-	-	-	-	35	70	-	-
Exposure	-	-	-	-	-	-	-	-	-	-	-	-	-	-	-	-	-	-	-	-	-	-	-	-	-	-	-	E	SE	-	-
Chiliadenus bocconei	.	.	.	.	.	.	.	.	.	.	.	.	.	.	.	.	.	.	.	.	.	.	.	.	.	.	+	.	.	.	.
Centaurea sicula	.	.	.	.	.	.	.	.	.	.	.	.	.	.	.	.	.	.	.	.	.	.	.	.	.	+	.	.	.	.	.
Foeniculum piperitum	.	.	.	.	.	.	.	.	.	.	.	.	.	.	.	.	.	.	.	.	.	.	.	+	.	.	.	.	.	.	.
Lobularia maritima	.	.	.	.	.	.	.	.	.	.	.	.	.	.	.	.	.	.	.	.	.	.	.	.	.	.	.	+	.	.	.
Rhamnus oleoides subsp. *oleoides*	.	.	.	.	.	.	.	.	.	.	.	.	.	.	.	.	.	.	.	.	.	.	.	+	.	.	.	.	.	.	.

Rel. 1–3: Comino, 15.4.1987; Rel. 4–5: Comino, 27.9.1985; Rel. 6–8: Malta, Qrejten Point, 25.6.1973; Rel. 9–12: Malta, near Qrejten Point, 25.6.1973; Rel. 13–15: Malta, Marfa, 26.6.1973; Rel. 16–17: Malta, Ramla tar-Torri, 26.6.1973; Rel. 18: Malta, Ramla tal Madonna, 26.6.1973; Rel. 19–20: Malta, Wied Babu, 26.6.1973; Rel. 21: Malta, Ghar Lapsi, 27.6.1973; Rel. 22: Malta, Ta Lippija Tower, 28.6.1973; Rel. 23: Gozo, Ta Cenc, 29.6.1973; Rel. 24–25: Malta, Ramla Tal-Bir (Marfa), 29.4.2007; Rel. 26: Malta, Ramla Tal-Torri (Marfa), 29.4.2007; Rel. 27: Malta, Mnajdra (presso Qrendi), 1.5.2007; Rel. 28–29: Malta, Wadi Babu, near Blue Grotto (Zurrieq), 1.5.2007; Rel. 30–31: Malta, Ghar Lapsi (Siggewi), 1.5.2007

hermanniae s.l., it is frequent also in the scrubland of the *Cisto eriocephali-Ericion multiflorae*, where, however, it shows low coverage value, assuming simply a role of territorial characteristic, respect to other associations of this alliance, occurring in other localities of the Mediterranean countries. Instead, in the coastal places *Anthyllis hermanniae* subsp. *melitensis* seems to have its optimum, as can be deduced by its remarkable abundance and constant presence in this vegetation. Besides, this species, together with the other shrubs of this community, assumes a characteristic pulvinate habit, giving to it a peculiar physiognomy, which is typical of the all dwarf shrublands linked to very windy environments, occurring along the coastal places (see Pignatti et al. 1980). Therefore, these species, widespread in all islands of the Maltese archipelago, for the significant role that they play in this coastal habitat, are considered as characteristics of an endemic association, proposed as *Antllyllido melitensis-Euphorbietum melitensis*.

Syntaxonomically, this association falls for its floristic set and ecological requirements in the *Crithmo maritimi-Limonietea* class. In fact, it is closely linked to the marine aerosol that affects directly the coastal environments, as emphasized by the occurrence of numerous halophytes of this class, among them *Daucus gingidium*, *Reichardia picroides* var. *maritima*, *Lotus cytisoides*, *Plantago macrorhiza*, *Jacobea bicolor*, *Frankenia hirsuta*, etc. As concerns instead *Crithmum maritimum* and *Limonium* sp.pl., they are here quite rare because their optimum is in the belt closer the sea. Among the *Cisto eriocephali-Ericion multiflorae* shrubs occurring in the association at issue, the most reppresetative one for its constant presence and abundance is *Thymbra capitata*. In fact, the *Anthyllido melitensis-Euphorbietum melitensis* sometimes is in contact with the *Erico multiflorae-Antlyllidetum melitensis,* but limitedly to the more rocky stands, as wadis or otherwise protected slopes, unsuitable to the establishment of *Periplocion angustifoliae* maquis. For the occurrence of *Crucianella rupestris*, *Cichorium spinosum*, *Hypericum aegyptiacum* subsp. *webbii* and *Daucus rupestris*, the association clearly belongs to the *Crucianellion rupestris*. From the physiognomic-structural and floristic-ecological viewpoint, this association shows remarkable relation with the so-called nanophanerophitic and chamaephitic belts described for various coastal countries of central and western Mediterranean. Among these associations must be mentioned the *Thymelaeo-Helichrysetum conglobati* Brullo & Furnari 1990 from Cyrenaica, the *Chiliadenetum lopadusani* Bartolo et al. 1990 from Lampedusa, both belonging to the *Crucianellion rupestris*. These associations are replaced in the Tyrrhenian territories (Sicily, Italian Peninsula, Sardinia, Corsica, S. France) from other ones, physiognomically and ecologically very similar but belonging to the *Crithmo maritimi-Limonion* and in particular to the suballiance *Plantagini-Tlymelaeenion hirsutae* Bartolo, Brullo & Marcenò 1982 (cfr. Bartolo et al. 1982; Brullo and Marcenò 1983; Bartolo and Brullo 1993). Plant communities very similar to *Anthyllido melitensis-Euphorbietum melitensis* was described also in the eastern Mediterranean area by Brullo et al. (2017a), which belong to different alliance, such as *Cichorio spinosi-Limonion roridi* and *Crithmo maritimi-Elytrigion athericae*.

7.5 *Crucianello rupestris-Helichrysetum melitensis* ass. nov. hoc loco (Table 7.5)

Holotypus: rel. 2, hoc loco.

The rocky plateaux that overhang the high limestone cliffs of Dwejra Point (Gozo) are partly colonized by a low shrub vegetation dominated by *Helichrysum melitense* (Fig. 7.1c), endemic of this coastal tract of the island. It is a community showing a low value of coverage, since it grows on surfaces due to the wind and rains are smooth and leached, with soil accumulating only in the rocky crevices.

In these stands are frequent several species of the *Crucianellion rupestris* and *Crithmo maritimi-Limonietea* such as *Crucianella rupestris*, *Hypericum aegyptiacum* subsp. *webbii*, *Daucus rupestris*, *Crithmum maritimum*, *Lotus cytisoides*, *Limonium melitense*, *Plantago macrorhiza*, *Frankenia hirsuta*, *Allium commutatum*, etc. Apart from *Helichrysum melitense*, other endemic chasmophytes are here frequent, as *Hyoseris frutescens* and *Darniella melitensis*, which are to be considered characteristic of a new association, proposed as *Crucianello rupestris-Helichrysetum melitensis*. It represents an ecological vicariant of the *Antllyllido melitensis-Euphorbietum melitensis,* which occurs in the low rocky reefs.

Table 7.5 *Crucianello rupestris-Helichrysetum melitensis*

Number of relevé	1	2	3	4	5	6	7	8	9	10	11	12
Plot size (m^2)	50	50	100	100	100	50	40	40	20	20	20	50
Plant cover (%)	50	80	70	70	80	70	90	85	5	70	70	70
Fall	5	5	5	10	5	5	10	10	10	10	15	10
Exposure	W	SW	SW	W	NW	W	W	SW	SW	SW	SW	SW
Char. Association												
Helichrysum melitense	3	4	4	3	4	2	4	3	1	+	1	2
Darniella melitensis	1	+	2	3	3	1	2	1	.	.	1	.
Hyoseris frutescens	.	+	1	1	+	1	.	.	.	.	.	1
Char. All. (*Crucianellion rupestris*)												
Crucianella rupestris	2	2	2	2	2	3	2	2	3	3	3	1
Daucus rupestris	3	3	2	2	1	3	2	1	3	2	2	.
Hypericum aegyptiacum subsp. *webbii*	2	2	2	3	2	2	1	+	.	.	+	2
Cichorium spinosum	.	.	.	.	.	.	.	.	+	+	.	+
Char. Cl. (*Crithmo maritimi-Limonietea*)												
Lotus cytisoides	3	2	2	2	2	3	1	2	2	1	3	2
Limbarda crithmoides subsp. *longifolia*	2	2	2	1	2	2	1	1	1	2	2	2
Limonium melitense	+	+	+	+	+	+	+	+	1	1	+	.

(continued)

Table 7.5 (continued)

Number of relevé	1	2	3	4	5	6	7	8	9	10	11	12
Plot size (m^2)	50	50	100	100	100	50	40	40	20	20	20	50
Plant cover (%)	50	80	70	70	80	70	90	85	5	70	70	70
Fall	5	5	5	10	5	5	10	10	10	10	15	10
Exposure	W	SW	SW	W	NW	W	W	SW	SW	SW	SW	SW
Crithmum maritimum	2	1	2	2	1	2	.	.	1	1	+	1
Frankenia hirsuta	+	.	+	+	+	+	.	.	.	.	.	.
Plantago macrorhiza	+	+	+	+	+	.	.	.	.	.	.	.
Reichardia picroides var. *maritima*	1	1	+	+	.	.	.	.	.	.	.	.
Allium commutatum	.	.	.	+	.	+	.	.	.	.	.	.
Other species												
Capparis spinosa	+	+	1	1	+	1	+	+	.	+	+	1
Asparagus aphyllus	2	1	+	1	+	1	.	.	1	+	.	1
Charybdis pancration	+	1	+	.	.	+	1	+	.	.	.	+
Euphorbia pinea	+	+	+	+	.	1	.	.	.	.	.	.
Silene vulgaris subsp. *angustifolia*	+	+	.	.	+	+	.	.	.	.	.	.
Arthocnemum macrostachyum	.	.	.	.	.	.	.	+	+	+	+	.
Euphorbia dendroides	1	.	+	.	.	.	2	1	.	.	.	.
Senecio leucanthemifolius	.	.	.	.	.	.	+	+	+	.	.	1
Romulea melitensis	.	.	.	.	.	.	.	.	+	+	.	+
Matthiola incana subsp. *melitensis*	.	.	.	.	.	.	1	+	.	.	.	2
Carlina corymbosa	1	.	+	.	.	.	.	.	.	.	.	.

Rel. 1–6: Gozo, Dwejra Point, 30.4.2007; Rel. 7–11: Gozo, Dwejra Point, 9.3.2014; Rel. 12: Dwejra Point, 9.4.1984

Chapter 8
Psammophilous Vegetation

Along the coasts of the Maltese Islands the sandy beaches are not very spread. Currently, some examples can be observed only in bays or small coves, where the sandy deposits tend to form small dunes. These surfaces usually are colonized by a paucispecific psammophilous vegetation, represented by perennial communities of the *Ammophilion australis*. Sometimes, also annual halo-nitrophilous communities, belonging to the *Cakiletea maritimae,* occur in stands near the shore. Unfortunately, many of these dune environments for a long time have been destroyed by urbanization and tourist activities, remaining only a few sites, some of which are luckily designated as natural protected areas.

8.1 *Eryngio maritimi-Sporoboletum arenarii* (Arènes ex Géhu & Biondi 1994) Rivas Martinez & Cantò in Rivas Martinez et al. 2002 (Table 8.1, ril. 15)

At the base of the embryonic dunes, sometimes a vegetation dominated by *Sporobolus arenarius* occurs. It represents a marked pioneer community, linked to flat surfaces, usually affected by heavy seas, which is included in the *Sporobolenion arenarii* (Géhu & Géhu-Frank ex Géhu and Biondi 1994) Rivas Martinez & Cantò in Rivas Martinez et al. 2002, suballiance of the *Ammophilion australis* Br.-Bl. 1933, belonging to *Euphorbio paraliae-Ammophiletea australis* Géhu & Rivas Martinez in Rivas Martinez et al., 2011. For the occurrence of *Eryngium maritimum*, it can be attributed to *Eryngio maritimi-Sporoboletum arenarii*, association distributed mainly in the West Mediterranean.

S. Brullo et al., *The Vegetation of the Maltese Islands*, Geobotany Studies,
https://doi.org/10.1007/978-3-030-34525-9_8

Table 8.1 *Cypero capitati-Agropyretum juncei* (rel. 1–13), *Eryngio maritimi-Sporoboletum arenarii* (rel. 14)

Number of relevé	1	2	3	4	5	6	7	8	9	10	11	12	13	14	15
Plot size (m^2)	100	100	50	20	30	30	50	50	20	50	50	10	10	10	5
Plant cover (%)	70	80	70	80	60	60	50	70	80	100	100	90	100	90	50
Char. Association and All. (*Ammophilion australis*)															
Elytrigia juncea	4	4	4	3	3	3	3	3	4	5	3	5	5	3	1
Sporobolus arenarius	1	2	1	1	2	1	2	+	2	.	1	1	2	2	3
Cyperus kallii	.	.	.	.	.	.	1	2	+	.	.	.	.	.	.
Char. Ord. (*Ammophiletalia australis*) and Cl. (*Euphorbio paraliae-Ammophiletea australis*)															
Pancratium maritimum	+	1	2	2	1	2	2	2	1	1	2	1	2	3	.
Eryngium maritimum	.	2	.	1	+	1	2	1	2	2	1	3	4	1	1
Scolymus hispanicus	2	+	+	1	1	1	+	1	+	.	+	.	.	+	.
Medicago marina	+	.	1	2	1	.	2	+	1	.	3	.	.	.	.
Lotus cytisoides	+	+	.	.	.	.	.	.	.	2	.	1	+	3	.
Sixalix maritima	+	+	.	.	.	.	.	.	.	1	.	.	1	3	.
Echinophora spinosa	.	.	.	.	.	2	1	1	1	.	1	.	.	.	.
Euphorbia terracina	1	.	.	.	.	1	.	2	1	.	1	.	.	.	.
Ononis ramosissima	.	+	.	.	.	.	1	1	+	.	1	.	.	.	.
Euphorbia paralias	.	.	.	.	.	+	+	.	.	1	.	.	.	.	.
Calystegia soldanella	.	.	.	.	.	+	1	1	.	.	2	.	.	.	.
Other species															
Cutandia maritima	+	+	1	1	.	+	1	+	+	1	+	+	.	+	.
Echium arenarium										1	.	.	.	.	.
Sonchus oleraceus										+	.	.	.	.	.
Reichardia picroides var. *picroides*										1	+	.	.	.	.
Medicago littoralis	1	+	1	+	+	.	+	1	1	.	+	.	.	.	.
Salsola tragus	+	.	.	+	1	+	.	.	+	.	.	1	+	1	.
Silene colorata	+	.	.	+	+	.	+	+	1	.	+	.	.	.	.

Cakile maritima	.	.	.	+	+	.	.	+	.	.	.	1	+	2	.
Erodium laciniatum	+	+	.	.	.	+	.	1	.	+	.	.	.	.	.
Anisantha rigida	.	.	.	+	+	.	1	+	.	.	.	.	.	.	.
Polygonum maritimum	.	.	.	.	.	.	.	.	.	.	.	+	1	+	1
Pseudorlaya pumila	.	.	.	.	.	+	+	.	+	.	.	.	.	.	.
Matthiola tricuspidata	.	.	.	1	2	.	.	.	.	.	.	.	.	.	.
Glebionis coronaria	+	.	.	.	.	+	.	.	.	.	.	.	.	.	.
Lagurus ovatus subsp. *ovatus*	.	.	+	.	+	.	.	.	.	.	.	.	.	.	.
Polycarpon diphyllum	.	.	.	.	.	+	+	.	.	.	.	.	.	.	.
Ononis variegata	.	.	.	.	.	.	.	2	+	.	1	.	.	.	.
Vulpia fasciculata	+	.	.	.	.	.	.	.	.	.	.	.	.	.	.

Rel. 1–2: Malta, Mellieha Bay, 26.6.1973; Rel. 3: Malta, Golden Bay, 26.6.1973; Rel. 4–5: Malta, Golden Bay, 12.4.1984; Rel. 6–9: Gozo, il Ramla Bay, 13.4.1984; Rel.10: Malta, Ramla Tat-Torri (Marfa), 29.04.2007; Rel. 11: Gozo, il Ramla Bay, 26.3.2016; 12–15: Malta, Ramla Tat-Torri (Marfa), 22.5.2017

Fig. 8.1 Psammophilous vegetation: (**a**) *Cypero capitati-Agropyretum juncei* at Armier Beach, (**b**) Vegetation with *Ononis ramosissima* at Ramla Bay

8.2 *Cypero capitati-Agropyretum juncei* (Kuhnholtz-Lordat 1923) Br.-Bl. 1933 (Table 8.1, ril. 1–14)

Along the shoreline, mainly in correspondence of inlets or bays, are frequent sand deposits colonized by a typical psammophilous vegetation belonging to *Euphorbio paraliae-Ammopliletea australis*. The most common plant community frequent in these stands is the *Cypero capitati-Agropyretum juncei* (Fig. 8.1a), association linked to the embryonic dunes of the Mediterranean coast (Brullo et al. 2001b). It is a pioneer vegetation characterized by the dominance of *Elytrigia juncea*, which usually grows together with *Sporobolus arenarius* and *Cyperus kallii*, as well as various other psammophytes as *Pancratium maritimum, Medicago marina, Eryngium maritimum*, *Echinophora spinosa, Calystegia soldanella*, etc. This association, that until a few decades ago it was quite common to these islands, gradually it went disappearing in many localities of the Maltese island, due to the progressive expansion of the tourist settlements and the road network.

8.3 *Medicagini marinae-Ammophiletum australis* Br.-BI. 1921 corr. Prieto & Diaz 1991 (Table 8.2)

This association, which replaces the *Cypero capitati-Agropyretum juncei* on the highest dunes, is characterized by the dominance of *Ammophila australis*, caespitose grass growing usually on the top of the dunes (Brullo et al. 2001b). This species, in the past already very rare, observed only in some northern sandy places of Malta, seems now all but disappeared. In fact, recent surveys carried out in the few places where occurred the last examples of this vegetation, no trace of it have been found. The cause of this is probably due to the leveling of the dunes, as a result of anthropic action.

Table 8.2 *Medicagini marinae-Ammophiletum australis*

Number of relevé	1	2
Plot size (m^2)	10	5
Plant cover (%)	25	40
Char. Association and All. *(Ammophilion australis)*		
Ammophila australis	2	3
Echinophora spinosa	1	+
Char. Cl. (*Euphorbio paraliae-Ammophiletea australis*)		
Elytrigia juncea	1	2
Eryngium maritimum	1	+
Pancratium maritimum	1	.
Medicago marina	+	.
Other species		
Cutandia maritima	+	1
Cynodon dactylon	.	1
Chamaesyce peplis	+	.
Hedypnois cretica	+	.
Salsola tragus	+	.

Rel. 1: Malta, Ramla Tat Torri, 26.6.1973; Rel. 2: Malta, Ghadira Bay, 26.6.1973

8.4 *Centaureo sphaerocephalae-Ononidetum ramosissimae* Br.-Bl. & M. Frei in M. Frei 1937 (Table 8.3)

In the innermost stands of the sandy coast, it is possible to found a psammophilous vegetation more mature than previous ones, referable to the *Centaureo spaerocephalae-Ononidetum ramosissimae*. Among the characteristic species of this syntaxon occurring in Maltese islands there is only *Ononis ramosissima*, while *Centaurea sphaerocephala* is absent altogether (Fig. 8.1b). As concerns the species of higher rank, they are represented by *Euphorbia terracina*, *Sixalis maritima*, *Lotus cytisoides*, *Scolymus hspanicus*, *Pancratium maritimum*, *Calystegia soldanella*, etc. This association, linked to rather stabilized retrodunal places, was previously known only from Sicily, where it was recorded by various authors (Frei 1937, Brullo and Furnari 1970, Bartolo et al. 1982, Brullo et al. 1988a, 2001b, etc.). In the Maltese archipelago it probably does not finds its optimal conditions, since here there are not dune formations very large inward, but the coastal sandy environmental are mostly represented by narrow belts in contact with the rock or the salt marshes.

8.5 *Salsolo tragi-Cakiletum maritimae* Costa & Mansanet 1981, nom. corr. hoc loco (Table 8.4)

Syn.: *Salsolo kali-Cakiletum aegyptiacae* Costa & Mansanet 1981.

Table 8.3 *Centaureo sphaerocephalae-Ononidetum ramosissimae*

Number of relevé	1	2	3	4
Plot size (m^2)	50	50	30	50
Plant cover (%)	50	90	80	100
Char. Association and All. (*Ononidion ramosissimae*)				
Ononis ramosissima	2	4	4	4
Euphorbia terracina	1	2	2	2
Sixalix maritima	2	2	+	+
Lotus cytisoides	1	2	.	+
Scolymus hispanicus	.	.	2	3
Char. Cl. (*Euphorbio paraliae-Ammophiletea australis*)				
Sporobolus arenarius	+	1	+	+
Pancratium maritimum	+	1	1	3
Cyperus kallii	+	.	1	.
Elytrigia juncea	.	.	1	1
Medicago marina	+	.	.	1
Eryngium maritimum	.	.	1	+
Echinophora spinosa	.	.	+	.
Calystegia soldanella	.	.	.	+
Other species				
Erodium laciniatum	1	1	+	+
Cutandia maritima	+	1	+	1
Cakile maritima	+	+	.	+
Lagurus ovatus subsp. *ovatus*	+	+	.	.
Rumex bucephalophorus subsp. *bucephalophorus*	1	1	.	.
Hedypnois cretica	1	+	.	.
Anisantha rigida	.	1	+	2
Silene colorata	1	.	1	+
Salsola tragus	+	.	.	+
Ononis variegata	.	.	+	+
Medicago littoralis	.	.	+	.
Euphorbia pinea	+	.	.	2
Pseudorlaya pumila	.	.	+	+

Rel. 1–3: Malta, Mellieha Bay, 26.6.1973; Rel. 4: Gozo, il Ramla Bay, 26.3.2016

Along the sandy coastline in the belt nearest the shore, where there is accumulation of organic material washed up by the waves, there is usually an annual psammophilous community of subnitrophilous type, floristically characterized by succulent terophytes having a summer phenology. The plants linked to this vegetation are very specialized and represented in particular by taxa belonging to the *Cakile maritima* cycle and by annual species of the *Salsola kali* group, previously ascribed to *Kali* genus. In the Maltese islands, as in the all Mediterranean territories, *Cakile maritima* is represented by subsp. *maritima*, which is replaced in the Atlantic coasts mainly by the subsp. *integrifolia* (Hornem) Greuter & Burdet (cfr. Ortiz 1993), while

Table 8.4 *Salsola tragi-Cakiletum maritimae*

Number of relevé	1	2	3	4	5	6
Plot size (m^2)	10	15	10	10	5	5
Plant cover (%)	70	50	40	70	60	60
Char. Association						
Cakile maritima	3	2	3	1	4	2
Char. All. (*Euphorbion peplidis*) and Cl. (*Cakiletea maritimae*)						
Salsola tragus	3	3	1	4	2	3
Polygonum maritimum	.	.	.	.	1	2
Chamaesyce peplis	+	.	.	.	.	.
Atriplex littoralis	.	.	.	1	.	.
Other species						
Beta maritima	.	.	.	1	+	+
Lolium rigidum subsp. *rigidum*	.	.	.	.	.	+
Elytrigia juncea	.	.	.	.	.	1
Soda inermis	.	+	.	.	.	.
Parapholis incurva	.	.	.	+		

Rel. 1–2: Malta, Mellieha Bay, 26.6.1973; Rel. 3: Malta, Golden Bay (Ghajn Tuffieha), 28.6.1973; Rel. 4: Gozo, Marsalforn, 13.4.1987; Rel. 5–6: Malta, Ramla Tat-Torri (Marfa), 22.5.2017

the genus *Salsola* is represented by *Salsola tragus* (=*Kali tragus*), species widespread mainly in the Mediterranean coasts, vicaried in the Atlantic shoreline by *S. kali* L. (=*K. turgidum*) and in the Black Sea by *Salsola pontica* (*K. ponticum*), as underlined by Brullo et al. (2015a, b).

From the nomenclatural point of view, this vegetation can be referred to *Salsolo kali-Cakiletum aegyptiacae*, association described by Costa and Mansanet (1981), but according to art. 45 this name was modified by Rivas-Martinez et al. (2002) in *Salsolo kali-Cakiletum maritimae* nom. mut. Unfortunately, the last name is uncorrect because *Salsola kali* is usually replaced in the Mediterranean area by *Salsola tragus*, therefore the name of the syntaxon proposed by Costa & Mansanet (1981), basing on the art. 43, must be correct in *Salsolo tragi-Cakiletum maritimae.*

It is an ephemeral association widespread in the Mediterranean, which represents the first step in the process of colonization of sandy coasts, adapting very well to the mobile substrates, subject to periodic flooding, especially during storm surges. It is in fact a purely pioneer association, which tolerates a certain edaphic salinity, having its optimum in the summer-autumn period. As concerns the Maltese islands, this association is currently very rare, since the sandy beaches are markedly anthropized due to urbanisation and tourist activities. From taxonomical viewpoint, this vegetation is included in the *Euphorbion peplidis* R.Tx. ex Oberd. 1952, alliance of the *Cakiletea maritimae* R.Tx. & Preising ex Br.-Bl. & R.Tx. 1952.

Chapter 9
Salt Marshes Vegetation

The coastal lagoons, usually localized in the inner part of sandy belt, are colonized by hygro-halophilous plant communities, characterized by succulent chamaephytes and nanophanerophytes, as well as by perennial helophytes and therophytes. They are splitted in several associations well distinct from floristic and physiognomic point of view, which are linked to well-defined edaphic conditions. The shrubby communities belong to *Sarcocornietea fruticosae*, the helophytic ones to *Juncetea maritimi*, while that one prevalently dominated by annual species are referable to *Thero Salicornietea.* The salt marshes are characterized also by submerged communities differentiated by algae and *Ruppia* species.

9.1 *Arthrocnemo macrostachyi-Juncetum subulati* Brullo & Furnari 1976 (Table 9.1)

Along the coast there are, especially in Malta, wetland depressions often very extended periodically flooded by rain and sea water. These salt marshes are colonized by markdly specialized halophilous communities, although not very rich floristic like those ones occurring in other countries of the Central Mediterranean. One of the most frequent associations linked to these habitat is the *Arthrocnemo glauci-Juncetum subulati* belonging to *Arthrocnemion macrostachyi* Rivas-Martínez & Costa 1984, alliance of *Sarcocornietea fruticosae* Br.-Bl. & R.Tx. ex A. & O. Bolòs 1950. This vegetation, characterized by *Arthrocnemum glaucum* and *Juncus subulatus*, was described initially from Sicily by Brullo and Furnari (1976) and after recorded from Spain (Rivas-Martinez et al. 1980, 1999; Rivas-Martinez and Costa 1984), Greece (Géhu et al. 1986a), Cyprus (Géhu et al. 1984b) and Tunisia (Barbagallo et al. 1990). In Malta this association is represented mainly by subass. *juncetosum subulati* Brullo & Furnari 1976, which is differentiated by the dominance of *Juncus subulatus* forming an dense populations on the surfaces subject

S. Brullo et al., *The Vegetation of the Maltese Islands*, Geobotany Studies,
https://doi.org/10.1007/978-3-030-34525-9_9

Table 9.1 *Arthrocnemo macrostachyi-Juncetum subulati*

Number of relevé	1	2	3	4	5	6	7
Plot size (m^2)	30	30	50	20	30	10	15
Plant cover (%)	100	100	100	80	90	90	90
Char. Association							
Juncus subulatus	5	5	5	1	2	+	+
Limonium virgatum	.	.	.	+	.	.	.
Char. All. (*Arthrocnemion macrostachyi*) and Cl. (*Sarcocornietea fruticosae*)							
Arthrocnemum macrostachyum	.	1	+	3	4	5	5
Limbarda crithmoides subsp. *longifolia*	1	.	2	1	+	2	2
Triglochin barrelieri	2	+	+	1	.	.	.
Other species							
Parapholis filiformis	3	+	2	3	1	+	+
Phragmites australis	1	1	2	.	.	.	.
Polypogon monspeliensis	.	.	+	+	+	.	.
Soda inermis	.	+	+	.	+	.	.
Cressa cretica	1	.	+	+	.	.	.
Spergularia marina	.	.	.	+	+	.	.
Salicornia patula	.	.	+	+	.	.	.
Symphyotrichum squamatum	.	.	+	.	.	.	.
Sonchus oleraceus	.	.	+	.	.	.	.
Suaeda vera	.	.	.	.	.	.	1
Juncus maritimus	.	.	.	.	.	.	1

Rel. 1–3: Malta, Mellieha Bay, 26.6.1973; Rel. 4–5: Malta, Salina Bay, 26.9.1985; Rel. 6–7: Malta, Marsaxlook, 16.11.1986

to a very long submersion. More rare is instead the subass. *arthrocnemetosum* Brullo & Furnari 1976, that replaced the previous one in the more arid stands having an edaphic higher salt concentration mainly in the summertime.

9.2 *Agropyro scirpei-Inuletum longifoliae* Brullo in Brullo et al. 1988 nom. corr. hoc loco (Table 9.2)

Syn.: *Agropyro scirpei-Inuletum crithmoidis* Brullo in Brullo et al. 1988.

This association as the previous one is a typically halophylous vegetation linked always to salt marshes, limited to soils not subject to flooding. It is characterized by the dominance of *Limbarda crithmoides* subsp. *longifolia* which grows usually together with *Elytrigia scirpea* (=*Agropyron scirpeum*). When Brullo (see Brullo et al. 1988a) described this syntaxon, used *Inula crithmoides* for give the nome, but according to Greuter (2003) the current name of this species is *I. crithmoides* subsp. *longifolia* (=*Limbarda crithmoides* subsp. *longifolia*). Therefore, the name "*Agropyro scirpei- Inuletum crithmoidis*"

Table 9.2 *Agropyro scirpei-Inuletum longifoliae*

Number of relevé	1	2	3	4
Plot size (m^2)	20	10	5	5
Plant cover (%)	90	100	70	80
Char. Association				
Elytrigia scirpea	+	3	1	2
Char. All. (*Inulion longifoliae*) and Cl. (*Sarcocornietea fruticosae*)				
Limbarda crithmoides subsp. *longifolia*	4	2	4	4
Arthrocnemum macrostachyum	.	4	2	.
Triglochin barrelieri	1	.	.	.
Other species				
Suaeda vera	.	2	1	1
Parapholis filiformis	3	.	.	2
Salicornia patula	1	.	.	2
Limonium virgatum	.	.	1	.
Juncus bufonius	1	.	.	.
Soda inermis	1	.	.	.
Cressa cretica	1	.	.	.
Juncus maritimus	.	+	.	.

Rel. 1: Malta, Mellieha Bay, 26.6.1973; Rel. 2–3: Malta, Marsaxlokk, 16.11.1986; Rel. 4: Malta, Salina Bay, 26.9.1985

must be corrected in *Agropyro scirpei-Inuletum longifoliae* (Art. 43). As emphasized by Brullo et al. (1988a), the association must be considered as the vicariant in the Central Mediterranean of the *Agropyro elongati-Inuletum longifoliae* Br.-Bl. in Br.-Bl., Roussine & Negre 1952 nom. corr. hoc loco (=*Agropyro elongati-Inuletum crithmoidis* Br.-Bl. in Br.-Bl., Roussine & Negre 1952), distributed in the West Mediterrannean. On the whole, the two associations are ecologically closely related, but diverging essentially for the occurrence of a different species of *Agropyron* (=*Elytrigia*). They were included by Brullo et al. (1988a) in a distinct alliance of *Sarcocornietea fruticosae,* represented by the *Inulion crithmoidis* Brullo & Furnari 1990, name that for the above reasons must be corrected in *Inulion longifoliae* Brullo & Furnari 1990 nom. corr. hoc loco.

9.3 *Halimiono portulacoidis-Suaedetum verae* (Molinier & Tallon 1970) Gèhu in Gèhu et al. 1984 (Table 9.3)

In the salt marsh where the surfaces are enriched by organic material grows a halophilous shrubby vegetation with subnitrophilous exigences, which is dominated by *Suaeda vera,* growing usually together with *Halimione portulacoides*. This plant community for its ecology and floristic set is referable to *Obioneto-Suaedetum fruticosae*, association described by Molinier and Tallon (1970) from southern

Table 9.3 *Halimiono portulacoidis-Suaedetum verae*

Number of relevé	1	2	3	4	5
Plot size (m^2)	10	10	5	10	10
Plant cover (%)	100	90	70	100	90
Char. Association and All. (*Suaedion verae*)					
Suaeda vera	5	5	4	5	4
Char. Cl. (*Sarcocornietea fruticosae*)					
Limbarda crithmoides subsp. *longifolia*	1	1	1	2	3
Halimione portulacoides	.	.	1	.	1
Elytrigia scirpea	.	.	1	.	.
Arthrocnemum macrostachyum	1	.	.	.	.
Other species					
Hordeum marinum	.	.	.	1	+
Lygeum spartum	.	.	.	.	+
Soda inermis	+	.	.	.	.

Rel. 1: Malta, Marsaxlokk, 16.11.1986; Rel. 2–5: Malta, Salina Bay, 26.9.1985

France and referred to *Staticion gallo-provincialis*. Although this syntaxon is validly described, its name must be corrected (Art. 43), since it is based on the misidentification of *Suaeda vera*. In fact, the authors attributed it to *S. fruticosa*, which is a species with a South-eastern Mediterranean distribution, absent in the West Mediterranean territories. The correct name for this association was proposed by Géhu et al. (1984a), sub *Halimiono-Suaedetum verae*, including it in the *Arthrocnemion glauci*. According to Brullo and Furnari (1990), this association must be more correctly included in *Suaedion verae*, alliance belonging to *Sarcocornietea fruticosae,* which ranges plant community with halo-nitrophilous requirements, dominated by *Suaeda vera*.

9.4 *Inulo longifoliae-Juncetum maritimi* Brullo in Brullo et al. 1988 nom. corr. hoc loco (Table 9.4)

Syn.: *Inulo crithmoidis-Juncetum maritimi* Brullo in Brullo et al. 1988.

The wet depressions characterized by sandy-clay-silty feebly salty soils are usually colonized by a community dominated by *Juncus maritimus*. It is a vegetation with halo-hygrophilous requirements referable to *Inulo crithmoidis-Juncetum maritimi*, association described by Brullo (see Brullo et al. 1988a) from Sicily and southern Italy. Floristically, it is differentiated by *Limbarda crithmoides* subsp. *longifolia* and therefore the name of this syntaxon must be corrected in *Inulo longifoliae-Juncetum maritimi* (art. 43). It is an association of *Juncetea maritimi* Br.Bl. in Br.-Bl., Roussine & Nègre 1952 closely related to *Puccinellio festuciformis-Juncetum maritimi* (Pignatti 1953) Géhu in Géhu et al. 1984a,

Table 9.4 *Inulo longifoliae-Juncetum maritimi*

Number of relevé	1	2	3	4	5
Plot size (m^2)	20	10	20	5	10
Plant cover (%)	100	100	100	70	90
Char. Association					
Limbarda crithmoides subsp. *longifolia*	+	1	+	+	1
Char. All. (*Juncion maritimi*) and Cl. (*Juncetea maritimi*)					
Juncus maritimus	5	4	3	4	5
Juncus acutus	+	1	.	+	+
Carex extensa	+	+	3	.	.
Centaurium tenuiflorum	.	1	+	.	.
Schenkia spicata	.	.	+	.	+
Other species					
Dittrichia viscosa	.	1	1	+	.
Hordeum marinum	.	+	+	.	+
Schedonorus arundinaceus subsp. *arundinaceus*	.	3	3	.	.
Polypogon monspeliensis	.	+	1	.	.
Suaeda spicata	1	.	.	+	.
Helminthotheca echioides	.	+	+	.	.
Euphorbia pinea	.	+	.	.	.
Parapholis filiformis	1	.	.	.	.
Phragmites australis	.	1	.	.	.
Soda inermis	+	.	.	.	.

Rel. 1: Malta, Mellieha Bay, 26.6.1973; Rel. 2–3: Malta, Ghajn Hadid Tower, 26.6.1973; Rel. 4–5: Malta, Marsaxlokk, 16.11.1986

association distributed in the northern Mediterranean salt marshes. Both associations was previously referred to *Juncetum maritimi* s.l. (cfr. Brullo and Furnari 1976).

9.5 *Juncetum maritimo-acuti* Horvatic 1934 (Table 9.5)

The *Inulo longifoliae-Juncetum maritimi* is replaced on sandy soils in correspondence of interdunal depressions by a vegetation physiognomically dominated by *Juncus acutus*, usually growing with *Juncus maritimus*. For its ecology and floristic set, this vegetation can be attributed to the *Juncetum maritimo-acuti*, psammophilous association linked to periodically submerged surfaces. It can be considered a transition community between the typical psammophilous associations of the *Euphorbio paraliae-Ammophiletea australis* and those halophilous ones of *Sarcocornietea fruticosae*. This association, previously known to the Adriatic and the Sicilian coasts, (see Horvatić 1934, 1963; Brullo et al. 1988a), shows some floristis and ecological affinities with other syntaxa, always characterized by

Table 9.5 *Juncetum maritimo-acuti*

Number of relevé	1	2	3	4
Plot size (m^2)	20	30	20	50
Plant cover (%)	80	90	90	80
Char. Association				
Juncus acutus	3	5	4	4
Char. All. (*Juncion maritimi*) and Cl. (*Juncetea maritimi*)				
Juncus maritimus	+	1	.	+
Centaurium tenuiflorum	.	.	+	+
Carex extensa	.	1	.	.
Other species				
Parapholis incurva	2	1	2	+
Limbarda crithmoides subsp. *longifolia*	3	+	2	2
Suaeda spicata	+	+	.	+
Sporobolus arenarius	1	1	+	.
Salicornia patula	+	1	.	1
Polypogon monspeliensis	.	.	+	+
Juncus bufonius	+	+	.	.
Symphyotrichum squamatum	+	.	1	.
Salsola tragus	.	.	+	.
Melilotus massanensis	.	.	.	1

Rel. 1: Malta, Dragunara, 24.6.1973; Rel. 2–4: Malta, Mellieha Bay, 26.6.1973

dominance of *Juncus acutus*, among them the *Limonio-Juncetum acuti* Brullo & Di Martino 1974 and the *Juncetum acuti* Molinier & Tallon 1970, all belonging to *Juncion maritimi*.

9.6 *Caricetum divisae* Br. -Bl. in Br.-Bl., Roussine & Nègre 1952 (Table 9.6)

In the mouth of some streams of Malta, another helophytic vegetation showing feebly halophilous exigences was surveyed. It is dominated by *Carex divisa*, stoloniferous species tending to constitute small populations, referable to *Caricetum divisae,* belonging to *Juncion maritimi.* This association, widespread in the Mediterranean territories (Braun-Blanquet et al. 1952; Valsecchi and Diana Corrias 1973; Biondi 1999; Brullo and Siracusa 2000; Rivas-Martínez et al. 2001; Biondi and Bagella 2005; etc.), occurs in the damp depressions with silty-clay soils. It differs from *Inulo longifoliae-Juncetum maritimi*, with which shows some relations especially in ecological terms, for the dominance of *Carex divisa* and for the less halophily and more marked hygrophily.

Table 9.6 *Caricetum divisae*

Number of relevé	1	2
Plot size (m^2)	20	30
Plant cover (%)	100	100
Char. Association		
Carex divisa	4	3
Char. All. (*Juncion maritimi*) and Cl. (*Juncetea maritimi*)		
Juncus maritimus	+	1
Centaurium tenuiflorum	1	+
Juncus acutus	.	1
Carex extensa	1	.
Schenkia spicata	1	.
Other species		
Polypogon monspeliensis	1	+
Hordeum marinum	2	+
Schedonorus arundinaceus subsp. *arundinaceus*	3	3
Dittrichia viscosa	1	1
Phragmites australis	.	+
Mentha pulegium	+	.
Narcissus tazetta	1	.
Juncus bufonius	.	1

Rel.1: Malta, Ghajin Hadid Tower, 26.6.1973; Rel.2: Malta, Salina Bay, 26.9.1985

9.7 *Suaedo spicatae-Salicornietum patulae* Brullo & Furnari ex Géhu & Géhu-Franck 1984 corr. Alcaraz, Ríos, De la Torre, Delgado & Inocencio 1998 (Table 9.7)

In the central part of the salt marshes subject to very long submersion periods, it is sometimes localized an annual vegetation, showing its optimum in the summertime. This is a marked halophilous community characterized by *Salicornia patula*. This succulent therophyte usually constitutes monophytic populations, giving a characteristic red-purple color to them.

The community characterized by this species can be attributed to *Suaedo spicatae-Salicornietum patulae*, association of *Salicornion patulae* Géhu & Géhu-Franck ex Rivas-Martinez 1990 widespread in the Mediterranean area, but quite rare in Malta (Brullo and Furnari 1976; Géhu et al. 1984a, b; Rivas-Martínez et al. 2001). According to literaure (Mucina et al. 2016), this vegetation is included in the class *Thero-Salicornietea* R. Tx. in R. Tx. et Oberd. 1958.

Table 9.7 *Suaedo spicatae-Salicornietum patulae*

Number of relevé	1	2	3	4	5
Plot size (m^2)	20	20	10	50	50
Plant cover (%)	80	70	70	50	50
Char. Association and All. (*Salicornion patulae*)					
Salicornia patula	5	4	4	3	5
Char. Cl. (*Thero-Salicornietea*)					
Suaeda spicata	+	+	.	1	+
Soda inermis	.	+	.	+	+
Cressa cretica	.	.	+	.	.
Atriplex littoralis	.	.	.	.	+
Other species					
Hordeum marinum	.	.	.	+	.
Juncus acutus	.	.	.	.	+
Sporobolus arenarius	.	.	.	+	.

Rel. 1–3: Malta, Salina Bay, 26.9.1985; Rel. 4–5: Malta, Mellieha, 26.6.1973

9.8 *Salsoletum sodae* Pignatti 1953 (Table 9.8)

In the sandy places near the salt marshes of Malta, on the deposit of the organic material beached by the waves it is frequent the *Salsoletum sodae*. This association is dominated by *Salsola soda* (= *Soda inermis* Fourr.), succulent therophyte, which grows together with other annual halo-nitrophytes, such as *Suaeda spicata* and *Atriplex prostrata*. The *Salsoletum sodae*, belonging to *Thero-Suaedion* alliance of *Thero-Salicornietea,* was recorded from various localities of the Italian peninsula, Sicily and Greece (cf. Pignatti 1953, 1966; Brullo and Furnari 1976; Géhu et al. 1984a, b).

9.9 *Suaedetum spicatae* Pignatti 1953 corr. hoc loco (Table 9.9)

Syn. *Suaedetum maritimae* Pignatti 1953.

In the salt marshes of Malta, the surfaces characterized by heavy silty-clay soils, rather humid in the summertime, are colonized by a vegetation dominated by *Suaeda spicata* (= *Suaeda maritima* auct. fl. medit.). This vegetation, likewise the previous one, shows halo-nitrophilous exigences too, since linked to slightly anthropized stands or however to soils with a significant nitrogenous component. From literature data, several associations characterized by *Suaeda maritima* s.l. was described in various country of Europe (Soò 1927, 1933; Wendelberger 1943; Tüxen 1974; Loidi et al. 1999; Ditè et al. 2015), Mediterranean (Braun-Blanquet 1933; Braun-Blanquet et al. 1952; Pignatti 1953; de Bolós and Vigo 1984; Rivas-Martínez et al. 2001, 2002; Tomaselli and Sciandrello 2016) and also North America (Conard 1935). Previously, the annual species of *Suaeda* were usually attributed to *S. maritima*,

Table 9.8 *Salsoletum sodae*

Number of relevé	1	2	3	4	5	6	7	8	9	10
Plot size (m^2)	10	25	10	5	10	10	5	20	10	10
Plant cover (%)	90	70	80	70	80	70	80	90	80	90
Char. Association										
Soda inermis	4	4	4	4	5	4	5	4	4	5
Char. All. (*Thero-Suaedion splendentis*) and Cl. (*Thero-Salicornietea*)										
Suaeda spicata	2	1	1	+	1	1	1	.	+	1
Atriplex littoralis	2	.	.	1	+	2	1	3	2	2
Salicornia patula	.	1	1	.	.	.	.	.	.	.
Cressa cretica	.	.	1	.	.	.	.	.	+	.
Other species										
Parapholis incurva	+	.	.	2	2	1	1	+	1	1
Spergularia marina	+	.	+	.	.	.	.	.	1	1
Symphyotrichum squamatum	.	.	.	.	.	+	+	.	.	+
Parapholis filiformis	.	+	1	.	.	.	.	.	.	.
Hordeum marinum	.	.	.	.	.	.	.	.	+	2
Atriplex halimus	.	.	.	1	.	.	+	.	.	.
Phragmites australis	.	.	.	.	.	.	.	.	.	+
Limbarda crithmoides subsp. *longifolia L.*	.	.	.	+	.	.	.	.	.	.
Sonchus oleraceus	.	.	.	.	.	.	+	.	.	.

Rel. 1–2: Malta, Mellieha Bay, 26.6.1973; Rel. 3–6: Malta, Marsaxlokk, 26.11.1986; Rel. 7: Malta, St. Paul Bay, 26.6.1973; Rel. 8–10: Malta, Salina Bay, 29.9.1985

Table 9.9 *Suaedetum spicatae*

Number of relevé	1	2	3	4
Plot size (m^2)	8	5	5	10
Plant cover (%)	80	90	90	70
Char. Association				
Suaeda spicata	4	5	5	4
Char. All. (*Thero-Suaedion splendentis*) and Cl. (*Thero-Salicornietea*)				
Soda inermis	2	1	2	1
Atriplex prostrata var. *salina*	2	1	2	+
Other species				
Limbarda crithmoides subsp. l*ongifolia*	.	1	1	1
Parapholis incurva	1	.	+	1
Symphyotrichum squamatum	.	1	1	.
Arthrocnemum macrostachyum	+	.	.	.

Rel. 1–3: Malta, Marsaxlokk, 16.11.1986; Rel. 3–4: Malta, Salina Bay, 29.9.1985

species complex currently splitted in several taxa, having a well-defined distribution (Pedrol and Castroviejo 1990; Mile and Walter 2003). In particular, *Suaeda maritima* is widespread in the Atlantic coast of Europe and North America, *S. spicata* replaces the previous one in the Central-western Mediterranean area, while *S. pannonica*, *S. prostrata* and *S. salsa* occurs in the East Europe and Euxinic countries. The Maltese coummunity for its ecology and floristic set can be attributed to *Suaedetum maritimae*, association described by Pignatti (1953) from the Venetian lagoon and later recorded by Brullo and Furnari (1976) from Sicily. According to art. 43 (ICPN), the name of this syntaxon must be corrected in *Suaedetum spicatae*. As concerns the *Atriplici salinae-Suaedetum spicatae* Bolòs & Vigo 1984 corr. Rivas-Martinez et al. 2002 (=*Atriplici hastatae-Suaedetum maritimae*), it represents a synonym of the previous association. The last syntaxon, which is clearly an illegitimate name, apart from the Iberian peninsula is reported also in Sicily and southern Italy by Sciandrello (2007) and Tomaselli and Sciandrello (2016).

9.10 *Cressetum creticae* Brullo & Furnari 1976 (Table 9.10)

This association, which is quite rare in Malta, replaces the *Suaedetum spicatae* in the muddy surfaces, quite wet even during the summer. The characteristic species of this association is *Cressa cretica*, halophilous therophyte linked to soils submerged for long time, usually placed in the central part of the salt marshes. The *Cressetum creticae*, belonging to *Thero-Suaedion*, initially described from Sicily by Brullo and Furnari (1976), was surveyed in other country of the Mediterranean, such as Spain (Galán de Mera et al. 1997), the Italian Peninsula (Tomaselli et al. 2010, 2011), Sardinia (Biondi et al. 2001), Corsica (Géhu and Biondi 1994), Tunisia (Barbagallo et al. 1990) and Croatia (Jasprica et al. 2015).

Table 9.10 *Cressetum creticae*

Number of relevé	1	2	3
Plot size (m^2)	50	30	50
Plant cover (%)	50	60	70
Char. Association			
Cressa cretica	3	4	4
Char. All. (*Thero-Suaedion splendentis*) and Cl. (*Thero-Salicornietea*)			
Salicornia patula	2	1	1
Soda inermis	+	+	.
Suaeda spicata	.	+	1

Rel. 1–3: Malta, Mellieha Bay, 26.6.1973

9.11 *Lamprothamnietum papulosi* Corillion 1957

According to Casha (2015), in the Ghadira Natural Reserve (North Malta) it is frequent *Lampothamnium papulosum*, annual Characea growing in retrodunal salt marshes, subject to summer drying. It is a pioneer and heliophilous species, linked to brackish, hyperhaline, subneutral to basic water, slightly or moderately deep (0.50–1 m), very light and rapidly warming, with sandy to sandy-loamy substrata. This halophilous hydrophyte usually grows together with other submerged cormophytes, such as *Ruppia drepanensis* and *Althenia orientalis*. The last species is recently reported for the same locality by Casha and Mifsud (2013). The populations dominated by *Lampothamnium papulosum* were described by Corillion (1957) as *Lamprothamnietum papulosi*, association with winter phenology, having a Mediterranean-Atlantic distribution. Besides, this author described a mixed association with *Lamprothamnium papulosum*, *Althenia filiformis* (=*A. orientalis*) and *Ruppia* sp.pl., considering it a transition community between the *Lamprothamnietum papulosi* and the *Ruppia* meadows. This association, belonging always to *Charetea intermediae*, is included in the *Charion canescentis* Krausch 1964, alliance regarding the stonewort swards occurring in the brackish waters. Previously, it was recorded from France (Corillion 1957; Delassus and Magnanon 2014), Iberian Peninsula (Rivas-Martinez et al. 2001), Sicily (Brullo and Furnari 1976), Italian Peninsula (Corbetta 1970; Caniglia et al. 1984), Sardinia (Corbetta and Lorenzoni 1976; Valsecchi 1983).

9.12 *Enteromorpho intestinalidis-Ruppietum maritimae* Westhoff ex R. Tx. & Böckelmann 1957 (Table 9.11)

In the brackish waters of coastal marshes of Malta is frequent a halo-hygrophilous vegetation, characterized by the occurrence of *Ruppia maritima*. This hydrophite usually colonizes the surfaces not subject to summer drying, sumberged by fairly deep water, not very salty. It constitute a dense populations having a winter-spring

Table 9.11 *Enteromorpho intestinalidis-Ruppietum maritimae*

Number of relevé	1	2
Plot size (m^2)	10	10
Plant cover (mq)	90	90
Char. Association		
Enteromorpha intestinalis	2	+
Char. All. (*Ruppion maritimae*) and Cl. (Ruppietea maritimae)		
Ruppia maritima	4	5
Other species		
Cheatomorpha linum	1	+

Rel. 1–2: Malta, Salina Bay, 9.4.1984

Table 9.12 *Ruppietum drepanensis*

Number of relevé	1	2
Plot size (m^2)	30	20
Plant cover (mq)	90	100
Char. Association and Cl. (*Ruppietea maritimae*)		
Ruppia drepanensis	5	5
Other species		
Enteromorpha intestinalis	.	+

Rel. 1: Malta, Marsaxlokk, 16.11.1986; Rel. 2: Malta, Salina Bay, 29.9.1985

cycle, often growing with some green seaweed, such as *Enteromorpha instestinalidis* and *Chaetomorpha linum*. This vegetation can be referred to *Enteromorpho intestinalidis-Ruppietum maritimae*, association widespread along the coasts of the Mediterranean territories (Rivas-Martinez et al. 2001; Biondi and Bagella 2005; Tomaselli et al. 2010; Pisanu et al. 2014; Jasprica et al. 2015). It is a kind of submerged rooted herbaceous vegetation of brackish waters belonging to *Ruppion maritimae* Br.-Bl. ex Westhoff in Bennema et al. 1943, alliance of *Ruppietea maritimae* J. Tx. ex Den Hartog et Segal 1964.

9.13 *Ruppietum drepanensis* Brullo & Furnari 1976 (Table 9.12)

The previous association is replaced in the salt marshes with surfaces dried in summer, by a vegetation with *Ruppia depranensis*. It represents also a monophytic population, referable to *Ruppietum drepanensis*, association having markedly thermophilous requirements, linked to water with high salt concentrations. It is recorded from various countries of the Mediterranean area and especially in Sicily, Sardinia and Spain (Brullo and Furnari 1976; Marchioni Ortu 1982; Rivas-Martinez et al. 1980, 2001; Martinez Parras and Peinado Lorca 1993).

Chapter 10
Hygrophilous Vegetation

In the Maltese Archipelago although it is represented by small islands with almost absent water basins and very limited watercourses, the freshwater plant communities are quite frequent and well diversified. Within the hygrophilous vegetation it is possible to distinguish perennial and annual associations linked to stagnant or slowly flowing waters, as well as to running waters. They are mostly differentiated by helophytes, pleustophytes and submerged hydrophytes, which belong to various phytosociological classes, such as *Magnocarici elatae-Phragmitetea australis*, *Molinio-Arrhenatheretea*, *Potametea* and *Lemnetea*.

10.1 *Phragmitetum australis* Savič 1926 (Table 10.1)

In the Maltese Islands, along the damp depressions, flooded by brackish or fresh water, there are often monophytic populations of *Phragmites australis*. This hydrophytic vegetation, linked to submerged soils, which also fits into short periods of soil drying, can be ascribed to *Phragmitetum australis*, which is the correct name of this association (Savič 1926; Šumberová 2011). According to literature (Bartolo et al. 1982), this vegetation, belonging to *Phragmition australis* Koch 1926, is very common in the Euro-Mediterranean territories.

10.2 *Typhetum domingensis* Brullo, Minissale & Spampinato 1994 (Table 10.2)

Along the watercourses limitedly to swamped stands, the *Phragmitetum australis* is replaced by the *Typhetum domingensis*. This association dominated by *Typha domingensis*, as well as the previous one, constitute a dense almost monophytic

S. Brullo et al., *The Vegetation of the Maltese Islands*, Geobotany Studies,
https://doi.org/10.1007/978-3-030-34525-9_10

Table 10.1 *Phragmitetum australis*

Number of relevé	1	2	3	4	5
Plot size (m^2)	20	10	20	15	10
Plant cover (%)	100	100	100	100	100
Char. Association and All. (*Phragmition australis*)					
Phragmites australis	5	5	5	5	5
Char. Cl. (*Magnocarici elatae-Phragmitetea australis*)					
Bolboschoenus maritimus subsp. *compactus*	+	.	+	.	.
Polygonum salicifolium	.	1	.	+	.
Schedonorus arundinaceus subsp. *arundinaceus*	.	.	+	.	1
Typha latifolia .	.	.	.	1	.
Other species					
Dittrichia viscosa	1	+	+	.	+
Symphyotrichum squamatum	.	.	+	1	.

Rel. 1: Malta, Salina Bay, 9.4.1984; Rel. 2: Malta, Bahrija Walley, 25.6.1973; Rel. 3: Malta, Melllieha Bay, 26.6.1973; Rel. 4: Malta, Bahrija Walley, 14.11.1984; Rel. 5: Malta, Salina Bay, 29.9.1985

Table 10.2 *Typhetum domingensis*

Number of relevé	1	2	3
Plot size (m^2)	10	10	5
Plant cover (%)	90	100	70
Char. Association			
Typha domingensis	5	5	4
Char. All. (*Phragmition australis*) and Cl. (*Magnocarici elatae-Phragmitetea australis*)			
Phragmites australis	1	1	2

Rel. 1–3: Malta, Girgenti Valley, 16.4.1987

population tied to muddy soils inundated with fresh water. Basing on literature data, this vegetation, apart from Malta where it is quite rare, occurs in various countries of the southern Europe (Brullo et al. 1994; Biondi et al. 2004; Biondi and Bagella 2005; Landucci et al. 2013).

10.3 *Polygono salicifolii-Phragmitetum australis* Barbagallo, Brullo & Furnari 1979 (Table 10.3)

In Malta along the perennial streams, such as that one of Bahrija Valley, a peculiar hygrophilous vegetation, quite rich floristically, occurs. A dominant role is played by *Polygonum salicifolium*, which grows together with *Phragmites australis*, *Veronica anagallis-aquatica*, *Rumex conglomeratus*, *Cyperus longus*, *Apium nodiflorum*, etc. In this places it was observed also *Colocasia esculenta*, naturalized species, which is linked to these wet environments. For its ecology and floristic set, this vegetation can

Table 10.3 *Polygono salicifolii-Phragmitetum australis*

Number of relevé	1	2	3	4	5	6
Plot size (m^2)	50	30	30	30	50	20
Plant cover (%)	100	100	100	100	100	100
Char. Association						
Polygonum salicifolium	3	4	4	4	5	4
Char. All. (*Phragmition australis*) and Cl. (*Magnocarici elatae-Phragmitetea australis*)						
Phragmites australis	1	1	1	2	1	2
Veronica anagallis-aquatica	2	1	+	1	+	+
Rumex conglomeratus	.	+	+	.	1	1
Cyperus longus subsp. *longus*	1	1	1	2	.	.
Apium nodiflorum	+	.	.	.	+	+
Nasturtium officinale	.	.	.	.	1	1
Schedonorus arundinaceus subsp. *arundinaceus*	.	.	+	.	.	.
Scrophularia auriculata	.	.	.	.	.	1
Other species						
Colocasia antiquorum	1	.	2	1	1	2
Scirpoides holoschoenus subsp. *australis*	3	1	2	1	.	.
Dittrichia viscosa	+	1	+	.	.	1
Cynodon dactylon	+	+	+	.	+	.
Agrostis stolonifera	1	2	2	2	.	.
Foeniculum piperitum	1	+	+	.	.	.
Arum italicum	.	+	1	+	.	.
Potentilla reptans	+	.	+	.	+	.
Panicum repens	+	.	+	.	.	.
Plantago major	.	.	1	+	.	.

Rel. 1–2: Malta, Bahrija Valley, 25.6.1973; Rel. 3–4: Malta, Bahrija Valley, 14.11.1984; Rel. 5–6: Malta, Bahrija Valley, 11.4.1987

be referred to *Polygono salicifolii-Phragmitetum australis*, association described from the South-eastern Sicily by Barbagallo et al. (1979). It is tied to the semi-submerged riverbanks of the short watercourses, affected by slightly flowing and clear fresh waters. Usually, it tends to constitute a dense and intricate vegetation in contact with the *Helosciadietum nodiflori*. The *Polygono salicifolii-Phragmitetum australis*, which until now was known only from Sicily (Brullo and Spampinato 1990; Minissale and Spampinato 1990) and Calabria (Maiorca et al. 2007), in the Maltese islands is very rare and only in the stands where it was surveyed, shows a certain spread.

Table 10.4 *Cyperetum distachyi*

Number of relevé	1	2
Plot size (m^2)	2	3
Plant cover (%)	100	90
Char. Association		
Cyperus distachyos	5	5
Other species		
Melilotus messanensis	+	.

Rel. 1–2: Gozo, Wied il-Mielah, 13.4.1987

10.4 *Cyperetum distachyi* O. Bolós & R. Molinier 1984 (Table 10.4)

Another hygrophilous community is represented by populations of *Cyperus distachyos* colonizing small dampy surfaces near the sea. It is a very rare vegetation, observed only in the northern coast of Gozo, where it is linked to muddy submerged soils, places near the sources. For the dominance of *Cyperus distachyos*, it can be referred to *Cyperetum distachyi*, association of *Scirpion compacti*, Dahl & Hadac 1941 corr. Rivas-Martínez et al. 1980 alliance of *Phragmito-Magnocaricetea* Klika in Klika & Novák 1941, described by de Bolós and Molinier (1984) from Balearic Islands. This community, showing sub-halophilous requirements, was recorded also in Tunisia by Barbagallo et al. (1990), in Sicily by Licandro et al. (2011), in Spain by de Bolós (1996) and Alcaraz Ariza (2011).

10.5 *Bolboschoenetum compacti* Van Langend. 1931 corr. Bueno & F. Prieto in Bueno 1997 (Table 10.5)

Along watercourses, particularly in the flat surfaces near their mouth, was observed a vegetation characterized by the dominance of *Bolboschoenus maritimus* subsp. *compactus*. It shows sub-halophilous requirements, occurring on silty-clay soils, which tolerates also short emersion periods during the summertime. This community for its floristical and ecological features is referable to *Bolboschoenetum compacti*, association widespread in the Euro-Mediterranean territories, mainly near the coast (cfr. Brullo et al. 1988a; Bueno 1997; Rivas-Martínez et al. 2011).

10.6 *Helosciadietum nodiflori* Maire 1924 (Table 10.6)

This association, quite widespread in the Maltese islands, grows along the streams with perennial fresh waters and is characterized by *Apium nodiflorum* (= *Helosciadium nodiflorum*) usually mixed with *Nasturtium officinale*. It is an

Table 10.5 *Bolboschoenetum compacti*

Number of relevé	1	2	3	4	5
Plot size (m^2)	20	30	20	50	50
Plant cover (%)	100	100	100	100	100
Char. Association and All. (*Scirpion compacti*)					
Bolboschoenus maritimus subsp. *compactus*	4	4	5	5	5
Char. Cl. (*Magnocarici elatae-Phragmitetea australis*)					
Schedonorus arundinaceus subsp. *arundinaceus*	2	2	1	.	.
Phragmites australis	2	1	.	.	.
Rumex conglomeratus	.	1	2	.	.
Veronica anagallis-aquatica	.	.	.	+	.
Epilobium tetragonum	.	.	.	.	1
Alisma lanceolatum	.	.	1	.	.
Other species					
Scirpoides holoschoenus subsp. *australis*	2	2	.	.	.
Dittrichia viscosa	1	1	.	.	.
Polypogon monspeliensis	.	+	.	.	+
Lythrum junceum	1	2	.	.	.
Beta maritima	1	.	.	.	.
Bromus hordeaceus	1	.	.	.	.
Ranunculus macrophyllus	.	.	.	.	1

Rel. 1–2: Malta, Gnejna Valley, 12.4.1984; Rel. 3: Malta, Gnejna Valley, 11.4.1987; Rel. 4–5: Malta, Wied I-Arkata, 16.4.1987

amphibious community often favored by a certain pollution or however eutrophication of water, due to the nitrogenous constituent coming from the sewage. It usually is localized in the central part of the watercourse, taking contact outwards with the community of the *Phragmition australis*. From the syntaxonomical point of view, it belongs to *Nasturtion officinalis* Gèhu & Gèhu-Frank 1987, alliance of the *Nasturtio officinali-Glyceritalia fluitantis* Pignatti 1953. The association is distributed in the all Mediterranean territories (cf.. Maire 1924; Braun-Blanquet et al. 1952; Brullo and Spampinato 1990; Venanzoni and Gigante 2000; Pirone et al. 2003; Venanzoni et al. 2003; Biondi et al. 2004; Pedrotti 2008; Lastrucci et al. 2010; Mereu et al. 2010; Rivas-Martínez and coauthores 2011; Corbetta and Pirone 1989).

10.7 *Apio nodiflori-Glycerietum plicatae* Brullo & Spampinato 1988 (Table 10.7)

In the river places with deep and calm water the *Helosciadietum nodiflori* is replaced by *Apio nodiflori-Glycerietum plicatae* association linked to submerged surfaces with clay or marl substrates. The association is characterized by *Glyceria plicata*, which usually constitutes a dense tangle partially floating, mixed to other

Table 10.6 *Helosciadetum nodiflori*

Number of relevé	1	2	3	4	5	6	7	8	9	10	11	12
Plot size (m²)	20	3	10	5	6	5	5	10	15	10	10	4
Plant cover (%)	100	70	100	100	70	70	70	80	90	90	100	100
Char. Association and All. (*Nasturtion officinalis*)												
Apium nodiflorum	4	3	2	5	3	1	1	3	2	3	3	3
Nasturtium officinale	2	3	4	2	3	4	4	3	4	2	2	1
Char. Cl. (*Phragmito-Magnocaricetea*)												
Rumex conglomeratus	2	.	2	1	.	1	2	1	1	2	1	2
Cyperus longus subsp. longus	+	.	.	.	.	.	1	1	+	1	1	.
Veronica anagallis-aquatica	1	.	.	.	.	.	.	.	+	3	.	3
Polygonum salicifolium	.	.	.	.	.	.	.	2	2	2	3	.
Phragmites australis	+	.	1	.	.	.	.	2	.	.	.	.
Scrophularia auriculata	.	.	.	1	.	.	.	.	.	+	2	.
Schedonorus arundinaceus subsp. *arundinaceus*	.	.	.	.	.	.	.	.	1	+	1	.
Other species												
Arum italicum	.	.	.	.	.	.	.	+	1	1	+	.
Scirpoides holoschoenus subsp. *australis*	.	.	.	.	.	1	.	+	.	1	2	.
Colocasia antiquorum	.	.	.	.	.	.	.	+	.	1	1	.
Lythrum junceum	.	.	.	.	.	+	.	.	.	.	.	2
Verbena officinalis	.	+	.	.	.	+	.	.	.	.	.	.
Agrostis stolonifera	2	1	.	.	.	.	.	.	.	.	.	.
Samolus valerandi	.	1	.	.	1	.	.	.	.	.	.	.
Galium aparine	.	.	2	+	.	.	.	.	.	.	.	.
Ranunculus sardous	.	.	1	+	.	.	.	.	.	.	.	.
Dittrichia viscosa	+	.	.	.	.	.	.	.	.	.	.	.
Polypogon monspeliensis	1	.	.	.	.	.	.	.	.	.	.	.
Potentilla reptans	.	.	.	.	.	.	.	.	.	.	1	.
Mentha pulegium	+	.	.	.	.	.	.	.	.	.	.	.
Juncus bufonius	.	.	.	.	.	.	.	.	.	.	.	1

Rel. 1–2: Gozo, Xlendi Valley, 29.6.1973; Rel. 3–4: Gozo, Xlendi Valley, 11.4.1984; Rel. 5: Gozo, Dwejra, 11.4.1984; Rel. 6: Malta, St. Paul Bay, 11.4.1987; Rel. 7–11: Malta, Bahrrija Valley, 11.4.1987; Rel. 12: Malta, Fiddien Valley, 16.4.1987

amphibious plants, such as *Apium nodiflorum*, *Nasturtium officinale*, *Veronica anagallis-aquatica*, *Alisma lanceolatum*, *Eleocharis palustris*, *Schedonorus arundinaceus*, etc. This vegetation is quite rare in Malta, replacing at the edge of the watercourses the *Ranunculetum trichophylli* submerged vegetation. Previously the *Apio nodiflori-Glycerietum plicatae* was only recorded from Sicily by Brullo and Spampinato (1990) and Fici and Gianguzzi (2011).

Table 10.7 *Apio nodiflori-Glycerietum plicatae*

Number of relevé	1	2	3	4	5
Plot size (m^2)	5	5	4	5	5
Plant cover (%)	70	0	90	0	70
Char. Association					
Glyceria plicata	4	3	5	3	3
Char. All. (Nasturtion officinalis)					
Apium nodiflorum	+	+	.	2	2
Nasturtium officinale	.	2	+	.	.
Char. Cl. (Magnocarici elatae-Phragmitetea australis)					
Veronica anagallis-aquatica	1	2	1	.	+
Alisma lanceolatum	1	.	1	2	3
Rumex conglomeratus	+	.	+	1	1
Eleocharis palustris	.	.	1	1	+
Cyperus longus subsp. *longus*	.	.	1	2	3
Schedonorus arundinaceus subsp. *arundinaceus*	.	.	+	1	+
Epilobium tetragonum	.	.	+	.	+
Other species					
Ranunculus macrophyllus	.	1	+	.	2

Rel. 1–5: Malta, Fiddien Valley, 16.4.1987

10.8 *Cyperetum longi* Micevski 1957 (Table 10.8)

In Malta along the outer riversides of watercourses, mainly in the stands subject to short periods of submersion, occurs a helophitic vegetation physiognomically characterized by *Cyperus longus*. It must be referred to *Cyperetum longi* Micevski 1957, association of *Magnocaricion elatae* Koch 1926, very rich in hygrophites, such as *Rumex conglomeratus*, *Phragmites australis*, *Schedonorus arundinaceus*, *Carex cuprina*, *Alisma lanceolatum*, *Eleocharis palustris*, etc. The association can be considered as a termophilous southern vicariant of *Caricetum elatae* W. Koch 1926, widespread in the rivers of central and northern Mediterranean. The *Cyperetum longi* was recorded from Sicily and several Euro-Mediterranean and Asiatic countries by several authors (Tansley 1965; Barbagallo et al. 1979; Brullo and Spampinato 1990; Venanzoni and Gigante 2000; Venanzoni et al. 2003; Jasprica et al. 2003; Mehrvarz et al. 2016).

10.9 *Cypero longi-Caricetum cuprinae* R. Tx. ex Díaz Gonzalez & Fernández-Prieto 1994 (Table 10.9)

Along the canals of Malta, which are submerged during the winter period, the *Cyperetum longi* is replaced by another helophitic vegetation dominated by *Carex cuprina* (=*C. otrubae*). For its ecological requirements and floristic set, it can be

Table 10.8 *Cyperetum longi*

Number of relevé	1	2	3	4	5	6	7	8	9	10
Plot size (m^2)	30	10	20	20	10	15	15	20	10	10
Plant cover (%)	100	100	100	100	100	100	100	100	100	100
Char. Association										
Cyperus longus subsp. *longus*	4	4	4	3	5	4	3	3	4	4
Char. All. (*Magnocaricion elatae*) and Cl. (*Magnocarici elatae-Phragmitetea australis*)										
Rumex conglomeratus	2	+	2	2	+	3	3	1	2	.
Schedonorus arundinaceus subsp. *arundinaceus*	.	.	.	.	1	1	2	3	2	2
Carex cuprina	.	.	.	.	.	.	+	1	1	.
Alisma lanceolatum	.	.	.	.	.	.	1	2	2	.
Phragmites australis	.	2	1	3	.	.	.	.	.	.
Epilobium tetragonum	.	.	.	.	.	.	.	.	+	2
Eleocharis palustris	.	.	.	.	.	.	.	.	+	.
Veronica anagallis-aquatica	2	.	.	.	.	.	.	.	.	.
Apium nodiflorum	+	.	.	.	.	.	.	.	.	.
Polygonum salicifolium .	1	.	.	.	.	.	.	.	.	.
Other species										
Scirpoides holoschoenus subsp. *australis*	1	2	1	2	.	.	.	.	.	.
Ranunculus macrophyllus	.	.	.	.	1	1	2	.	.	.
Potentilla reptans	1	.	.	.	+	1	.	.	.	.
Arum italicum	.	1	+	1	.	.	.	.	.	.
Galium aparine	.	.	2	1	.	.	.	.	.	.
Carex divisa	1	.	.	.	.	.	.	.	.	.
Dittrichia viscosa	1	.	.	.	.	.	.	.	.	.
Geranium dissectum	.	.	.	.	.	.	.	+	.	.
Panicum repens	.	.	.	.	1	.	.	.	.	.
Juncus acutus	1	.	.	.	.	.	.	.	.	.
Agrostis stolonifera	2	.	.	.	.	.	.	.	.	.
Calocasia antiquorum	2	.	.	.	.	.	.	.	.	.
Populus alba	.	.	+	.	.	.	.	.	.	.

Rel. 1: Malta, Bahrija Valley, 25.6.1973; Rel. 2: Malta, Gnejna Valley, 11.4.1987; Rel. 1: Malta, Bahrija Valley, 11.4.1987; Rel. 3–4: Malta, Bahrija Valley, 25.6.1973; Rel. 5–9: Malta, Fiddien Valley, 25.6.1973; Rel. 10: Malta, Wied l-Arkata, 16.4.1987

attributed to *Cypero longi-Caricetum cuprinae*, association described as provisional from the Iberian Peninsula by Tüxen in Tüxen and Oberdorfer (1958) and named *Cypero- Caricetum otrubae*. It was later examined by de Bolós 1967 and Díaz-Gonzales 1975, and validated by Díaz Gonzalez and Fernández-Prieto (1994). It was also surveyed in Sicily by Barbagallo et al. (1979) and Brullo & Spampinato (1990), always on the bottom of canals near the rivers in the eastern part of the island. In this vegetation are frequent several other helophytes, among them *Cyperus longus*, *Schedonorus arundinaceus*, *Phragmites australis*, *Rumex conglomeratus*, *Carex*

Table 10.9 *Cypero longi-Caricetum cuprinae*

Number of relevé	1	2	3	4	5	6	7	8	9	10	11	12
Plot size (m^2)	20	20	20	25	30	30	15	10	10	20	30	10
Plant cover (%)	100	100	100	100	100	100	100	100	100	100	100	100
Char. Association												
Carex cuprina	4	5	5	5	4	3	3	4	3	4	2	2
Cyperus longus subsp. *longus*	+	.	.	.	.	1	2	.	+	.	3	3
Char. All. (*Magnocaricion elatae*) and Cl. (*Magnocarici elatae-Phragmitetea australis*)												
Schedonorus arundinaceus subsp. *arundinaceus*	2	3	+	1	+	4	3	2	2	2	2	3
Phragmites australis	1	.	+	1	1	1	1	.	2	1	.	.
Rumex conglomeratus	2	.	.	+	+	.	1	2	2	2	3	2
Carex distans	.	.	.	.	.	.	.	.	.	+	+	1
Epilobium tetragonum	.	.	.	.	.	.	.	.	.	1	.	1
Alisma lanceolatum	.	.	.	.	.	.	.	.	.	.	.	1
Epilobium parviflorum	.	.	.	.	2	.	.	.	.	.	.	.
Typha latifolia	.	.	.	.	1	.	.	.	.	.	.	.
Eleocharis palustris	.	.	.	.	.	.	.	.	.	.	.	+
Other species												
Scirpoides holoschoenus subsp. *australis*	1	.	+	.	2	.	2	1	1	.	.	.
Lythrum junceum	+	.	.	1	1	+	.	.	.	.	.	.
Dittrichia viscosa	1	+	.	2	1	.	.	.	.	.	.	.
Polypogon monspeliensis	+	+	.	+	+	.	.	.	.	.	.	.
Carex divisa	.	.	.	.	.	.	+	+	1	.	.	1
Agrostis stolonifera	.	.	.	.	.	+	1	1	.	.	.	.
Bromus hordeaceus	1	+	.	.	.	.	.	.	.	+	.	.
Ranunculus macrophyllus	.	.	.	.	.	.	.	.	.	+	+	+
Schenkia spicata	.	+	.	+	+	.	.	.	.	.	.	.
Helminthotheca echioides	.	+	.	1	1	.	.	.	.	.	.	.
Narcissus tazetta	.	.	.	1	+	.	.	.	.	.	.	.
Mentha pulegium	.	.	.	+	+	.	.	.	.	.	.	.

(continued)

Table 10.9 (continued)

Number of relevé	1	2	3	4	5	6	7	8	9	10	11	12
Plot size (m²)	20	20	20	25	30	30	15	10	10	20	30	10
Plant cover (%)	100	100	100	100	100	100	100	100	100	100	100	100
Panicum repens	.	.	.	1	.	.	.	.	.	+	.	.
Plantago lanceolata	.	1	.	.	.	.	.	.	.	.	.	.
Pulicaria dysenterica	.	.	.	.	.	.	.	.	.	1	.	.
Arum italicum	.	.	.	.	.	.	.	.	.	+	.	.
Juncus acutus	.	.	.	.	+	.	.	.	.	.	.	.

Rel. 1: Malta, Gnejna Bay, 12.4.1984; Rel. 2–3: Malta, Il Fawwara, 27.6.1973; Rel. 4–5: Malta, Gnejna Bay, 28.6.1973; Rel. 7: Malta, il-Karraba, 12.4.1984; Rel. 6: Malta, il-Karraba, 12.4.1984; Rel. 7–9: Malta, St. Paul Bay, 11.4.1987; Rel. 10: Malta, Girgenti, 16.4.1987; Rel. 11–12: Malta Fiddien Valley, 16.4.1987

Table 10.10 *Caricetum hispidae*

Number of relevé	1	2
Plot size (m²)	10	20
Plant cover (%)	100	100
Char. Association		
Carex hispida	5	3
Char. All. (*Magnocaricion elatae*) and Cl. (*Magnocarici elatae-Phragmitetea australis*)		
Schedonorus arundinaceus subsp. *arundinaceus*	2	4
Phragmites australis	1	1
Carex cuprina	1	.
Epilobium parviflorum	+	.
Other species		
Scirpoides holoschoenus subsp. *australis*	2	1
Dittrichia viscosa	.	+
Limbarda crithmoides subsp. *longifolia*	.	+
Foeniculum piperitum	.	+

Rel. 1: Malta, Mtahleb. 26.9.1985; Rel. 2: Malta, il Fawwara, 27.6.1973

distans, etc. With the decreasing of edaphic humidity it is normally replaced by communities of the *Potentillion anserinae*.

10.10 *Caricetum hispidae* Brullo & Ronsisvalle 1975 (Table 10.10)

In the dampy places, represented by riverside with slightly salty waters or otherwise rich in minerals, the *Cyperetum longi* is replaced by the *Caricetum hispidae*. This association is a strictly coastal helophytic community, occurring mainly near the mouth of the river or in the canals. It is characterized by *Carex hispida* species physiognomically dominant, which grows together with few species of the *Phragmitetea*, such as *Schedonorus arundinaceus*, *Phragmites australis*, *Carex cuprina* and *Epilobium parviflorum*. The association, quite rare in Malta, occurs also in Sicily, where it was described by Brullo and Ronsisvalle (1975), in southern Spain and Italy (Brullo and Furnari 1976; Cirujano 1980; Géhu and Biondi 1988).

10.11 *Festuco arundinaceae–Caricetum distantis* J. Duvign. 1967 (Table 10.11)

It is a relatively rare association, observed only in some streams of Malta, where it grows in periodically flooded stands with clay soil, limitedly to the more shady and fresh surfaces. It is characterized by *Carex distans* and *Schedonorus arundinaceus*, which grow together other species of *Potentillion anserinae* R.Tx. 1947 and *Potentillo-Polygonetalia avicularis* R.Tx. 1947, such as *Rumex conglomeratus*, *Potentilla reptans*, *Carex cuprina*, *Epilobium tetragonum*, *Agrostis stolonifera* and *Carex divisa*, as well as of the *Molinio-Arrhenatheretea* R.Tx. 1937, such as *Ranunculus macrophyllus*, *Scirpoides australis*, *Trifolium pratense* and *Pulicaria dysenterica*. According to literature (Duvigneaud 1967, Billy 2000, de Foucault and Catteau 2012), this association occurs also in France and Germany.

10.12 *Phalarido coerulescentis-Schedonoretum arundinacei* ass. nov. hoc loco (Table 10.12)

Holotypus: ril. 4 hoc loco.

The canals and wetlands periodically submerged by running water, especially on outcrops of clay and clay-marl, habitat quite spread in the Maltese Islands, are colonized mainly in the less sloped surfaces by a perennial hygrophilous vegetation physiognomically characterized by *Schedonorus arundinaceus* and *Phalaris caerulescens*. Like the previous one, this community for its floristic set belongs to the *Potentillion anserinae*. Among the species of this alliance and higher order occurs *Potentilla reptans*, *Mentha pulegium*, *Plantago major*, *Scirpoides australis* and *Poa trivialis*. It differs from the *Festuco arundinaceae-Caricetum distantis*, apart from its ecology and the absence of *Carex distans*, also for the occurrence of *Phalaris coerulescens*, sometimes dominant, which allows to differentiate a new association, proposed as *Phalarido coerulescentis-Schedonoretum arundinacei*.

Table 10.11 *Festuco arundinaceae-Caricetum distantis*

Number of relevé	1	2	3	4	5	6	7
Plot size (m^2)	10	10	5	10	4	5	5
Plant cover (%)	100	100	100	80	100	80	70
Char. Association							
Carex distans	3	3	2	3	4	3	3
Char. All. (*Potentillion anserinae*) and Ord. (*Potentillo anserinae-Polygonetalia avicularis*)							
Rumex conglomeratus	2	2	2	1	1	2	+
Schedonorus arundinaceus subsp. *arundinaceus*	1	1	3	.	1	.	1
Potentilla reptans	.	.	.	2	2	1	2
Carex cuprina	1	+	1	.	.	.	.
Epilobium tetragonum	+	1	1	.	.	.	.
Agrostis stolonifera	.	.	.	.	1	1	+
Carex divisa	.	.	.	.	.	.	1
Char. Cl. (*Molinio-Arrhenatheretea*)							
Ranunculus macrophyllus	1	+	1	3	.	.	.
Oenanthe globulosa	.	.	.	1	2	2	2
Trifolium pratense subsp. *pratense*	.	.	.	+	2	1	1
Pulicaria dysenterica	.	1	1	.	.	.	.
Other species							
Ranunculus sardous	+	.	1	+	1	1	.
Carex spicata	2	3	3	.	.	.	.
Juncus bufonius	.	.	.	.	+	1	1
Arum italicum	+	2	+	.	.	.	.
Geranium dissectum	+	+	1	.	.	.	.
Phragmites australis	2	1	1	.	.	.	.
Bromus hordeaceus	1	.	+	.	.	.	.
Acanthus mollis	1	+	.	.	.	.	.
Trifolium resupinatum	.	.	.	.	+	.	+
Alisma lanceolatum	.	.	.	.	.	+	.

Rel. 1–3: Malta, Girgenti Valley, 14.4.1987; Rel. 4–7: Malta, il Wied tal-Fiddien (Rabat), 16.4.1987

10.13 *Schedonoro arundinacei-Caricetum divisae* ass. nov. hoc loco (Table 10.13)

Holotypus: ril. 6 hoc loco.

On surfaces submerged for short periods, such as those ones near the watercourses and in the damp depressions characterized by clay soils, a hygrophilous subnitrophilous vegetation grows, which is dominated by *Schedonorus arundinaceus* and *Carex divisa*. It is a community linked to habitats more or less humid for all the year, where there is a certain occurrence of organic materials, as a result of decantation by eu-hypertrophic waters. The submentioned species are usually associated with species of *Potentillion anserinae* and *Molinio-*

Table 10.12 *Phalarido coerulescentis-Schedonoretum arundinacei*

Number of relevé	1	2	3	4	5
Plot size (m^2)	10	10	15	8	10
Plant cover (%)	100	100	100	100	100
Char. Association					
Phalaris coerulescens	3	2	4	3	1
Char. All. (*Potentillion anserinae*) and Ord. (*Potentillo anserinae-Polygonetalia avicularis*)					
Schedonorus arundinaceus subsp. *arundinaceus*	2	5	2	3	5
Potentilla reptans	+	.	1	+	.
Mentha pulegium	.	+	1	2	.
Plantago major	.	.	.	+	+
Ranunculus macrophyllus	+	.	.	.	.
Char. Cl. (*Molinio-Arrhenatheretea*)					
Scirpoides holoschoenus subsp. *australis*	+	1	1	.	+
Poa trivialis	.	+	.	.	1
Other species					
Avena barbata	2	+	.	1	+
Ranunculus muricatus	2	+	.	+	.
Phragmites australis	1	.	2	2	.
Medicago arabica	1	.	2	2	.
Ononis mitissima	+	1	.	.	.
Phalaris paradoxa	+	+	.	.	.
Trifolium resupinatum	1	.	1	.	.
Geranium dissectum	+	.	.	+	.

Rel. 1–2: Malta, Girgenti Valley, 16.4.1987; Rel. 3–5: Gozo, il- Harrax, 14.4.1987

Arrhenatheretea, such as *Scirpoides australis*, *Agrostis stolonifera*, *Mentha pulegium*, *Dactylis glomerata*, *Potentilla reptans, Rumex conglomeratus* etc. The high covering values of *Carex divisa* emphatized a certain edaphic salinity, approaching this vegetation to *Caricetum divisae* Br.-Bl. in Br.-Bl., Roussine & Nègre 1952, halo-hydrophilous association of the *Juncetea maritimi*, surveyed in Malta too, but in the salt marshes. On the whole, the vegetation at issue is ecologically and floristically well differentiated from the latter, therefore it is proposed as a new association, named *Schedonoro arundinacei-Caricetum divisae*. It can be considered as a transition association between the strictly halophilous communities of *Juncion maritimi* and halo-subnitrophilous those ones of the *Potentillion anserinae*.

10.14 *Potentillo reptantis-Panicetum repentis* ass. nov. hoc loco (Table 10.14)

Holotypus: ril. 4 hoc loco.

Table 10.13 *Schenodoro arundinacei-Caricetum divisae*

Number of relevé	1	2	3	4	5	6	7	8	9	10	11	12	13	14
Plot size (m^2)	30	50	20	20	30	10	10	25	30	50	50	10	20	10
Plant cover (%)	100	100	100	90	100	100	100	100	100	100	100	100	100	90
Char. Association														
Carex divisa	4	2	4	3	4	3	5	5	5	5	4	4	5	4
Char. All. (*Potentillion anserinae*) and Ord. (*Potentillo anserinae-Polygonetalia avicularis*)														
Schedonorus arundinaceus subsp. *arundinaceus*	3	4	.	2	2	2	1	2	2	+	.	+	3	1
Agrostis stolonifera	.	+	1	.	1	2	.	.	+	1	1	.	+	+
Rumex conglomeratus	.	.	2	1	1	2	+	+	.	2	1	.	.	.
Mentha pulegium	+	.	.	.	+	1	.	.	+	+	1	.	.	+
Potentilla reptans	.	.	.	+	.	1	.	.	.	1	1	+	.	.
Plantago major	.	.	.	.	.	.	+	+	.	.	.	.	+	.
Ranunculus macrophyllus	.	.	.	.	.	.	.	.	.	.	.	.	+	1
Char. Cl. (*Molinio-Arrhenatheretea*)														
Scirpoides holoschoenus subsp. *australis*	.	.	2	3	2	3	1	1	.	+	2	2	.	.
Lythrum junceum	.	+	.	.	.	.	.	.	.	1	1	1	.	.
Oenanthe globulosa	.	.	.	.	.	.	.	1	.	2	2	1	.	.
Poa trivialis	1	.	.	.	.	.	.	.	.	.	.	.	+	2
Other species														
Trifolium fragiferum	.	+	.	.	.	.	+	+	+	.	.	.	.	.
Ranunculus muricatus	.	.	.	.	.	.	+	.	.	2	1	2	.	.
Juncus maritimus	+	.	+	.	1	.	.	.	.	.	.	.	.	.
Geranium dissectum	.	.	.	.	.	.	.	.	.	2	1	2	.	.
Phragmites australis	.	1	1	2	.	.	.	.	.	.	.	.	.	.
Centaurium tenuiflorum	1	+	.	.	.	.	.	.	.	.	.	.	.	.
Polypogon monspeliensis	1	+	.	.	.	.	.	.	.	.	.	.	.	.
Dittrichia viscosa	1	1	.	.	.	.	.	.	.	.	.	.	.	.

Helminthotheca echioides	+	+	.	.	.	.	.	.	.	.	.	.	.	.
Juncus acutus	.	1	.	.	.	.	.	.	.	.	.	.	.	.
Hordeum marinum	1	.	.	.	.	.	.	.	.	.	.	.	.	.
Sonchus oleraceus	.	.	.	.	.	.	.	.	+	.	.	.	.	.
Daucus carota	+	.	.	.	.	.	.	.	.	.	.	.	.	.
Bromus racemosus	1	.	.	.	.	.	.	.	.	.	.	.	.	.

Rel. 1–2: Malta, Mistra Valley, 11.4.1987; Rel. 3–6: Malta, St. Paul Bay, 11.4.1987; Rel. 7–8; Malta, Fiddien (Rabat), 16.4.1987; Rel. 9; Gozo, Dwejra, 13.4.1987; Rel. 10–12; Malta, Fiddien (Rabat), 16.4.11987; Rel. 13–14; Malta, Wied Arkata (Mosta) 16.4.11987

Table 10.14 *Potentillo reptantis-Panicetum repentis*

Number of relevé	1	2	3	4	5
Plot size (m^2)	8	10	5	10	10
Plant cover (%)	100	100	100	100	100
Char. Association					
Panicum repens	5	4	4	2	2
Char. All. (*Paspalo distichi-Polypogonion viridis*) and Ord. (*Potentillo anserinae-Polygonetalia avicularis*)					
Potentilla reptans	2	2	1	3	3
Plantago major	+	1	+	1	2
Agrostis stolonifera	.	+	.	3	4
Rumex conglomeratus	+	1	1	.	.
Lythrum junceum	.	+	+	.	.
Verbena officinalis	.	1	1	.	.
Schedonorus arundinaceus subsp. *arundinaceus*	.	.	.	.	+
Carex cuprina	.	+	.	.	.
Mentha pulegium	1	.	.	.	.
Epilobium tetragonum	.	.	+	.	.
Char. Cl. (*Molinio-Arrhenatheretea*)					
Scirpoides holoschoenus subsp. *australis*	+	2	1	1	+
Poa trivialis	.	1	+	.	.
Other species					
Helminthotheca echioides	+	+	2	.	+
Rumex pulcher	.	.	+	2	1
Trifolium resupinatum	+	.	.	1	1
Equisetum ramosissimum	+	1	2	.	.
Convolvulus arvensis	.	+	+	+	.
Anagallis arvensis	+	.	+	.	.
Arum italicum	.	.	+	+	.
Symphyotrichum squamatum	.	.	+	.	.

Rel. 1–3: Malta, Mtahleb, 26.9.1985; Rel. 4–5: Malta, Wied ir-Rum, 26.9.1985

The damp stands near the watercourses, with incoherent and well drained soils, are colonized by a hygro-nitrophilous community having its vegetative peak at the end of the summer. Significant here is the dominance of *Panicum repens*, which grows together with different species of the *Paspalo distichi-Polypogonion viridis* Br.-Bl. in Br.-Bl., Roussine & Nègre 1952 and *Potentillo-Polygonetalia avicularis* as *Potentilla reptans*, *Plantago major*, *Agrostis stolonifera*, *Verbena officinalis*, *Rumex conglomeratus*, *Lythrum junceum*, *Schenodorus arundinaceus*, *Carex cuprina*, etc. This vegetation shows close relations with *Lippio-Panicetum repentis* O. de Bolós 1957, but it differs for the ecology and also for the floristic set. In fact, the *Lippio-Panicetum repentis* is a coastal association linked to sandy soils, differentiated by *Lippia nodiflora* (cfr. de Bolós 1967; Brullo and Marcenò 1985b; Brullo and Sciandrello 2006). Therefore, the Maltese vegetation is proposed as *Potentillo*

Table 10.15 *Ranunculetum trichophylli*

Number of relevé	1	2	3	4	5
Plot size (m^2)	5	8	5	10	6
Plant cover (%)	90	100	70	90	100
Char. Association and All. (*Ranunculion fluitantis*)					
Ranunculus trichophyllus	4	5	4	5	5
Char. Cl. (*Potametea*)					
Zanichellia melitensis	2	2	1	+	2
Other species					
Chara vulgaris	1	1	.	+	1

Rel. 1–5: Malta, il Wied tal-Fiddien (Rabat), 16.4.1987

reptantis-Panicetum repentis ass. nova, having *Panicum repens* as characteristic species.

10.15 *Ranunculetum trichophylli* Melendo, Cano & Valle 2003 (Table 10.15)

In some rivers affected by fairly deep and clear waters, the submerged vegetation is represented by a community physiognomically characterized by *Ranunculus trichophyllus*. This species, which often tends to constitute a dense and intricate vegetation, is considered as the differential of a peculiar association, named *Ranunculetum trichophylli*, described from Iberian Peninsula by Melendo et al. (2003). Usually, it is linked to more or less flowing waters and is ascribed to *Ranunculion fluitantis* Neuhäusl 1959, alliance of *Potametea* Klika in Klika & Novak 1941 gathering the submerged communities spread in the watercourses, with more or less dynamic waters and characterized by the dominance of water buttercups. The association was recorded from Spain, Italy and Sicily (Buchwald 1994; Brullo et al. 2002b; Melendo et al. 2003), where it occurs in shallow, oligotrophic streams of calcareous habitats.

10.16 *Lemnetum minoris* Oberd. ex Müller & Görs 1960

The *Lemnetum minoris* consists essentially of floating unilayered mats of *Lemna minor*, tiny species usually aggregated in dense populations on the water surfaces (Fig. 10.1). This ephemeral community is linked of stagnant or very slow-moving, mesotrophic to eutrophic waters. It occurs over the surfaces of ponds, edges of streams, ditches and canals, also in disturbed stands, with neutral or basic fresh water. This association belongs to *Lemnion minoris* R.Tx. ex O. de Bolós and Masclans 1955, in the Maltese Islands it is not very frequent and can by observed only in the winter-spring period. As concerns its distribution, this pleustophytic

Fig. 10.1 Floating vegetation with *Lemna minor*

association is widespread in the Euro-Mediterranean territories (de Bolós and Masclans 1955; Scoppola 1982; Sburlino et al. 2004; Felzines 2012). No releves are reported on this association since it is monophytic with dense coverage.

Chapter 11
Seagrass Vegetation

The coasts around the Maltese Islands are characterized by seagrass meadows, linked mainly to more or less sandy bottoms. These communities are mainly represented by monophytic populations of *Posidonia oceanica*, *Cymodocea nodosa* or *Halophila stipulacea*, which occur prevalently in the infralittoral zone from the seashore up to 30–40 m deep. For this vegetation the authors not carried out original releves, due to their localization in deep marine habitats.

11.1 *Posidonietum oceanicae* Molinier 1960

Seagrass meadows dominated by *Posidonia oceanica* are frequent in the all Mediterranean coasts (Den Hartog 2003, 2016). This vegetation is common on the sea bottom of Maltese Archipelago as emphasized by Devillers and Devillers-Terschuren (2001), Borg and Schembri (1995, 2002) and Borg et al. (2009), where grows usually at a depth of 10–40 m on sandy substrata. It is linked to warm-temperate waters, tolerating very variable hydrodynamism conditions. This perennial submerged vascular vegetation was described by Feldmann (1937) as *Posidonietum* nom. nud. and also by Braun-Blanquet et al. (1952), while it was validly published by Molinier (1960), since based on a phytosociological table. This monophytic association was attributed to *Posidonion oceanicae* by several authors (Braun-Blanquet et al. 1952; Molinier 1960; Den Hartog 1976, 2003; Bardat et al. 2004), that according to Mucina et al. (2016) must be referred to *Posidonietalia oceanicae,* order of the *Zosteretea*.

S. Brullo et al., *The Vegetation of the Maltese Islands*, Geobotany Studies,
https://doi.org/10.1007/978-3-030-34525-9_11

11.2 *Cymodoceetum nodosae* Giaccone & Pignatti 1967

In the Maltese islands mainly in bays and harbours it is possible to observe seagrass meadows characterized by *Cymodocea nodosa*. The species is distributed in the Mediterranean sea and in warmer coasts of the Atlantic Sea from southern Portugal to the northwestern African coast, Canary Islands included (Borum and Greve 2004; Den Hartog 2003, 2016). It likes sea bottom with fine sands not oxidized, often mixed with slime and sheltered waters, deep up to 6–7 m, rarely more (45–48 m). This species was considered as characteristic of a peculiar vegetation, named "Associàtion à *Cymodocea nodosa*" by Feldmann (1937), but it is an invalid name (art. 2b, 7), while it was validly described by Giaccone and Pignatti (1967). In this association, often monophytic, *Cymodocea nodosae* grows together with other marine cormophytes, such as *Zostera marina*, *Z. noltii*, *Ruppia cirrhosa* and *Halophila stipulacea* (Feldmann 1937; Giaccone and Pignatti 1967; Licandro et al. 2011), while in Malta sometimes was observed sometimes together with *Posidonia oceanica*, Previously, the occurrence in Malta of this association, was recorded by Devillers and Devillers-Terschuren (2001) and Borg and Schembri (2002) As concerns its syntaxonomical arrangement, the *Cymodoceetum nodosae* is included in *Cymodoceion nodosae* Den Hartog ex Mucina in Mucina et al. 2016, alliance of the *Halodulo wrightii-Thalassietea testudinum* Den Hartog ex Rivas Mart. et al. 1999.

11.3 *Halophiletum stipulaceae* Augier ex Brullo ass. nov. hoc loco

Syn.: *Halophiletum stipulaceae* Augier 1982 ass. provv.; *Halophiletum stipulaceae* Giaccone 1988, nom. nud.; *Cymodocietum nodosae* facies a *Halophila stipulacea* Licandro et al. 2011, nom. inval.

Holotypus: rel. 3, tab. 11, Licandro et al. (2011), hoc loco.

The alien seagrasse *Halophila stipulacea* were recorded in Malta by Lanfranco (1970). It is a tropical species whose penetration into the Mediterranean from the Red Sea took place through the Suez Canal (Lessepsian migration). In the Mediterranean initially this marine phanerogam was distributed on the eastern part of the basin, reaching in the last decades also the coasts of the central part (Schembri 1995; Gambi et al. 2009; Sghaier et al. 2011). Usually, it colonizes the sandy sediments, often enriched with fine particles, mainly near the harbours, constituting discontinuous meadows often monophytic or mixed with *Cymodocea nodosa*, *Posidonia oceanica* and *Caulerpa* sp.pl. This vegetation was described initially by Augier (1982) as *Halophiletum stipulaceae* ass. provv. and later by Giaccone (1968) as *Halophiletum stipulaceae* nom. nud., since the author does not refer to any phytosociological relevé. Therefore, this syntaxon is here validated, choosing as a holotype a relevè published by Licandro et al. (2011) from Sicily and attributed to

Cymodocietum nodosae facies *Halophila stipulacea*. This vegetation, as the previous association, must be included in the *Cymodoceion nodosae* and is recorded in the Mediterranean basin by various authors (Augier 1982; Meinesz et al. 1983; Giaccone et al. 1994; Licandro et al. 2011).

Chapter 12
Rocky Pool Vegetation

The Maltese Islands are characterized by large limestone outcrops, where are very frequent rocky pools usually submerged during winter-spring period. These habitats have a relevant phytosociological role, since a specialized flora of small hydrophytes, is here localized. It is represented by several submerged and amphibious species, some of them endemic, which differentiate various peculiar plant communities. They belong mainly to the *Isoeto-Nanojuncetea* class, which are included in one of the most important priority habitat (Cod. 3170, Mediterranean temporary ponds). Community of *Charetea* and *Potametea* are also frequent in these habitats.

12.1 *Charetum vulgaris* Corillion 1957

According to literature (Lanfranco 1969a, 2002; Henwood 2014; Casha 2015; etc.), some species of Characeae were recorded from Maltese Island, among them *Chara vulgaris*, *C. fragilis*, *C. globularis*, *Tolypella glomerata* and *Lamprothamnium papulosum*. In particular, *Chara vulgaris* is the more spread species in these islands and usually localized in the calcareous rocky pools with freshwater. The species likes neutral or slightly alkaline eutrophic and meso-eutrophic shallow water on silty deposits, from 0.2 to 1 m in depth. It characterizes a monospecific vegetation or sometimes paucispecific, growing sometimes with other species of *Chara* sp., *Cladophora* sp. and *Zannichellia melitensis*. From the phytosogiological point of view, this hygrophilous plant community can be referred to *Charetum vulgaris* Corillion 1957, association often recorded on temporary pools on flat limestone outcrops (Corillion 1957; Krausch 1964; Krause and Lang 1977; Hrivnák et al. 2005). For this plant community, represented usually by monospecific populations, no phytosociological releves were carried out. As concern its syntaxonomical arrangement, the *Charetum vulgaris* is framed in the *Charion vulgaris* (Krause ex Krause & Lang 1977) Krause 1981, alliance of *Charetea intermediae* Fukarek 1961 (see Mucina et al. 2016).

S. Brullo et al., *The Vegetation of the Maltese Islands*, Geobotany Studies,
https://doi.org/10.1007/978-3-030-34525-9_12

12.2 *Zanichellietum melitensis* ass. nov. hoc loco (Table 12.1)

Holotypus: rel. 9, hoc loco

In the streams with muddy bottoms and shallow water, as well as in the temporarily water-filled small depressions in the rock, it is frequent a dense submerged, often monophytic community dominated by *Zannichellia melitensis* (Fig. 12.1b), endemic hydrophyte described from Maltese Archipelago by Brullo et al. (2001a). This vegetations shows its optimum in the winter-spring period, enduring quite well periods of soil desiccation. It is proposed as a new association, named *Zanichellietum melitensis*, which shows some relations with the *Zanichellietum obtusifoliae* Brullo & Spampinato 1990, described from Sicily. Both associations can be included in the *Ranunculion fluitantis*, alliance of the *Potametea*.

12.3 *Damasonio bourgaei-Ranunculetum saniculifolii* ass. nov. hoc loco (Table 12.2)

Holotypus: ril. 2. hoc loco.

The rocky pools are very frequent in the flat limestone outcrops of the Maltese Islands, which during the autumn and winter period are filled with rainwater, that sometimes persist until the end of spring. The deeper pool are usually colonized by a very peculiar hydrophitic community, characterized by submerged species rooting on the muddy bottom and forming a dense and intricate vegetation. Physiognomically it is dominated by few species represented by *Ranunculus saniculifolius*

Table 12.1 *Zanichellietum melitensis*

Number of relevé	1	2	3	4	5	6	7	8	9
Plot size (m^2)	2	2	1	1.5	1	1	0.6	0.3	1
Plant cover (%)	50	90	90	100	100	0	100	100	90
Char. Association									
Zanichellia melitensis	3	5	5	5	5	3	5	5	5
Char. All. (*Ranunculion aquatilis*) and Cl. (*Potametea*)									
Ranunculus saniculifolius	.	.	.	1	.	.	.	.	.
Callitriche truncata	.	.	.	.	.	.	.	.	1
Other species									
Chara vulgaris	.	.	.	.	.	2	1	.	.
Elatine gussonei	.	.	.	+	.	.	.	.	.

Rel. 1–2: Gozo, Tal Blata, 9.4.1984; Rel. 3: Malta, Ghallis, 8.4.1984; Rel. 4–5; Gozo, Xlendi Valley, 13.4.1987; Rel. 6; Gozo, Dwerja, 13.4.1987; Rel. 7: Gozo. Il-Mielah, 13.4.1987; Rel. 8: Malta, Wadi Babu, near Blue Grotto (Zurrieq), 01.05.2007; Rel. 9: Malta, Mistra, 8.3.2014

Fig. 12.1 Rocky pool vegetation: (**a**) *Crassulo vaillantii-Elatinetum gussonei* with *Ranunculus saniculifolius*, (**b**) *Zanichellietum melitensis*

and *Damasomium bourgaei*, often associated with *Callitriche truncata* and, more rarely, with *Zanichellia melitensis*. This very peculiar vegetation is proposed as a new association of the *Ranunculion fluitantis* and named *Damasonio bourgaei-Ranunculetum saniculifolii*. In particular, *Ranunculus saniculifolius* and *Damasomium bourgaei* are proposed as characteristic species. From the floristic and ecological point of view, this associations approaches to *Ranunculo saniculifolii-Callitrichetum brutiae*, association described by Brullo et al. (1976) from the southern Sicily. On the whole, it belongs to a group of vicariant associations widespread in the pools of many Mediterranean territories, which are differentiated

Table 12.2 *Damasonio bourgaei-Ranunculetum saniculifolii*

Number of relevé	1	2	3	4	5	6	7	8	9	10	11	12	13	14
Plot size (m^2)	1	1	1	2	3	2	2	2	1	2	1	1	1	0.5
Plant cover (%)	100	100	70	50	90	70	40	40	50	70	100	100	70	80
Char. Association														
Ranunculus saniculifolius	5	3	1	2	3	2	3	2	3	3	3	2	4	.
Damasonium bourgaei	.	4	2	3	1	.	1	.	.	3	2	4	.	.
Char. All. (*Ranunculion aquatilis*) and Cl. (*Potametea*)														
Callitriche truncata	.	1	3	.	.	.	2	4	2	.	4	3	.	4
Zanichellia melitensis	.	.	.	+	1	.	.	.	+	.	.	.	1	.
Other species														
Lythrum hyssopifolia	1	.	.	1	1	2	1	.	.	+	.	.	+	.
Elatine gussonei	+	.	.	.	.	.	.	.	1	1	1	.	+	+
Juncus bufonius	.	.	.	.	2	3	1	.	.	+	.	.	.	.
Polypogon subspathaceus	.	.	.	.	1	1	+	.	.	.	.	.	.	.
Mentha pulegium	.	.	.	.	2	1	1	.	.	.	.	.	.	.
Trifolium tomentosum	.	.	.	1	1	+	.	.	.	.	.	.	.	.
Hypericum pubescens	.	.	.	.	1	1	.	.	.	.	.	.	.	.

Rel. 1–5: Malta, Ghallis, 8.4.1984; Rel. 6–7: Malta, Mistra, 14.4.1983; Rel. 8–12: Malta, Naxxar, 18.4.1987; Rel. 13: Gozo, Xlendi Valley, 13.4.1987; Rel. 14: Malta, Mistra 8.3.2014

by species of *Ranunculus* subgen. *Batrachium* and *Callitriche* sp. pl. (Bergmeier and Raus 1999; Bergmeier 2001; Deil 2005; Rivas-Martinez and coauthores 2011; Lumbreras et al. 2011). The *Damasonio-Ranunculetum saniculifolii* replaces in the pools with fairly deep water (generally more than 15–20 cm) the *Zanichellietum melitensis*, association with more hydric requirements.

12.4 *Crassulo vaillantii-Elatinetum gussonei* Bartolo, Brullo, Minissale & Spampinato 1990 (Table 12.3)

The temporary freshwater pools localized in the more or less large hollows of the inland plateaux are very frequent in the Maltese islands. These habitats, occurring exclusively on the limestones rock, are periodically flooded by rainwater, mainly during the winter-spring period. Several annual hydrophytes, rooting on the muddy bottom, colonize these rocky pools, reaching the water surface, forming a dense and intricate micro-vegetation. It is a very specialized community belonging to *Isoetion durieui* Br.-Bl. 1935, alliance of *Isoeto-Nanojuncetea*, Br.-Bl. & R.Tx. in Br.-Bl. et al. 1952, well represented in the all Mediterranean area (Brullo and Minissale 1998). For the occurrence and often dominance of *Elatine gussonei* and *Buillardia vaillantii* (=*Crassula vaillatii*), this vegetation must be referred to *Crassulo vaillantii-Elatinetum gussonei* (Fig. 12.1a), association described from the neighboring island of Lampedusa (Pelagian Archipelago) by Bartolo et al. (1990).

Table 12.3 *Crassulo vaillantii-Elatinetum gussonei*

Number of relevé	1	2	3	4	5	6	7	8	9	10	11	12	13	14	15	16	17	18	19	20	21	22
Plot size (m^2)	1	2	1	1	2	1	0.5	0.5	1	1	0.8	0.5	1	1	1	2	1	0.5	0.6	0.5	1	0.6
Plant cover (%)	80	80	80	100	100	70	70	70	80	80	100	90	90	80	50	50	90	100	100	50	85	85
Char. Association																						
Elatine gussonei	+	1	4	4	4	3	2	4	2	3	5	5	5	4	2	1	3	4	2	2	4	4
Char. All. (*Isoetion*) and Cl. (*Isoeto-Nanojuncetea*)																						
Crassula vaillantii	3	4	+	1	1	2	3	1	+	+	1	.	1	1	3	3	3	3	2	1	.	.
Polypogon subspathaceus	2	+	1	3	2	2	1	+	+	+	2	+	.	+	.	+	1	.	1	.	.	+
Lythrum hyssopifolia	3	+	1	.	2	2	2	1	.	2	.	+	.	2	1	2	2	+	2	.	.	+
Mentha pulegium	1	.	.	1	+	.	1	+	2	3	.	+	.	.	.	.	.	.	3	.	.	.
Juncus bufonius	.	.	+	.	.	1	+	.	3	+	.	.	1	.	.	1	2	1	+	.	.	.
Juncus capitatus	1	.	+	.	.	+	.	1	.	.	.	.	.	.	.	.	.	.	.	.	.	.
Ochlopoa infirma	.	.	.	.	.	.	.	.	.	.	.	.	.	1	.	+	.	.	.	.	.	.
Other species																						
Ranunculus saniculifolius	.	+	.	.	.	.	.	1	.	.	1	.	1	+	.	.	.	+	.	.	.	.
Triglochin laxiflorum	+	.	.	.	.	.	.	.	1	1	.	.	.	.	+	1	.	.	.	.	.	.
Bellis annua	.	.	.	.	.	.	+	.	.	+	.	.	.	.	.	.	.	.	.	.	.	.
Damasonium bourgaei	.	.	.	.	.	.	.	.	1	+	.	.	.	.	.	.	.	.	.	.	.	.
Bromus hordeaceus	.	.	.	.	.	1	.	.	.	.	.	.	.	.	.	.	.	.	.	.	.	.
Callitriche truncata	.	.	.	.	.	.	.	.	.	.	.	.	.	.	.	.	.	.	.	.	2	.
Plantago coronopus	.	.	.	.	.	.	.	.	.	1	.	.	.	.	.	.	.	.	.	.	.	.

Rel. 1: Malta, Ghallis, 8.4.1984; Rel. 2–4: Malta, Tal Blata, 9.4.1984; Rel. 5–8: Malta, Ghallis, 8.4.1984; Rel. 9–10: Malta, Naxxar, 12.4.1987; Rel. 11–18: Gozo, Xlendi Valley, 13.4.1987; Rel. 19: Malta, Wied Incita (Attard), 13.4.1987; Rel. 20–22: Malta, Mistra, 8.3.2014

According to Brullo et al. (1990) and Molnar et al. (2014), *Elatine gussonei* is endemic of Lampedusa, Maltese Archipelago and Sicily, where it occurs always in the calcareous rocky pools. As concerns the distribution of this association, it was surveyed also in southern and western Sicily by Minissale and Sciandrello (2016). It can be considered as a thermophilous vicariant of the *Elatinetum macropodae* Br.-Bl 1936, association described from South France (Braun-Blanquet 1936; de Foucault 2013a), but probably occurring also in other Mediterranean territories. Other associations characterized by *Elatine macropoda* are represented by *Crassulo-Elatinetum macropodae* Brullo & Minissale 1998, from Algeria, *Apio crassipedis-Elatinetum macropodae* Bagella et al. 2009 from Sardinia and *Elatino macropodae-Lindernietum procumbentis* Ubriszy 1948 from Hungary.

12.5 *Lolio lepturoidis-Spergularietum marinae* ass. nov. hoc loco (Table 12.4)

Holotypus: ril. 1, hoc loco

In the small rocky pools near the sea, that are filled by weakly salty waters, it can be observed a therophytic hygrophilous vegetation characterized by *Spergularia marina* and *Lolium rigidum* subsp. *lepturoides*. In these habitats are frequent

Table 12.4 *Lolio lepturoidis-Spergularietum marinae*

Number of relevé	1	2	3
Plot size (m^2)	5	5	2
Plant cover (%)	60	70	70
Char. Association			
Spergularia marina	2	3	1
Lolium rigidum subsp. *lepturoides*	1	1	2
Char. All. (*Isoetion*) and Cl. (*Isoeto-Nanojuncetea*)			
Polypogon subspathaceus	+	1	2
Lythrum hyssopifolia	3	3	2
Mentha pulegium	3	4	3
Romulea ramiflora	+	+	1
Other species			
Phalaris paradoxa	1	1	.
Bellis annua	+	+	2
Plantago lagopus	1	1	2
Symphyotrichum squamatum	+	+	+
Moraea sisyrinchium	+	+	+
Anthemis urvilleana	1	+	1
Cichorium pumilum	.	+	+
Centaurium pulchellum	.	.	+
Plantago coronopus	1	1	1

ril.1–3: Malta, Pembroke (St. Andrews), 29.04.2007

some species of *Isoetion durieui* and *Isoeto-Nanojuncetea*, such as *Polygonum subspathaceus*, *Lythrum hyssopifolia*, *Mentha pulegium* and *Romulea ramiflora*. For its floristic and ecological peculiarities, it is described as a new association and named *Lolio lepturoidis-Spergularietum marinae*. Based on our field investigations, it seems quite rare in Malta.

12.6 *Triglochino laxiflori-Romuleetum melitensis* ass. nov. hoc loco (Table 12.5)

Holotypus: ril. 8, hoc loco

The rocky pools of the limestone rock filled with quite shallow water, drying already by the end of winter season, are colonized by a vegetation showing less hygrophilous requirements. Physiognomically, this community is characterized by *Romulea melitensis*, endemic to Maltese archipelago, and *Triglochin laxiflorum*, not strictly hydrophilous species, but well adapted to moderate edaphic humidity with short flooding periods. These two species allow to differentiate a new association of the *lsoetion durieui*, proposed as *Triglochino laxiflori-Romuleetum melitensis*. This association can be considered as a less hygrophilous vicariant of *Crassulo vaillantii-Elatinetum gussonei*, which is linked to more humid edaphic conditions. In fact, in the floristic set of the *Triglochino laxiflori-Romuleetum melitensis*, it is possible

Table 12.5 *Triglochino laxiflori-Romuletum melitensis*

Number of relevé	1	2	3	4	5	6	7	8
Plot size (m^2)	2	1	2	0.5	1	1	1	0.5
Plant cover (%)	50	50	40	40	70	100	100	90
Char. Association								
Romulea melitensis	3	3	3	2	1	2	2	3
Triglochin laxiflorum	.	+	2	1	2	2	4	2
Char. All. (*Isoetion*) and Cl. (*Isoeto-Nanojuncetea*)								
Lythrum hyssopifolia	2	2	1	2	2	3	3	2
Polypogon subspathaceus	1	1	+	2	3	3	+	1
Juncus bufonius	.	.	1	+	.	+	2	1
Mentha pulegium	.	.	.	.	3	3	1	3
Crassula vaillantii	.	.	+	+	.	.	.	.
Juncus capitatus	.	.	+	+	.	.	.	.
Ochlopoa infirma	.	.	.	.	.	.	1	.
Other species								
Cichorium pumilum	+	.	.	.	.	.	.	.
Rumex conglomeratus	.	.	.	.	.	1	.	.
Bellis annua	.	.	.	.	.	.	.	+

Rel. 1–4: Malta, Mistra Valley, 9.4.1984; Rel. 5–6: Malta, Wied il-Ghasel (Mosta), 9.4.1984; Rel. 7: Gozo, Xlendi Valley, 13.4.1987; Rel. 8: Malta, Mosta, 9.4.1984

observed the lack or rarity of the most hygrophilous species, among them *Crassula vaillantii* and *Elatine gussonei*, while the less hygrophilous tend to dominate, as *Lythrum hissopifolia*, *Polypogon subspathaceus*, *Mentha pulegium*, etc.

12.7 *Damasonio bourgaei-Crypsietum aculeatae* Rivas-Martínez & Costa in Rivas-Martínez et al. 1980 corr. V. Silva & J.C. Costa in Costa et al. 2012 (Table 12.6)

Syn.: *Damasonio alismatis-Crypsietum aculeatae* Rivas-Martínez & Costa in Rivas-Martínez et al. 1980.

In the small marshes characterized by eutrophic water, persistent for long periods, an annual hygrophilous vegetation with subnitrophilous requirements occurs. It is a very peculiar community, characterized by *Crypsis aculeata*, species very rare in Malta, which was observed in some coastal places during the summer. From the floristic and ecological point of view, it can be referred to *Damasonio bourgaei-Cypsietum aculeatae*, association belonging to *Verbenion supinae* Slavnic 1951, alliance of *Nanocyperetalia* Klika 1935, described by Rivas-Martínez et al. (1980)

Table 12.6 *Damasonio bourgaei-Crypsietum aculeatae*

Number of relevé	1	2	3
Plot size (m^2)	3	3	5
Plant cover (%)	90	80	80
Char. Association and All. (*Verbenion supinae*)			
Crypsis aculeata	3	3	2
Damasonium bourgaei	.	.	1
Char. Cl. (*Isoeto-Nanojuncetea*)			
Lythrum hyssopifolia	1	+	2
Polypogon subspathaceus	1	+	1
Mentha pulegium	.	1	1
Juncus bufonius	1	.	.
Other species			
Symphyotrichum squamatum	3	2	2
Atriplex littoralis	2	1	1
Plantago coronopus	+	+	1
Cressa cretica	+	+	1
Parapholis incurva	+	.	+
Suaeda spicata	1	+	.
Anagallis arvensis	.	+	+
Spergularia marina	.	+	1
Polypogon monspeliensis	+	1	.
Centaurium tenuiflorum	+	.	+

Rel. 1–2: Malta, Dragunara, 24.6.1973; Rel. 3: Malta, Ghallis, 24.6.1973

from Spain sub *Damasonio alismatis-Crypsietum aculeatae* (cf. Brullo and Minissale 1998), name corrected by Silva and Costa (see Costa et al. 2012), since *Damasonium alisma* was been mistaken with *D. bourgaei*. Previously, this association was reported, apart from the Iberian peninsula (Costa et al. 2012), also from northern Tunisia by Barbagallo et al. (1990), and basing on unpublished data occurs in Sicily too.

Chapter 13
Woody Riparian Vegetation

The riparian woodland in the Maltese Islands are very rare and represented by scattered and isolated patches, which are very depauperated and degraded due to intense anthropic activities. Besides, it is to be noted that in this territory the watercourses are short with a limited flow and usually dried in summertime. Currently, this type of vegetation is represented just by two associations, one more thermophilous belonging to *Nerio oleandri-Tamaricetea* and another more mesic of the *Alno glutinosae-Populetea albae*.

13.1 *Tamarici gallicae-Viticetum agni-casti* ass. nova hoc loco (Table 13.1)

Holotypus: ril. 2, hoc loco.

In the Maltese islands along the pebbly bed of some streams was surveyed a riparian scrub vegetation characterized by *Vitex agnus-castus*, usually associated with *Tamarix gallica*. This plant community, belonging to *Nerio oleandri-Tamaricetea* Br.-Bl. & O. 1958, is floristically well differentiated from the other associations of the *Rubo ulmifolii-Nerion oleandri* O. Bolòs 1985 physiognomically dominated by *Vitex agnus-castus*. In particular, it differs from the *Tamarici africanae-Viticetum agni-casti* Brullo & Spampinato 1997, which is distributed in Calabria (Brullo and Spampinato 1997), Sicily (Licandro et al. 2011) and Sardinia (Angius and Bacchetta 2009), for the lack of *Nerium oleander* and especially of *Tamarix africana*, that is here replaced by *T. gallica*, while *Rubus ulmifolius* is rare. It is also floristically well differentiated from the *Leucojo pulchelli-Viticetum agni-casti* O. Bolós & Molinier 1958 and *Vinco majoris-Viticetum agni-casti* O. Bolós 1956, associations described from Spain (see Rivas-Martinez and coauthores 2011). Therefore, the Maltese vegetation is described as new association and named

S. Brullo et al., *The Vegetation of the Maltese Islands*, Geobotany Studies,
https://doi.org/10.1007/978-3-030-34525-9_13

Table 13.1 *Tamarici gallicae-Viticetum agni-casti*

Number of relevé	1	2	3	4	5	6
Plot size (m^2)	20	20	30	10	20	20
Plant cover (%)	100	100	100	90	100	90
Char. Association and all. (*Rubo ulmifolii-Nerion oleandri*)						
Vitex agnus-castus	2	3	5	2	5	3
Rubus ulmifolius	.	.	.	.	.	3
Char. Cl. (*Nerio oleandri-Tamaricetea*)						
Tamarix gallica	3	3	.	3	.	1
Punica granatum	.	.	.	1	.	.
Other species						
Arundo donax	3	3	2	.	+	.
Atriplex halimus	3	2	1	.	.	.
Foeniculum piperitum	.	.	.	1	1	1
Parietaria judaica	+	+	.	.	+	.
Limbarda crithmoides subsp. *longifolia*	.	.	.	2	1	.
Phragmites australis	.	.	.	2	1	.
Calystegia sepium	1	+	.	.	.	.
Charybdis pancration	.	.	.	.	2	.
Oxalis pes-caprae	.	.	.	.	1	.
Dittrichia viscosa	.	.	.	1	.	1
Smilax aspera	.	.	.	.	.	1
Euphorbia dendroides	.	.	.	.	.	2
Erica multiflora	.	.	.	.	.	1
Ficus carica	.	.	.	.	.	1
Ceratonia siliqua	.	.	.	.	.	2
Jacobaea mariitma subsp. *bicolor*	.	.	.	.	.	1

Rel. 1–2: Gozo, Ramla Bay, 25.9.1985; Rel. 3: Gozo, Dwerja, 25.9.1985; Rel. 4: Gozo, Marsalforn, 13.4.1987; Rel. 5: Gozo, Dwerja, 13.4.1987; Ril. 6: Gozo, Triq Kalate Punic, 19.5.2017

Tamarici gallicae-Viticetum agni-casti, having *Tamarix gallica* as characteristic species.

13.2 *Ulmo canescentis-Salicetum pedicellatae* Brullo & Spampinato, 1990 (Table 13.2)

In former times, the watercourses of the Maltese Islands were probably more widespread and characterized by the occurrence of riparian woodlands of the *Alno glutinosae-Populetea albae* P. Fukarek & Fabijanic 1968. Currently, this vegetation has been unfortunately almost completely destroyed by human activities, due to the progressive cutting of the trees, as well as for the total or partial drying up of streams for the taking of water for urban or agricultural uses. Some remains of this vegetation

Table 13.2 *Ulmo canescentis-Salicetum pedicellatae*

Number of relevé	1	2	3	4	5
Plot size (m^2)	30	20	30	50	20
Plant cover (%)	100	100	100	100	100
Char. Association					
Ulmus canescens	+	5	5	5	+
Char. All. (*Populion albae*) and Ord. (*Populetalia albae*)					
Arum italicum	2	3	1	+	1
Populus alba	5	1	.	.	.
Ficus carica	1	+	.	.	.
Salix pedicellata	.	.	.	.	3
Char. Cl. (*Alno glutinosae-Populetea albae*)					
Hedera helix	2	1	2	1	.
Rosa gallica	+	.	1	+	.
Dioscorea communis	.	.	1	.	3
Brachypodium sylvaticum	1	.	.	.	+
Other species					
Phragmites australis	2	2	.	.	2
Acanthus mollis	2	1	.	.	+
Rubus ulmifolius	2	1	.	.	1
Laurus nobilis	2	+	.	.	.
Rubia peregrina	2	2	.	.	.
Lathyrus annuus	+	.	.	.	.
Carex cuprina	+	.	.	.	.
Asparagus aphyllus	1	.	.	.	.
Carex hispida	.	.	.	.	+

Rel. 1–4: Malta, Girgenti Valley, 16.4.1987; Rel. 5: Malta, Mtahleb, 28.9.1985

can be observed in inaccessible or poorly anthropized stands. Among the tree occur only rare individuals of *Salix pedicellata*, *Populus alba*, *Ficus carica*, *Salix alba* and *Ulmus canescens*, while the lianes are represented by *Hedera helix*, *Rosa gallica*, *Tamus communis*, *Rubus ulmifolius* and *Rubia peregrina*. The occurrence of *Salix pedicellata* and *Ulmus canescens*, species in the past probably much more widespread in the Maltese territory, allows to attribute this riparian vegetation to the *Ulmo canescentis-Salicetum pedicellatae*, association described by Brullo and Spampinato (1990) for the rivers of north-western Sicily, belonging to *Populion albae*. Br.-Bl. ex Tchou 1948. The Maltese community is clearly more impoverished floristically than Sicily one, but overall it shows a certain affinity with the latter.

Chapter 14
Annual Xerophilous Vegetation

The marked environmental aridity featuring the Maltese Island has clearly a confirmation in the occurrence and spread mainly in rocky habitats of annual plant communities, linked to the edaphic xericity. Most of them show a pioneer character fitting well to shallow soils in contact with the rock, as well as to sands and clays. From the floristical viewpoint, they are differentiated by small therophytes and geophytes, showing a short biological life cycle, with their optimum in springtime. These communities must be referred to *Stipo capensis-Trachynietea distachyae*, while that ones linked to sandy soil belong to *Vulpietalia,* order of *Tuberarietea guttatae*.

14.1 *Thero-Sedetum caerulei* Brullo 1975 (Table 14.1)

The flat rocky surfaces and the niches of the rocky places, where there is a thin layer of soil, are colonized by a therophytic micro-vegetation characterized by the dominance of *Sedum coeruleum.* It is a purely pioneer community, which is very specialized and linked to a marked edaphic xericity. These ecological conditions allow the settlement only to small therophytes with a short vegetative cycle or sometimes to small geophytes. In these habitats, a part from *Sedum caeruleum*, other annual species of *Sedum*, such as *S. rubens*, *S. caespitosum* and *S. litoreum*, play a relevant physiognomical role, which grow usually with numerous microphytes of the *Trachynion distachyae* Rivas-Martinez 1978 and *Stipo capensis–Trachynietea distachyae* Brullo in Brullo, Scelsi & Spampinato 2001c. For its ecological and floristic features this vegetation must be referred to the *Thero-Sedetum coerulei* Brullo 1975, association described from the Hyblean territory (South Sicily), but widespread in Sicily and in some neighboring islands (Bartolo et al. 1982). In the Maltese islands within this association can be identified two variants, represented by the subass. *sedetosum caespitosi* Brullo 1975 and subass. *sedetosum litorei* Barbagallo, Brullo & Guglielmo 1980. The first one is distributed in the coastal

S. Brullo et al., *The Vegetation of the Maltese Islands*, Geobotany Studies,
https://doi.org/10.1007/978-3-030-34525-9_14

Table 14.1 *Thero-Sedetum coerulei*

Number of relevé	1	2	3	4	5	6	7	8	9	10	11	12	13	14	15	16	17	18	19
Plot size (m^2)	0.6	1	1	1.5	0.5	0.8	1	0.5	0.4	0.6	0.5	0.5	1	0.8	0.5	1	1	0.5	0.6
Plant cover (%)	60	70	40	50	40	50	20	60	70	70	50	30	70	70	50	80	80	50	50
Char. Association																			
Sedum coeruleum	4	4	3	2	3	3	2	4	3	4	2	2	3	3	3	4	4	3	+
Diff. Subassociation																			
Catapodium balearicum	1	+	1	2	+	2	1	1	3	2	1	1	1	2	.	+	.	.	.
Moraea sisyrinchium	+	1	+	1	+	.	.	1	2	+	2	1	+	.	.	.	.	.	+
Plantago coronopus	+	+	.	.	+	+	.	1	.	1	2	+	.	.	+	.	.	.	.
Parapholis incurva	.	.	+	.	.	+	.	+	+	.	.	+	1	1	.	.	.	.	+
Sedum caespitosum	.	.	+	.	.	.	.	.	.	.	.	2	3	2	.	.	.	.	.
Sedum litoreum	.	.	.	.	.	.	.	.	.	.	.	.	.	.	2	1	2	2	3
Char. All. (*Trachynion distachyae*) and Cl. (*Stipo-Trachynietea distachyae*)																			
Evax pygmaea	+	1	.	1	1	+	+	.	2	+	1	.	+	1	+	1	1	2	+
Romulea melitensis	.	+	1	2	+	1	+	2	1	+	1	.	+	1	2	1	+	.	1
Trifolium scabrum	+	.	+	2	1	1	+	+	+	.	1	.	.	1	+	1	1	1	+
Trachynia distachya	.	+	.	1	1	.	.	1	+	.	+	.	+	1	.	+	+	+	2
Sedum rubens	+	+	1	.	1	+	.	+	.	+	.	.	.	.	1	.	+	1	.
Linum strictum	.	.	1	.	+	+	.	.	+	+	.	+	.	.	+	.	+	+	1
Valantia muralis	.	+	.	.	.	+	+	.	.	.	1	+	1	.	+	+	.	1	1
Euphorbia exigua	+	+	.	.	+	+	.	.	+	+	.	.	+	.	.	.	1	1	.
Hypochoeris achyrophorus	.	.	.	.	1	.	+	.	.	+	+	.	.	.	+	+	.	+	+
Medicago minima	+	.	.	.	+	+	.	.	.	+	.	+	+	.	.	.	+	.	.
Stipella capensis	.	.	.	1	.	+	+	.	+	+	.	+	.	.	.	.	.	.	1
Arenaria leptoclados	+	.	.	.	.	.	.	.	.	.	.	.	.	.	.	.	.	.	.
Trigonella monspeliaca	.	.	.	.	.	+	.	.	.	.	.	.	.	.	.	.	.	.	.
Hyoseris scabra	.	.	.	.	.	.	.	+	.	.	.	.	.	.	.	..	.	.	.

Other species																			
Catapodium rigidum	.	+	.	+	.	+	1	+	1	.	.	.	+	1	.	.	1	+	+
Poa bulbosa	1	.	1	.	+	+	1	.	.	2	1	+	.	.	.	1	.	1	.
Bellis annua	.	.	.	+	+	+	.	+	.	+	.	.	+	1	.	.	+	.	.
Rostraria cristata	.	.	.	1	1	.	+	.	.	.	+	.	+	.	.	.	.	+	.
Hedypnois cretica	.	.	+	+	.	.	.	.	.	.	+	.	.	+	.	.	.	+	.
Trifolium nigrescens subsp. *nigrescens*	.	.	.	1	.	.	+	+	.	.	.	+	.	.	+	.	.	.	.
Lagurus ovatus subsp. *ovatus*	.	+	.	.	+	.	1	.	.	.	+	.	.	.	.	+	.	.	.
Anthemis urvilleana	.	.	.	.	.	.	.	+	.	.	+	.	+	.	+	.	.	+	.
Trisetaria aurea	.	.	.	.	.	.	.	1	.	.	+	+	+	+	.	.	.	.	.
Plantago afra subsp. *afra*	.	+	+	.	.	+	.	+	.	.	.	.	.	.	.	.	.	.	.
Medicago littoralis	.	.	.	.	.	.	.	+	.	+	1	.	.	.	.	.	.	.	+
Asteriscus aquaticus	.	.	.	.	.	.	.	+	.	.	1	.	.	.	+	.	.	.	.
Desmazeria pignatti	.	.	.	.	.	.	.	+	.	.	+	.	.	.	.	.	.	.	.
Medicago truncatula	.	.	.	.	.	.	.	.	.	.	.	.	+	.	.	.	.	.	.

Rel. 1–3: Malta, Wied il Gebel, 16.4.1984; Rel. 4–7: Malta, St. Andrews, 11.4.1984; Rel. 8–12: Gozo, Ta Cenc, 13.4.1987; Rel. 13: Gozo, Xlendi Valley, 13.4.1987; Rel. 14–16: Comino, 15.4.1987; Rel. 17–19: Malta, Wied Incita, 16.4.1987

and inland places and shows a marked xerophilous requirement, character confirmed by the occurrence of *Catapodium balearicum*, *Plantago coronopus*, *Parapholis incurva*, *Sedum caespitosum* and *Moraea sisyrinchium*. The second one is linked to fairly shaded and protected coastal stands and is differentiated by *Sedum litoreum*. From the syndynamic point of view, the *Thero-Sedetum coerulei* shows correlations with bryocoenosis of the *Barbuletea unguiculatae* Mohan 1978, which replace the association at issue in extreme edaphic conditions, as the almost total absence of soil. This factor causes an increase of the xericity, such that not to allow the settlement to phanerogamae, even is represented by ephemeral microphytes.

14.2 *Vulpio ligusticae-Trisetarietum aureae* Brullo 1975 (Table 14.2)

The *Thero-Sedetum caerulei* is replaced in the surfaces where the soil layer is significantly more deeper, by ephemeral grasslands physiognomically characterized by *Trisetaria aurea*. This vegetation, belonging always to *Trachynion distachyae*, is referrable to *Vulpio ligusticae-Trisetarietum aureae* Brullo 1975, association until now known only in Sicily (Bartolo et al. 1982). In Malta, among the characteristic species occur *Trisetaria aurea*, *Trifolium nigrescens* subsp. *nigrescens*, *Vulpia ciliata* and *Plantago lagopus*, which are associated with numerous annual species of the *Stipo capensis-Trachynietea distachyae*. This vegetation is linked to natural environments, such as the rocky ledges or the slight depressions of the calcareous rocky outcrops, where a soil layer of a few centimeters is deposited. In deeper depressions, where the rainwater persists for a long time, the association at issue is usually vicaried by communities of the *Isoeto-Nanojuncetea*.

14.3 *Ononido sieberi-Stipelletum capensis* ass. nova hoc loco (Table 14.3)

Holotypus: ril. 1, hoc loco.

The limestone outcrops characterized by large and flat surfaces, covered by a thick layer of soil, are sometimes colonized by dense grasslands dominated by *Stipella capensis* (=*Stipa capensis*). This vegetation is dynamically connected to the *Vulpio ligusticae-Trisetarietum aureae*, replacing it in the stands with deeper soils. Floristically, this plant community is characterized by the occurrence of *Ononis sieberi,* that allows to differentiate a new association proposed as *Ononido sieberi-Stipetum capensis*, which in the Maltese Islands is not very frequent. For its floristic and ecological peculiarity, this ephemeral community must be included in the *Trachynion distachyae*, alliance represented by numerous species, as well as those ones of the relative higher rank. For the dominance of *Stipella capensis* this association is closely related to *Ononido breviflorae-Stipetum capensis* Brullo, Guarino & Ronsisvalle 2000, association distributed in southern Sicily.

Table 14.2 *Vulpio ligusticae-Trisetarietum aureae*

Number of relevé	1	2	3	4	5	6	7	8
Plot size (m^2)	2	2	1	2.5	1	1	3	2
Plant cover (%)	90	70	90	80	100	70	70	70
Char. Association								
Trisetaria aurea	4	4	4	2	5	4	4	3
Trifolium nigrescens subsp. *nigrescens*	+	1	2	2	1	1	+	1
Vulpia ciliata	2	+	2	3	1	1	1	2
Plantago lagopus	+	1	2	2	2	2	1	.
Char. All. (*Trachynion distachyae*) and Cl. (*Stipo-Trachynietea distachyae*)								
Trachynia distachya	+	+	.	1	1	+	1	+
Moraea sisyrinchium	1	1	+	.	1	1	+	1
Stipella capensis	1	.	1	2	2	1	+	2
Evax pygmaea	+	1	+	.	1	+	+	.
Euphorbia exigua	+	+	+	+	.	1	.	1
Linum strictum	.	+	1	+	1	+	+	.
Hypochoeris achyrophorus	+	.	1	+	+	.	+	.
Trifolium scabrum	.	.	1	+	+	1	.	+
Medicago minima	+	.	+	.	.	+	.	+
Ranunculus paludosus	1	+	.	.	+	.	.	+
Hyoseris scabra	1	.	+	+	.	.	+	.
Romulea melitensis	.	.	+	.	1	+	+	.
Valantia muralis	.	.	+	+	.	.	+	.
Arenaria leptoclados	.	.	+	.	.	.	.	.
Trigonella monspeliaca	.	.	.	+	.	.	.	.
Other species								
Plantago afra subsp. *afra*	+	.	+	.	+	.	+	+
Silene colorata	.	+	+	+	1	.	+	.
Catapodium rigidum	.	+	+	+	.	+	+	.
Trifolium stellatum	+	+	.	1	1	.	.	1
Anisantha madritensis	.	.	1	2	+	.	+	.
Poa bulbosa	.	.	1	1	+	.	.	1
Lotus edulis	+	+	.	+	.	.	+	.
Medicago polymorpha	.	.	+	.	.	1	+	.
Bellardia trixago	.	.	+	+	.	.	+	.
Medicago littoralis	.	+	.	.	.	.	.	+
Bellis annua	.	.	.	1	.	.	+	.
Parapholis incurva	.	.	+	.	.	.	+	.
Scorpiurus muricatus	.	.	.	+	.	.	.	+
Rostraria cristata	.	.	.	.	.	+	.	+
Hedypnois cretica	.	.	.	.	.	+	+	.
Aegylops geniculata	.	.	.	1	.	.	.	.
Melilotus sulcata	.	.	.	.	+	.	.	.
Salvia verbenaca	.	.	.	.	.	+	.	.

Rel. 1–2: Malta, Mtahleb, 12.4.1987; Rel. 3–4: Malta, St. Andrews, 11.4.1987; Rel. 5: Gozo, Xlendi Valley, 13.4.1987; Rel. 6: Malta, Wied Incita, 16.4.1987; Rel. 7: Gozo, Ta Cenc, 13.4.1987; Rel. 8: Comino, 15.4.1987

Table 14.3 *Ononido sieberi-Stipelletum capensis*

Number of relevé	1	2
Plot size (m^2)	10	10
Plant cover (%)	90	90
Char. Association		
Ononis sieberi	2	1
Char. All. (*Trachynion distachyae*) and Cl. (*Stipo-Trachynietea distachyae*)		
Stipella capensis	5	4
Trachynia distachya	1	1
Trifolium scabrum	+	2
Euphorbia exigua	+	+
Ononis reclinata	1	+
Hedypnois rhagadioloides	+	1
Sideritis romana	+	+
Ammoides pusilla	+	+
Hypochoeris achyrophorus	1	1
Linum strictum	1	1
Medicago minima	.	1
Moraea sisyrinchium	.	+
Romulea melitensis	.	1
Lagurus ovatus subsp. *nanus*	.	+
Gastridium ventricosum	.	+
Anisantha rubens	+	.
Evax pygmaea	.	+
Euphorbia sommieriana	.	+
Other species		
Bellardia trixago	+	+
Galactites elegans	+	1
Blackstonia serotina	1	+
Trifolium stellatum	1	1
Vulpia ciliata	1	1
Centaurium pulchellum	1	1
Hippocrepis multisiliquosa	+	+
Plantago lagopus	1	2
Lotus ornithopodioides	1	+
Anacamptis coriophora	+	.
Anagallis arvensis	+	.
Medicago turbinata	.	+
Loncomelos narbonensis	1	.
Convolvulus elegantissimus	+	.
Daucus carota	+	.
Lotus edulis	+	.
Medicago littoralis	.	+
Aegylops geniculata	.	+

Rel.1–2: Malta, Pembroke (St. Andrews), 29.04.2007

14.4 *Allietum lojaconoi* Brullo 1985 (Table 14.4)

In the inner stands of the rocky coast and mainly in the clearings inside the *Anthyllido melitensis-Euphorbietum melitensis* or *Periploco-Euphorbietum dendroidis*, it is frequent an ephemeral vegetation characterized by the occurrence of *Allium lojaconoi*, small geophyte endemic to the Maltese archipelago (Brullo et al. 1982). This vegetation, described by Brullo (1985) and named *Allietum lojaconoi* (Fig. 14.1a), is very frequent in all islands. Both for the floristic set and ecology, this association must be included in the *Plantagini coronopi-Catapodion balearici* Brullo 1985 corr. Brullo nom. corr. hoc loco. Previously, this Mediterranean alliance was described by Brullo et al. (1985) as *Plantagini-Catapodion marini*, but *Catapodium marinum* L. used for name-giving taxon is incorrect, since according to Brullo et al. (2003) this species is distributed in the Atlantic coasts of Europe, while in the Mediterranean territories it is replaced mainly by *Catapodium balearicum* (Willk.) H. Scholz. Therefore, basing on Art. 43 and 48 of the ICPN (Weber et al. 2000) the name of this syntaxon must be corrected. In this association the characteristic species of the alliance are represented by *Catapodium balearicum*, *Plantago coronopus*, *Crepis pusilla*, *Bellis annua*, *Romulea melitensis*, *Lagurus ovatus* subsp. *nanus*, etc. The species of the order *Stipo-Bupleuretalia semicompositi* are frequent too, such as *Moraea sisyrinchium*, *Parapholis incurva*, *Medicago littoralis*, *Convolvulus lineatus*, *Asteriscus aquaticus*, *Hainardia cylindrica*, etc.

Apart from the subass. *typicum* (ril. 1–10), another two subassociations can be identified. The first one is proposed as *anthemidietosum urvilleanae* subass. nova hoc loco (ril 11–17, holotypus: ril. 16 hoc loco), which is linked to stands characterized by deeper soils and as differential species must be considered *Anthemis urvilleana*, endemic of Maltese islands; the second one is proposed as *linetosum trigyni* subass. nova hoc loco (ril 18–22, holotypus: ril. 19 hoc loco), localized in the small wet depressions of calcareous plateaux above the cliffs, which is differentiated by *Linum trigynum*. This association shows close relationship with other subhalophilous ephemeral micro-associations described in other Mediterranean territories, such as *Anthemido secundirameae-Desmazieretum siculae* Brullo 1985 from North Sicily, *Anthemido secundirameae-Allietum lehmannii* Brullo & Scelsi 1998 from South Sicily, *Filagini cossyrensis-Daucetum lopadusani* Brullo 1985 from Lampedusa, and *Oglifetum lojaconoi* Brullo 1985 from Linosa and Pantelleria.

14.5 *Silenetum melitensis* ass. nova hoc loco (Table 14.5)

Holotypus: ril. 1, hoc loco

In the plateaux above the cliffs characterized by marls or marly limestone with shallow soils, a procumbent vegetation dominated by annual xerophytes are frequent. These stands are often colonized by *Silene melitensis* (Fig. 14.1c), peculiar

Table 14.4 *Allietum lojaconoi*

Number of relevé	1	2	3	4	5	6	7	8	9	10	11	12	13	14	15	16	17	18	19	20	21	22
Plot size (m^2)	1	1	2	2	1	3	2	5	5	1	2	2	2	1	1	2	1	1	1	1	1	1
Plant cover (%)	50	70	40	50	50	50	40	50	50	50	50	50	50	50	50	50	50	40	50	40	50	30
Char. Association																						
Allium lojaconoi	1	1	1	1	+	2	1	1	2	1	2	+	1	+	1	2	1	2	2	1	2	1
Diff. Subassociation																						
Anthemis urvilleana	.	.	.	+	+	.	.	.	.	.	.	2	2	1	3	2	2	.	.	.	.	.
Linum trigynum	.	.	.	.	.	.	.	.	.	.	.	.	.	.	.	.	.	2	3	2	1	2
Char. All. (*Plantagini-Catapodion balearici*)																						
Catapodium balearicum	1	2	2	+	1	1	2	+	2	1	1	2	1	2	1	2	+	2	1	1	2	1
Romulea melitensis	1	+	+	+	1	+	1	1	.	2	2	+	1	1	.	1	1	1	1	+	1	1
Plantago coronopus	2	2	1	2	2	1	1	2	+	2	1	2	1	2	1	+	2	.	+	1	.	.
Bellis annua	1	+	.	1	+	.	+	.	.	.	.	.	+	+	+	1	+	.	.	.	.	.
Lagurus ovatus subsp. *nanus*	+	+	.	.	.	.	.	.	+	.	.	.	+	.	+	.	1	.	.	.	.	.
Echium parviflorum	1	.	.	.	.	.	.	.	+	.	+	+	+	.	.	+	.	.	.	.	.	.
Ononis sieberi	.	1	.	+	.	+	1	.	.	.	.	.	.	.	.	1	.	.	.	.	.	.
Euphorbia sommieriana	.	.	.	.	.	.	.	.	.	.	.	.	.	.	.	.	.	.	+	+	1	.
Crepis pusilla	.	.	.	.	.	.	.	.	.	.	2	1	.	.	.	.	.	.	.	.	.	.
Catapodium zwierleinii	.	.	.	.	.	.	.	.	.	.	.	.	.	.	.	.	.	+	+	.	.	.
Char. Ord. (*Stipo-Bupleuretalia semicompositi*)																						
Moraea sisyrinchium	3	3	1	2	2	1	2	1	1	2	3	2	1	1	2	3	2	1	2	2	2	2
Medicago littoralis	2	+	+	+	1	+	1	1	+	2	1	.	1	+	.	1	1	1	+	+	1	1
Parapholis incurva	+	1	+	1	+	+	1	+	1	+	.	2	+	+	.	+	+	.	.	.	.	.
Anisantha fasciculata	.	.	+	1	.	+	+	+	.	.	.	+	.	+	+	.	+	1	2	2	+	1

Convolvulus lineatus	2	1	+	1	+	.	1	+	+	.	.	1	2	+	1	1	.	.	.	.	.	.
Asteriscus aquaticus	1	+	.	+	+	.	+	.	.	.	.	+	.	.	.	1	+	+	.	1	+	.
Sulla capitata	.	+	.	+	1	+	+	.	.	.	.	.	.	+	1	+	.	.	.	.	.	.
Hainardia cylindrica	.	.	+	1	+	.	+	+	.	.	.	.	+	.	+	.	.	.	.	.	.	.
Trigonella monspeliaca	1	1	.	+	+	.	.	.	.	.	.	.	.	.	.	.	1	.	.	.	.	.
Astragalus sesameum	1	+	.	.	.	.	.	.	.	.	.	.	.	.	.	1	+	.	.	.	.	.
Crupina crupinastrum	.	.	+	.	.	.	+	+	.	.	.	.	.	.	.	.	.	.	.	.	.	.
Ononis ornithopodiodes	.	.	.	+	+	+	.	.	.	.	.	.	.	.	.	.	.	.	.	.	.	.
Bupleurum semicompositum	.	.	.	.	.	.	.	.	.	.	.	.	+	.	+	.	.	.	.	.	.	.
Char. Cl. (*Stipo-Trachynetea distachyae*)																						
Trachynia distachya	+	1	+	+	+	1	+	+	2	+	.	+	.	+	+	.	1	.	1	1	.	+
Evax pygmaea	1	2	.	+	+	.	.	+	1	1	1	+	2	1	1	+	.	1	+	.	+	1
Trifolium scabrum	+	1	+	+	+	1	+	+	+	1	+	.	.	1	+	1	.	.	.	.	.	.
Stipella capensis	1	.	1	+	.	+	1	.	.	.	.	.	+	+	.	1	+	2	2	1	2	1
Hypochoeris achyrophorus	1	.	+	.	+	+	+	.	1	.	.	.	.	.	+	+	.	.	+	+	.	+
Euphorbia exigua	1	+	.	.	+	+	+	.	2	.	.	.	1	.	1	+	2	.	.	.	.	.
Ranunculus paludosus	1	2	.	.	.	.	.	.	2	1	+	2	.	.	+	+	1	.	.	.	.	.
Linum strictum	+	+	+	+	.	+	.	+	.	.	.	.	.	.	+	.	.	+	.	+	.	.
Valantia muralis	+	.	.	+	+	.	1	+	+	.	.	1	.	+	.	.	.	.	.	.	.	.
Medicago minima	+	+	.	+	+	.	+	.	.	.	.	.	.	+	+	+	.	.	.	.	.	.
Hyoseris scabra	+	.	.	.	.	+	.	.	.	1	1	.	.	.	+	.	+	.	+	.	+	.
Sedum rubens	.	.	.	+	.	+	+	.	.	.	.	.	+	.	.	.	.	.	.	.	.	.
Arenaria leptoclados	.	.	.	+	+	+	.	+	.	.	.	.	.	.	.	.	.	.	.	.	.	.
Sideritis romana	+	.	.	.	.	.	.	.	.	.	.	+	.	.	+	+	.	.	.	.	.	.
Ononis reclinata	+	.	.	.	.	.	.	.	.	+	.	.	.	.	.	.	+	.	.	.	.	.
Silene nocturna	.	.	+	.	.	+	.	.	.	.	.	.	.	.	.	.	.	.	.	.	.	.

(continued)

Table 14.4 (continued)

Number of relevé	1	2	3	4	5	6	7	8	9	10	11	12	13	14	15	16	17	18	19	20	21	22
Plot size (m^2)	1	1	2	2	1	3	2	5	5	1	2	2	2	1	1	2	1	1	1	1	1	1
Plant cover (%)	50	70	40	50	50	50	40	50	50	50	50	50	50	50	50	50	50	40	50	40	50	30
Other species																						
Hedypnois cretica	+	1	+	1	+	+	+	1	.	+	.	.	.	+	+	.	.	.	1	2	+	.
Plantago serraria	+	1	+	+	.	+	.	.	.	2	1	.	.	.	.	1	+	.	1	1	.	+
Catapodium rigidum	.	.	1	+	1	1	+	1	.	+	.	.	.	.	.	+	1	.	.	.	.	.
Reichardia picroides var. *picroides*	+	.	.	1	+	.	.	+	+	.	.	+	.	+	+	+	.	.	.	.	.	.
Medicago truncatula	1	+	.	+	+	.	+	.	.	.	.	.	.	+	+	+	.	.	.	.	.	.
Lotus edulis	1	.	+	+	.	.	+	.	+	.	.	.	.	.	.	+	+	.	.	.	.	.
Trifolium stellatum	+	.	.	.	+	.	1	+	.	.	.	.	.	.	+	+	+	.	.	.	.	.
Rostraria cristata	1	.	.	.	1	.	+	+	.	.	+	.	.	.	+	+	.	.	.	.	.	.
Dactylis hispanica	+	1	+	.	.	.	.	+	.	.	.	.	1	.	+	1	.	.	.	.	.	.
Scorpiurus muricatus	.	1	.	.	+	+	+	+	.	.	.	.	.	.	.	.	+	.	.	.	.	.
Plantago lagopus	+	+	.	+	.	.	+	.	.	1	.	.	.	.	.	.	+	.	.	.	.	.
Poa bulbosa	+	1	.	.	.	.	.	.	.	2	1	.	.	.	+	+	.	.	.	.	.	.
Aegylops geniculata	.	.	+	+	1	.	+	+	.	.	.	.	.	.	.	.		.	.	.	.	.
Trisetaria aurea	.	+	.	.	.	.	.	.	.	+	+	.	.	.	.	+	+	.	.	.	.	.
Silene colorata	+	+	.	.	.	.	.	.	+	.	.	.	.	.	+	+	.	.	.	.	.	.
Filago pyramidata	.	.	+	1	+	+	.	1	.	.	.	.	.	.	.	.	.	.	.	.	.	.
Echium arenarium	+	.	.	.	.	.	.	.	.	.	.	.	+	.	+	1	+	.	.	.	.	.
Galium murale	1	.	.	.	.	.	.	.	.	.	.	.	2	2	+	1	.	.	.	.	.	.
Senecio leucanthemifolius	+	.	.	.	.	.	.	.	.	.	.	1	.	.	1	+	.	.	.	.	.	.
Anagallis arvensis	.	+	.	.	.	.	.	.	.	.	.	.	+	.	+	.	+	.	.	.	.	.

Centaurium pulchellum	.	.	.	+	+	+	.	+	.	.	.	.	.	.	.	.	.	.	.	.	.	.
Daucus carota	.	.	.	+	+	.	+	.	.	.	.	.	.	.	.	+	.	.	.	.	.	.
Rumex bucephalophorus subsp. *bucephalophorus*	+	+	.	.	.	.	.	.	.	.	.	.	.	+	.	+	.	.	.	.	.	.
Medicago polymorpha	.	.	+	.	.	.	+	+	.	.	.	.	.	.	.	.	.	.	.	.	.	.
Anisantha rubens	.	.	.	+	.	+	+	.	.	.	.	.	.	.	.	.	.	.	.	.	.	.
Trifolium nigrescens subsp. *nigrescens*	+	+	.	.	.	.	.	.	.	.	+	.	.	.	.	.	.	.	.	.	.	.
Spergularia marina	.	.	.	.	.	.	.	.	.	.	.	.	+	+	.	.	.	.	.	.	.	.
Frankenia hirsuta	+	.	.	.	.	.	.	.	.	.	.	.	+	+	.	.	.	.	.	.	.	.
Colchicum cupanii	+	+	.	.	.	.	.	.	.	.	.	.	.	.	.	.	.	.	.	.	.	.
Filago cf. *cosyrensis*	.	.	.	.	.	.	.	.	.	1	+	.	.	.	.	.	.	.	.	.	.	.
Sagina maritima	.	.	.	.	.	.	.	.	.	.	.	.	+	+	.	.	.	.	.	.	.	.
Avena barbata	.	.	.	.	+	+	.	.	.	.	.	.	.	.	.	.	.	.	.	.	.	.
Hordeum leporinum	.	.	.	.	.	+	+	.	.	.	.	.	.	.	.	.	.	.	.	.	.	.
Micromeria microphylla	.	.	.	+	.	.	+	.	.	.	.	.	.	.	.	.	.	.	.	.	.	.
Charybdis pancration	.	.	.	+	.	.	.	+	.	.	.	.	.	.	.	.	.	.	.	.	.	.
Leontodon tuberosus	.	+	.	.	.	.	.	.	.	+	.	.	.	.	.	.	.	.	.	.	.	.
Desmazeria pignattii	+	.	.	.	.	.	.	.	.	.	.	+	.	.	.	.	.	.	.	.	.	.
Cichorium spinosum	.	.	.	.	.	.	.	.	.	.	.	+	.	.	+	.	.	.	.	.	.	.
Bituminaria bituminosa	+	.	.	.	.	.	.	.	.	.	.	.	.	.	+	.	.	.	.	.	.	.
Anacamptis urvilleana	.	+	.	.	.	.	.	.	.	.	.	.	.	.	.	.	+	.	.	.	.	.

Rel. 1: Malta, Dragunara, 8.4.1984; Rel. 2: Malta, Dingli, 12.4.1984; Rel. 3: Gozo, Ta Cenc, 29.6.1973; Rel. 4–5: Malta, Wied Babu, 27.6.1973; Rel. 6: Malta, Dragunara, 24.6.1973; Rel. 7: Malta, St. Andrews, 25.6.1973; Rel. 8: Gozo, Dwejra, 29.6.1973; Rel. 9–10: Malta, Marfa, Madonna, 9.4.1984; Rel. 11–12: Malta, Dingli, 12.4.1984; Rel. 13–15: Malta, Mtahleb, 12.4.1987; Rel. 16–17: Malta, Mellieha, 12.4.1987; Ril. 18–22: Malta, Dingli, 21.5.2017

Fig. 14.1 Ephemeral xerophilous vegetation: (**a**) *Allietum lojaconoi*, (**b**) *Desmazerio pignattii-Senecionetum pygmaei*, (**c**) *Silenetum melitensis*, (**d**) *Catapodio balearici-Linarietum pseudolaxiflorae*

therophyte releted to *Silene crassiuscula* Brullo et al. (2017b), which is linked to sub-halophitic edaphic conditions. This species grows together with many others micro-xerophytes of the *Plantagini coronopi-Catapodion balearici* and higher rank, such as *Catapodium balearicum*, *Anthemis urvilleana*, *Plantago coronopus*, *Romulea melitensis*, *Bellis annua*, *Spergularia salina*, *Moraea sisyrinchium*, *Parapholis incurva*, *Medicago littoralis*, *Anisantha fasciculata*, etc. The association usually grows in the clearings localized in the flat surfaces, where are frequent the shrubs communities of *Suaedo verae-Darnielletum melitensis*.

14.6 *Phalaridetum paradoxae* ass. nov. hoc loco (Table 14.6)

Holotypus: ril. 1, hoc loco

In the clayey badlands markedly eroded by sliding water, colonized by *Lygeum spartum* grasslands, often it is possible observe on the small less steep surfaces an annual vegetation characterized by small therophytes. It is a very early community (March–April), localized in the clearings mixed with the perennial grassland, where several microphytes are frequent, as *Catapodium balearicum*, *Medicago*

Table 14.5 *Silenetum melitensis*

Number of relevé	1	2	3	4	5	6	7	8	9	10
Plot size (m^2)	2	4	2	2	2	2	2	3	2	3
Plant cover (%)	80	90	70	70	70	90	90	90	80	90
Char. Association										
Silene melitensis	3	2	3	3	2	2	3	2	4	3
Char. All. (*Plantagini-Catapodion balearici*)										
Catapodium balearicum	2	1	2	1	1	2	1	1	+	1
Anthemis urvilleana	3	2	2	1	1	3	3	3	2	3
Plantago coronopus	3	2	3	3	2	3	2	3	1	1
Romulea melitensis	3	2	1	2	1	2	1	2	1	1
Lagurus ovatus subsp. *nanus*	+	1	1	1	.	.	.	.	1	+
Bellis annua	2	1	1	1	2	.	.	.	.	.
Spergularia marina	2	2	1	+	2	.	.	.	.	.
Euphorbia sommieriana	.	.	.	.	.	+	.	+	.	.
Char. Ord. (*Stipo-Bupleuretalia semicompositi*)										
Moraea sisyrinchium	3	2	1	2	2	2	3	3	2	1
Medicago littoralis	2	4	2	2	2	3	3	2	2	3
Parapholis incurva	2	3	2	1	1	.	.	.	.	.
Anisantha fasciculata	2	1	1	+	1	.	.	.	.	.
Char. Cl. (*Stipo-Trachynetea distachyae*)										
Stipella capensis	2	2	3	2	2	1	2	1	1	2
Hypochoeris achyrophorus	1	1	.	+	+	+	1	+	+	1
Trachynia distachya	1	1	+	.	+	.	.	.	.	.
Linum strictum	+	+	+	.	.	.	.	.	.	.
Other species										
Mesembryanthemum nodiflorum	2	1	+	.	1	1	+	+	1	+
Hedypnois cretica	+	.	+	+	.	1	1	2	1	2
Lolium rigidum subsp. *rigidum*	1	1	+	+	.	.	.	.	.	.
Catapodium rigidum	.	+	+	.	+	.	.	.	.	.
Centaurium pulchellum	+	+	.	.	+	.	.	.	.	.
Sonchus oleraceus	.	.	.	.	.	.	+	+	.	1
Echium sabulicola	.	.	.	.	.	.	+	.	1	+

Ril. 1–5: Gozo, Xlendi Tower, 19.05.2017; Ril. 6–10: Malta, Migra i. Ferha, Rabat, 20.05.2017

littoralis, *Parapholis incurva*, *Senecio leucanthemifolius*, *Stipella capensis*, *Trifolium scabrum*, *Hypochoeris achyrophorus*, *Linum strictum*, *Romulea columnae*, *Trachynia distachya*, *Hedypnois cretica*, etc. For the floristic set and ecology, this vegetation can be included in the *Plantagini-Catapodion balearici*, alliance of the *Stipo-Trachynetea distachyae*. The occurrence of *Phalaris paradoxa* allows to attribute it to a new association, named *Phalaridetum paradoxae*, having as differential the species at issue.

Table 14.6 *Phalaridetum paradoxae*

Number of relevé	1	2	3
Plot size (m^2)	2	1	1
Plant cover (%)	80	80	80
Char. Association			
Phalaris paradoxa	2	1	3
Char. All. (*Plantagini-Catapodion balearici*) and Ord. (*Stipo-Bupleuretalia semicompositi*)			
Catapodium balearicum	1	2	1
Medicago littoralis	4	2	3
Parapholis incurva	1	2	2
Senecio leucanthemifolius	1	+	+
Plantago coronopus	.	1	1
Bellis annua	.	+	.
Char. Cl. (*Stipo-Trachynietea distachyae*)			
Stipella capensis	1	2	1
Trifolium scabrum	+	1	1
Hypochoeris achyrophorus	2	1	1
Linum strictum	+	.	1
Romulea columnae	2	+	1
Trachynia distachya	3	2	1
Hedypnois cretica	+	+	.
Arenaria leptoclados	.	+	+
Medicago minima	.	.	1
Ononis reclinata	.	.	+
Other species			
Lotus edulis	1	+	1
Scorpiurus muricatus	2	1	1
Anagallis arvensis	1	1	+
Aegilops geniculata	+	2	1
Beta maritima	+	1	.

Rel. 1–3: Gozo, Hondoq presso Qala, 19.05.2017

14.7 *Catapodio balearici-Linarietum pseudolaxiflorae* ass. nova hoc loco (Table 14.7)

Holotypus: ril. 2, hoc loco.

In the small rocky ledges of limestone outcrops localized in the upper parts of cliffs and hills, ephemeral plant communities characterized or sometimes dominated by *Linaria pseudolaxiflora* can be observed (Fig. 14.1d). It is a rare insular endemism currently known from Linosa in the Pelagian Archipelago and Maltese Islands, where it is more frequent in Gozo and very rare in Malta and Comino (Mifsud 2007, 2013; Sciberras et al. 2012; Mifsud et al. 2016a). Examples of this vegetation occur in coastal and inland places, covering small surfaces characterized by shallow soils.

Table 14.7 *Catapodio balearici-Linarietum pseudolaxiflorae*

Number of relevé	1	2	3	4	5
Plot size (m^2)	2	1	2	2	2
Plant cover (%)	70	80	80	70	70
Char. Association					
Linaria pseudolaxiflora	2	2	2	2	1
Char. All. (*Plantagini-Catapodion balearici*) and Ord. (*Stipo-Bupleuretalia semicompositi*)					
Catapodium balearicum	+	2	2	1	+
Echium parviflorum	+	+	1	1	+
Romulea melitensis	+	+	1	1	+
Plantago afra subsp. *zwerlenii*	1	+	+	+	+
Trigonella monspeliaca	+	1	+	1	+
Catapodium pauciflorum	+	1	+	+	.
Char. Cl. (*Stipo-Trachynietea distachyae*)					
Stipella capensis	2	2	3	3	3
Medicago minima	2	+	1	1	1
Trifolium scabrum	1	2	2	1	1
Valantia muralis	+	1	+	1	+
Hyoseris scabra	+	+	1	+	+
Sideritis romana	+	1	1	+	+
Hypochoeris achyrophorus	+	1	+	+	+
Evax pygmea	.	1	+	+	.
Hippocrepis biflora	.	.	+	+	.
Linum strictum	.	+	.	.	.
Other species					
Ferula melitensis	+	+	+	+	+
Medicago lupulina	1	+	+	+	+
Silene colorata	+	1	+	+	+
Hedypnois rhagadioloides	+	1	1	1	+
Erodium ciconium	1	1	+	+	+
Oxalis pes-caprae	1	+	+	+	1
Chamaesyce peplis	1	2	1	1	1
Bituminaria bituminosa	+	+	+	1	+
Prospero autunnalis	+	1	1	+	+
Salvia verbenaca	+	+	+	+	.
Colchicum cupanii	.	+	1	1	+
Sherardia arensis	.	+	+	1	+
Poa bulbosa	1	1	+	+	.
Hippocrepis multisiliquosa	+	+	.	+	+
Hedypnois cretica	.	1	+	+	+
Lotus edulis	.	+	+	+	.
Trifolium stellatum	+	+	+	.	.
Erodium chium	.	+	+	+	.
Trifolium cherleri	.	.	+	+	+

(continued)

Table 14.7 (continued)

Number of relevé	1	2	3	4	5
Plot size (m^2)	2	1	2	2	2
Plant cover (%)	70	80	80	70	70
Lotus cytisoides	+	.	.	+	.
Tordylium apulum	.	.	+	.	+
Senecio vulgaris	+	.	.	.	.

Rel. 1–5: Gozo, Ghammar, 9-3-14

From the syntaxonomical point of view, for the occurrence of *Catapodium balearicum*, *C. pauciflorum*, *Echium parviflorum*, *Romulea melitensis*, *Plantago afra* subsp. *zwerlenii* and *Trigonella monspeliaca*, this micro-phytocoenosis must be included in the *Plantagini coronopi-Catapodion balearici,* alliance of *Stipo-Bupleuretalia semicompositi.* The characteristics of the *Stipo-Trachynietea distachyae* are also well represénted, such as *Stipa capensis*, *Medicago minima*, *Trifolium scabrum*, *Valantia muralis*, *Hyoseris scabra*, *Sideritis romana*, *Hypochoeris achyrophorus*, etc. Floristically, this vegetation for the occurrence of *Linaria pseudolaxiflora* and some microphytes of the alliance and higher syntaxa shows some relations with *Plantagini zwierleinii-Erodietum linosae* Brullo 1985, association described for Linosa, where it is very widespread on incoherent volcanic soils mainly in the clearings of the maquis (Brullo 1985; Brullo and Siracusa 1996). This association differs from the Maltese one, apart from the ecology, also floristically, since are frequent *Erodium neuradifolium* subsp. *linosae*, *Catapodium zwierleinii*, *Silene neglecta* and *Reichardia tingitana*, besides *Linaria pseudolaxiflora* is more rare. Therefore, the latter is proposed as new association and named *Catapodio balearici-Linarietum pseudolaxiflorae*, with *Linaria pseudolaxiflora* as characteristic species.

14.8 *Cutandio maritimae-Pseudorlayetum pumilae* ass. nov. hoc loco (Table 14.8)

Holotypus: ril. 1, hoc loco

In the more developed and less disturbed dune systems, such as those currently noticeable in Gozo, an ephemeral psammophilous vegetation belonging to *Vulpietalia* Pignatti 1953 (= *Malcolmietalia* Rivas Goday 1958) was surveyed. It is localized mainly in the flat surfaces more distant from the shoreline and is differentiated by several therophytes linked to sandy soil, such as *Cutandia maritima*, *Erodium laciniatum*, *Pseudorlaya pumila*, *Echium sabulicola*, *Polycarpon diplyllum*, *Vulpia fasciculata*, etc. According to literature data (Brullo and Grillo 1985; Díez-Garretas et al. 2003; Minissale and Sciandrello 2013), this plant community can be ascribed to *Alkanno-Maresion nanae* Rivas Goday ex Rivas

Table 14.8 *Cutandio maritimae-Pseudorlayetum pumilae*

Number of relevé	1	2	3	4	5	6
Plot size (m^2)	4	5	5	2	5	3
Plant cover (%)	80	70	90	90	80	90
Char. Association						
Pseudorlaya pumila	2	+	2	2	2	+
Char. All. (*Alkanno-Maresion nanae*) and Cl. (*Tuberarietea guttatae*)						
Cutandia maritima	1	2	2	1	2	2
Medicago littoralis	3	2	3	2	1	3
Erodium laciniatum	1	1	+	1	2	1
Echium sabulicola	2	1	2	2	1	+
Silene colorata	2	2	2	+	2	1
Lagurus ovatus subsp. *ovatus*	1	.	+	+	+	+
Polycarpon diphyllum	2	2	1	1	2	.
Anisantha rigida	2	2	3	3	1	.
Vulpia fasciculata	2	+	.	2	.	4
Rumex bucephalophorus subsp. *bucephalophorus*	.	.	.	.	.	+
Other species						
Anagallis arvensis	1	+	.	+	+	.
Lobularia maritima	1	1	.	+	1	.
Ononis ramosissima	1	1	.	+	1	.
Hedypnois cretica	+	.	.	+	1	+
Reichardia picroides var. *picroides*	+	.	+	.	.	.
Euphorbia terracina.	.	.	.	.	.	1

Rel. 1–5: Gozo, Ramla Bay, 13.4.1984; Rel. 6: Malta, Mellieha Bay, 26.6.1973

Goday & Rivas-Martínez 1963 corr. Diaz-Garretas et al. 2001, alliance distributed mainly in the western Mediterranean area. For its floristic peculiarity, it is proposed as a new association named *Cutandio-Pseudorlayetum pumilae,* which is characterized by *Pseudorlaya pumila* and *Cutandia maritima.*

Chapter 15
Xerophilous Grasslands

The perennial herbaceous communities characterized by a marked edaphic xericity are widespread in all Maltese Islands. Physiognomically, these grasslands are differentiated by big hemicryptophytes showing a caespitose or stoloniferous habit, as *Lygeum spartum*, *Hyparrhenia hirta*, *Brachypodium retusum*, *Andropogon distachyos, Oryzopsis miliacea*, etc. Usually, they represent edaphophilous associations or can be considered as secondary aspects due to woodland degradation processes. These grasslands are included in *Lygeo sparti-Stipetea tenacissimae*, class with Circum-Mediterranean distribution.

15.1 *Leontodonto tuberosi-Lygeetum sparti* ass. nov. hoc loco (Table 15.1)

Holotypus: ril. 6, hoc loco

The acclived badland occurring in the coastal environments of the Maltese islands are colonized by dry grasslands dominated by *Lygeum spartum* (Fig. 15.1a). It is a very specialized vegetation, linked to clayey or more rarely to clayey-marly substrata, strongly eroded in the surface by sliding water. Basing on literature data (Brullo et al. 1990, 2002a, b, 2009c), this vegetation is referable to the *Lygeo sparti-Stipetea tenacissimae*, a Mediterranean class grouping the perennial termophilous dry grasslands. This syntaxon is here represented by *Dactylis hispanica*, *Asphodelus ramosus*, *Charybdis pancration*, *Reichardia picroides*, *Lobularia maritima*, *Euphorbia pinea*, *Pallenis spinosa*, etc. The Maltese grassland differs from the other Mediterranean communities dominated by *Lygeum spartum* for the occurrence of *Leontodon tuberosus*, which must be considered as a differential species of a new association, named *Leontodonto tuberosi-Lygetum sparti*. Within this association two different subassociations can be recognized, such as subass. *typicum*, which represents the community localized on surfaces with slope not

S. Brullo et al., *The Vegetation of the Maltese Islands*, Geobotany Studies,
https://doi.org/10.1007/978-3-030-34525-9_15

Table 15.1 *Leontodonto tuberosi-Lygeetum sparti*

Number of relevé	1	2	3	4	5	6	7	8	9	10	11	12	13	14	15	16	17	18
Plot size (m^2)	50	50	70	40	50	30	50	80	50	40	100	50	50	20	50	50	100	100
Plant cover (%)	90	90	100	90	90	100	100	100	90	90	90	90	90	30	50	50	100	100
Fall	–	–	–	–	–	–	–	–	–	–	30	45	35	–	–	–	25	25
Exposure	–	–	–	–	–	–	–	–	–	–	SW	SW	SW	–	–	–	SW	SW
Char. Association																		
Leontodon tuberosus	2	1	1	+	1	2	2	+	1	1	+	+	+	+	+	1	+	+
Diff. Subassociation																		
Fagonia cretica	.	.	.	.	.	.	.	.	.	.				2	1	2	+	+
Char. Ord. (*Lygeo sparti-Stipetalia tenacissimae*) and Cl. (*Lygeo sparti-Stipetetea tenacissimae*)																		
Lygeum spartum	4	5	5	5	4	5	5	5	4	5	4	4	4	1	1	2	5	5
Dactylis hispanica	+	1	1	1	2	1	2	2	2	1	1	1	1	.	1	1	2	2
Charybdis pancration	+	+	.	+	1	1	1	2	1	+	+	.	.	.	+	.	+	+
Reichardia picroides var. *picroides*	+	.	+	+	.	+	.	+	1	1	1	1	1	+	.	+	1	1
Euphorbia pinea	+	.	1	+	.	+	+	1	2	+	+	.	.	.	+	+	+	+
Asphodelus ramosus	.	+	1	1	+	2	+	2	2	1	.	.		.	.	.		.
Pallenis spinosa	1	+	.	+	+	.	+	1	+	.	.	.	.	.	.	.	.	.
Carlina corymbosa	.	+	.	1	+	1	.	1	1	+	.	.	.	.	.	.	.	.
Dittrichia viscosa	.	+	.	+	.	1	1	+	.	.	.	.	.	.	.	.	.	.
Lobularia maritima	+	+	.	.	+	.	.	1	+	.	.	.	.	.	.	.	.	.
Ferula melitensis	.	+	.	+	.	.	.	+	1	.	+	.	.	.	.	.	.	.
Centaurea sicula	.	.	.	.	.	.	.	.	.	.	+	+	+	.	.	.	1	1
Chamaeleon gummifer	.	.	.	.	.	.	.	.	.	.	+	+	.	.	.	.	+	+
Hyparrhenia hirta	.	.	.	.	.	.	.	.	.	.	+	.	.	.	.	.	1	1
Foeniculum piperitum	.	.	.	.	.	.	.	.	.	.	+	.	+	.	.	.	.	.
Anthyllis vulneraria subsp. *maura*	.	.	.	.	.	.	.	.	.	.	.	.	.	.	.	.	+	+
Other species																		
Sulla coronaria	1	1	2	.	1	+	2	1	+	1	1	+	.	1	2	2	1	+

Hedypnois cretica	1	1	1	1	1	1	+	.	.	+	.	.	.	.	+	+		.
Beta maritima	.	+	.	+	+	.	.	.	.	.	.	+	+	+	+	1	+	+
Moraea sisyrinchium	+	+	.	1	1	+	1	+	.	.	.	.	.	.	.	.	.	.
Anisantha madritensis	+	.	1	+	+	.	.	+	+	.	.	.	.	.	.	+	.	.
Cynara cardunculus	.	.	.	+	+	.	.	.	1	.	.	.	.	2	3	2	.	.
Parapholis incurva	.	.	+	.	.	+	+	+	.	.	.	.	.	.	+	+	.	.
Anisantha fasciculata	1	+	1	.	+	.	.	.	.	.	.	.	.	+	+	.	.	.
Asteriscus aquaticus	.	.	.	.	+	+	.	.	.	+	.	.	.	1	+	1	.	.
Stipella capensis	+	1	.	1	.	.	1	.	+	.	.	.	.	.	.	1	.	.
Melilotus sulcata	+	.	.	.	+	.	.	+	+	+				.	.	.		.
Scorpiurus muricatus	+	.	.	+	.	+	.	.	+	.	.	.	.	.	.	+	.	.
Mesembryanthemum nodiflorum	.	+	.	+	+	.	.	.	.	.	.	.	.	+	+	.	.	.
Linum strctum	+	+	.	+	+	.	+	.	.	.	.	.	.	.	.	.	.	.
Sonchus oleraceus	.	+	+	.	.	+	.	+	.	.	.	.	.	.	.	.	.	.
Trachynia distachya	+	.	.	.	+	.	.	+	+	.	.	.	.	.	.	.	.	.
Asparagus aphyllus	.	.	.	+	+	.	.	1	+	.	.	.	.	.	.	.	.	.
Limonium melitense	.	.	.	.	.	.	.	.	.	.	2	1	2	.	.	.	1	1
Phagnalon graecum	.	.	.	.	.	.	.	.	.	.	+	2	+	.	.	.	1	+
Avena barbata	.	.	.	.	.	.	.	.	.	.	+	+	+	.	.	.	+	+
Galactites elegans	.	.	.	.	.	.	.	.	.	.	+	+	+	.	.	.	1	1
Lotus edulis	.	.	.	.	.	.	.	.	.	.	+	+	+	.	.	.	.	.
Lotus cytisoides	.	.	1	.	.	.	.	1	+	.	.	.	.	.	.	.	.	.
Allium melitense	.	.	.	.	.	.	.	.	.	.	+	.	.	.	.	.	+	+
Daucus rupestris	.	.	.	.	.	.	.	2	.	1	.	.	.	.	.	.	.	.
Aegylops geniculata	.	.	.	+	.	.	+	.	.	.	.	.	.	.	.	.	.	.
Jacobaea maritima subsp. *bicolor*	.	.	.	.	.	.	.	.	.	.	.	.	+	.	.	.	.	.

Rel. 1–4: Malta, il-Karraba (Ghajn Tuffieha), 12.4.1984; Rel. 5–6: Malta, Ghajn Tuffieha Bay, 12.4.1984; Rel. 7–8: Malta, Ghajn Tuffieha Bay, 28.9.1985; Rel. 9–10: Malta, Mtahleb, 12.4.1987; Rel. 11–13: Malta, Ta' Zammitello (Gnejna Bay), 29.04.2007; Rel. 14–17: Malta, il-Karraba (Ghajn Tuffieha), 12.4.1984; Rel. 14–17: Malta, il-Karraba (Ghajn Tuffieha), 29.4.2007

Fig. 15.1 Perennial herbaceous communities: (**a**) *Leontodonto tuberosi-Lygeetum sparti* at Għajn Tuffieħa Bay, Malta (**b**) *Convolvulo oleifolii-Lygeetum sparti* at Qala, Gozo (**c**) *Feruletum melitensis* at White Rocks, Malta

exceeding 40°, while in the surfaces most inclined (40°–70°) the previous one is replaced by subass. *fagonietosum creticae* subass. nova hoc loco (holotypus: ril. 14 hoc loco), characterized usually by the low coverage of *Lygeum spartum* and by the presence of *Fagonia cretica*. This vegetation, although floristically rather poor, can be included from the phytogeographical point of view in the *Moricandio arvensis-Lygeion sparti* Brullo, De Marco & Signorello 1990, alliance with Central Mediterranean distribution. In particular, among the *Lygeum spartum* associations known in the literature, the Maltese association is quite related to *Phagnalo annotici-Lygeetum sparti* Biondi & Mossa 1992, distributed in southern Sardinia and Sicily (Biondi and Mossa 1992; Brullo et al. 2000; Bazan et al. 2006; Giusso del Galdo et al. 2008).

15.2 *Convolvulo oleifolii-Lygeetum sparti* ass. nov. hoc loco (Table 15.2)

Holotypus: ril. 8, hoc loco

On the marly or clayey-marly substrata, the *Leontodonto tuberosi-Lygeetum sparti* is usually replaced by a different *Lygeum spartum* grassland (Fig. 15.1b). The last vegetation is characterized by the occurrence of two peculiar shrubs, such as

Table 15.2 *Convolvulo oleifolii-Lygeetum sparti*

Number of relevé	1	2	3	4	5	6	7	8	9	10	11	12	13	14	15	16	17	18
Plot size (m^2)	10	10	20	20	10	10	20	20	20	20	20	20	50	50	50	50	50	5
Plant cover (%)	90	80	90	90	80	80	80	90	90	70	70	90	100	100	90	100	90	80
Fall	–	–	–	–	–	–	20	20	30	40	45	45	20	25	20	25	–	–
Exposure	–	–	–	–	–	–	S	S	S	S	S	S	SO	SO	SO	SO	–	–
Char. Association																		
Convolvulus oleifolius	1	+	1	1	+	1	2	3	2	2	1	+	2	2	1	1	1	2
Limbarda crithmoides subsp. *longifolia*	2	1	1	1	+	+	2	2	1	+	+	+	.	+	1	3	3	2
Pteranthus dichotomus	.	.	.	.	.	.	.	.	.	2	3	1	.	.	.	.	.	.
Char. Ord. (*Lygeo sparti-Stipetalia tenacissimae*) and Cl. (*Lygeo sparti-Stipetetea tenacissimae*)																		
Lygeum spartum	5	4	4	5	4	4	4	5	4	3	3	5	5	4	4	4	4	4
Reichardia picroides var. *picroides*	1	+	2	2	1	1	+	1	+	+	+	+	2	1	2	1	1	1
Asphodelus ramosus	1	1	1	2	1	1	2	2	2	1	.	+	1	2	1	2	2	2
Charybdis pancration	+	1	2	2	1	+	1	1	1	.	+	1	2	2	2	1	+	1
Dactylis hispanica	1	1	2	2	2	1	1	+	+	.	+	.	1	2	1	2	2	1
Phagnalon graecum	.	.	.	.	.	.	1	+	1	1	+	1	2	2	2	1	2	1
Carlina corymbosa	.	.	.	.	.	.	2	1	1	1	+	+	+	2	1	+	.	.
Centaurea sicula	.	.	+	+	1	1	.	.	.	.	.	.	+	.	+	+	+	.
Chamaeleon gummifer	+	1	1	2	2	1	.	.	.	.	.	.	.	.	.	.	.	2
Foeniculum piperitum	.	.	+	+	1	.	.	.	.	.			+	+	.	.	1	.
Convolvulus lineatus	.	.	.	.	.	.	+	1	2	2	+	.	.	.	.	.	.	.
Lathyrus articulatus	.	.	.	.	.	.	.	.	+	1	.	+	.	.	.	.	+	.
Euphorbia pinea	.	.	.	.	.	.	.	.	.	.	.	.	.	+	1	2	1	.
Hyparrhenia hirta	.	.	.	.	.	.	.	.	.	.	.	.	.	1	1	+	.	.
Bituminaria bituminosa	.	.	.	.	.	.	1	+	.	.	.	+	.	.	.	.	.	.
Pallenis spinosa	.	.	.	.	.	.	.	.	.	.	.	.	.	+	+	1	.	.

(continued)

Table 15.2 (continued)

Number of relevé	1	2	3	4	5	6	7	8	9	10	11	12	13	14	15	16	17	18
Plot size (m^2)	10	10	20	20	10	10	20	20	20	20	20	20	50	50	50	50	50	5
Plant cover (%)	90	80	90	90	80	80	80	90	90	70	70	90	100	100	90	100	90	80
Fall	–	–	–	–	–	–	20	20	30	40	45	45	20	25	20	25	–	–
Exposure	–	–	–	–	–	–	S	S	S	S	S	S	SO	SO	SO	SO	–	–
Sixalix maritima	.	.	.	.	.	.	.	.	.	.	.	.	.	.	.	.	+	.
Other species																		
Asparagus aphyllus	.	+	+	+	1	+	1	+	1	+	+	+	+	1	2	1	1	.
Lotus cytisoides	2	2	2	1	1	1	1	1	+	1	1	+	.	.	.	.	.	1
Ferula melitensis	.	.	.	.	.	.	1	2	2	+	.	+	2	1	1	1	.	.
Limonium melitense	+	+	+	.	.	.	+	.	+	+	+	.	.	.	.	.	+	.
Sulla coronaria	.	.	.	.	.	.	2	1	2	1	+	2	.	.	.	.	2	.
Euphorbia melitensis	1	2	1	1	1	+	.	.	.	.	.	.	.	.	.	.	.	.
Allium melitense	.	.	.	.	.	.	+	+	.	+	1	1	.	.	.	.	.	1
Cynara cardunculus	.	.	.	.	.	.	.	.	.	.	+	+	.	+	+	+	.	.
Sonchus oleraceus	.	.	.	.	.	.	.	.	.	.	.	.	1	+	1	+	.	.
Beta maritima	.	.	.	.	.	.	.	.	.	.	.	.	1	1	+	1	.	.
Daucus carota	.	.	1	2	1	+	.	.	.	.	.	.	.	.	.	.	.	.
Plantago serraria	.	.	+	+	.	+	.	.	.	.	.	.	.	.	.	.	.	.
Scorpiurus muricatus	+	+	+	.	.	.	.	.	.	.	.	.	.	.	.	.	.	.
Avena barbata	.	.	.	.	.	.	.	.	.	.	.	.	2	1	1	.	.	.
Allium commutatum	.	.	.	.	.	.	.	.	.	.	.	.	+	+	+	.	.	.
Daucus gingidium	.	.	.	.	.	.	.	.	.	.	.	.	.	.	.	.	1	+
Diplotaxis tenuifolia	.	.	.	.	.	.	.	.	.	.	.	.	.	.	.	.	1	.
Darniella melitensis	.	.	.	.	.	.	.	.	.	.	.	.	.	.	.	.	2	.

Rel. 1–6: Malta, Ghar Lapsi (Siggewi), 1.5.2007; Rel. 7–12: Gozo, Hondoq, presso Qala, 19.5.2017; Rel. 13–16: Malta, Migra I. Ferha, Rabat, 20.5.2017; ril. 17: Malta, Bahrija, Fommir-Rih, 20.05.2017; ril. 18: Malta, Ghar Lapsi (Siggewi), 21.05.2017

Convolvulus oleifolius and *Limbarda crithmoides* subsp. *longifolia*, which are quite constant in this community. It is here localized also *Pteranthus dichotomus*, small rare North African species, which together with the other two aforesaid taxa are proposed as characteristic of a new association, named *Convolvulo oleifolii-Lygeetum sparti*. This association is very common in the Maltese Islands and covers often wide surfaces, that can be flat or very acclived. The species of *Lygeo sparti-Stipetea tenacissimae* class are frequent, such as *Reichardia picroides*, *Asphodelus ramosus*, *Charybdis pancration*, *Dactylis hispanica*, *Phagnalon graecum*, *Carlina corymbosa*, *Centaurea sicula*, *Chamaeleon gummifer*, *Foeniculum piperitum*, *Convolvulus lineatus*, *Lathyrus articulatus*, *Euphorbia pinea*, *Hyparrhenia hirta*, *Bituminaria bituminosa*, etc.

15.3 *Hyparrhenietum hirto-sinaicae* A. & O. Bolos & Br.-Bl. in A. & O. Bolos 1950 (Table 15.3)

Another peculiar type of dry grasslands linked to rocky surfaces is characterized by the dominance of *Hyparrhenia hirta*, usually associated with *H, sinaica* and *Andropogon distachyos*. It is quite widespread in the islands of the archipelago, where it is usually localized on calcareous and calcarenitic outcrops. This vegetation is referable to the *Hyparrhenietum hirto-sinaicae* A. & O. Bolos & Br.-Bl. in A. & O. Bolos 1950 (=*Hyparrhenietum hirto-pubescentis*), association belonging to *Hyparrhenion sinaici* Br.-Bl., P. Silva & Rozeira 1956 corr. G.C. Costa et al. 2001, alliance of *Cymbopogono-Brachypodietalia ramosi* Horvatic 1963 (=*Hyparrhenietalia hirtae* Rivas-Martinez 1978). This association at first described from Spain, is currently recorded from many territories of Western Mediterranean area (de Bolós and de Bolós 1950; de Bolós and Molinier 1958; Brullo et al. 1977; Rivas-Martinez 1978a; Bartolo et al. 1982; Díez-Garretas and Asensi 1999; Costa et al. 2001; Brullo et al. 2010; etc.). In this grassland the perennial xerophytes of the *Lygeo sparti-Stipetea tenacissimae* and *Cymbopogono-Brachypodietalia ramosi* are very frequent. Within this association the subass. *feruletosum melitensis* subass. nova hoc loco (holotypus: ril. 10 hoc loco) is described. It is differentiated by *Ferula melitensis* and for its more nitrophilous requirements can be considered a transition aspect towards the *Feruletum melitensis*. The association represents usually a degradation stage of the maquis of *Periplocion angustifoliae*, due to fire or cutting of the shrub layer.

15.4 *Chamaeleo gummiferis-Brachypodietum retusi* C. Brullo, Brullo, Giusso & Tomaselli 2006 (Table 15.4)

In the Maltese islands limitedly to the surfaces covered by minute clastic material, constituted by Miocene limestone, with soil superficial or removed by the erosion, a peculiar vegetation dominated by *Brachypodium retusum* occurs. This is a perennial

Table 15.3 *Hyparrhenietum hirto-sinaicae*

Number of relevé	1	2	3	4	5	6	7	8	9	10	11	12	13
Plot size (m^2)	50	50	50	30	25	50	40	50	50	100	20	20	20
Plant cover (%)	90	100	100	100	100	100	100	100	90	100	90	100	80
Char. Association													
Andropogon distachyos	3	5	4	1	2	4	4	4	3	1	1	6	2
Char. Subassociation													
Ferula melitensis	.	+	.	.	+	.	.	+	.	2	1	1	1
Char All. (*Hyparrhenion hirtae)* and Ord. (*Cymbopogono hirti-Brachypodietalia ramosi*)													
Hyparrhenia hirta	3	1	2	4	3	1	2	1	4	4	4	5	4
Convolvulus althaeoides	1	1	1	+	+	+	1	+	.	1	.	.	.
Convolvulus elegantissimus	.	1	+	.	1	1	.	.	.	2	1	1	2
Phagnalon graecum	1	+	+	.	1	.	1	.	1	.	.	.	.
Lathyrus articulatus	1	.	+	1	1	.	+	.	.	.	.	.	.
Dittrichia viscosa	1	1	+	.	.	+	1	.	.	.	.	.	.
Lobularia maritima	.	+	+	.	1	.	.	+	.	.	.	.	.
Hyparrhenia sinaica	.	.	1	1	.	1	2	.	.	.	.	.	.
Oryzopsis miliacea	+	.	.	+	.	.	+	.	.	.	.	.	.
Char. Cl. (*Lygeo sparti-Stipetetea tenacissimae*)													
Carlina corymbosa	1	2	1	1	2	1	+	2	2	2	1	1	+
Dactylis hispanica	2	3	1	2	2	2	2	2	2	2	2	1	1
Foeniculum piperitum	1	1	1	1	+	+	1	1	1	2	2	1	1
Charybdis pancration	1	2	1	2	2	1	1	2	2	3	+	1	+
Bituminaria bituminosa	2	1	2	2	3	+	1	.	1	1	2	1	2
Reichardia picroides var. *picroides*	.	1	+	+	.	+	1	1	1	1	1	1	1
Asphodelus ramosus	2	2	2	2	+	+	.	.	2	3	2	1	1
Hyoseris radiata	.	+	2	1	2	2	1	+	+	+	.	.	.
Pallenis spinosa	.	1	.	1	1	.	.	+	+	+	+	1	+

Chamaeleon gummifer	.	+	.	.	1	+	.	1	1	2	2	1	1
Leontodon tuberosus	.	1	1	+	1	1	1	1	.	.	.	.	.
Euphorbia pinea	.	.	+	.	+	+	.	.	+	.	1	1	1
Sanguisorba minor	.	1	+	.	1	.	+	+	.	+	.	.	.
Sedum sediforme	+	1	.	1	+	.	.	1	.	.	.	.	.
Kundmannia sicula	+	.	+	1	+	.	.	1	.	.	.	.	.
Anthyllis vulneraria subsp. *maura*	.	+	1	+	.	+	.	.	.	+	.	.	.
Narcissus tazetta	.	.	+	2	2	.	+	.	.	.	.	.	.
Centaurea sicula	.	.	.	.	.	.	.	.	.	1	.	1	1
Sixalix maritima	.	.	.	+	+	.	.	.	.	.	.	.	.
Foeniculum vulgare	.	.	.	.	.	.	.	.	1	.	.	.	.
Other species													
Daucus carota	1	1	1	+	+	+	1	+	1	1	.	.	.
Asparagus acutifolius	2	1	1	1	+	1	1	.	+	+	.	.	.
Gladiolus italicus	+	.	+	1	1	.	+	.	+	.	.	.	.
Allium melitense	.	.	.	.	.	.	.	.	+	+	+	+	.
Anacamptis urvilleana	.	.	1	+	+	+	.	.	.	.	.	.	.
Leopoldia comosa	.	+	+	1	+	.	.	.	.	.	.	.	.
Loncomelos narbonensis	.	+	+	.	.	.	.	.	+	1	.	.	.
Anisantha rubens	.	1	.	+	.	.	.	1	.	.	.	.	.
Centaurium pulchellum	.	+	.	+	+	.	.	.	.	.	.	.	.
Lagurus ovatus subsp. *ovatus*	+	+	.	.	.	.	.	+	.	.	.	.	.
Micromeria microphylla	.	.	+	1	1	.	.	.	.	.	.	.	.
Teucrium flavum subsp. *flavum*	.	1	+	.	.	+	.	.	.	.	.	.	.
Aira cupaniana	.	+	.	+	.	.	.	.	.	.	.	.	.
Allium subhirsutum	1	.	.	+	.	.	.	.	.	.	.	.	.
Anacamptis pyramidalis	.	.	.	.	.	.	.	.	1	+	.	.	.
Arisarum vulgare	+	.	.	.	+	.	.	.	.	.	.	.	.

(continued)

Table 15.3 (continued)

Number of relevé	1	2	3	4	5	6	7	8	9	10	11	12	13
Plot size (m^2)	50	50	50	30	25	50	40	50	50	100	20	20	20
Plant cover (%)	90	100	100	100	100	100	100	100	90	100	90	100	80
Cynara cardunculus	.	.	.	.	.	.	.	.	+	1	.	.	.
Fumaria thymifolia	.	.	+	+	.	.	.	.	.	.	.	.	.
Galactites elegans	.	.	.	.	.	+	.	.	1	.	.	.	.
Lonicera implexa	.	.	.	+	1	.	.	.	.	.	.	.	.
Scorpiurus muricatus	1	.	.	+	.	.	.	.	.	.	.	.	.
Jacobaea maritima subsp. *bicolor*	.	.	.	.	.	1	1	.	.	.	.	.	.
Tetragonolobus purpureus	+	.	+	.	.	.	.	.	.	.	.	.	.
Trachynia distachya	+	.	.	.	.	.	.	+	.	.	.	.	.
Allium roseum	.	.	.	.	.	.	+	.	.	.	.	.	.
Asperula aristata	.	.	.	.	.	.	.	.	.	+	.	.	.
Thymbra capitata	.	.	.	.	.	.	.	.	+	.	.	.	.
Euphorbia melitensis	.	.	.	.	.	.	.	.	1	.	.	.	.
Stipella capensis	.	.	.	.	.	.	.	.	1	.	.	.	.
Lotus ornithopodioides.	.	.	.	.	.	.	.	+	.	.	.	.	.
Moraea sisyrinchium	.	.	.	.	.	.	.	.	+	.	.	.	.
Phlomis fruticosa	.	.	.	.	.	.	2	.	.	.	.	.	.
Medicago orbicularis	.	.	.	.	.	.	.	.	.	+	.	.	.
Poa bulbosa	.	.	.	.	.	.	.+	.	.	.	.	.	.
Ruta chalepensis	.	.	.	.	.	1	.	.	.	.	.	.	.
Sedum sediforme	.	.	.	.	.	.	.	.	+	.	.	.	.
Thesium humile	.	.	.	+	.	.	.	.	.	.	.	.	.

Rel. 1: Malta, Ghar Lapsi, 10.4.1984; Rel. 2: Malta, St. Andrews, 11.4.1987; Rel. 3–5: Malta, St. Paul Bay, 11.4.1987; Rel. 6: Malta, Bingemma, 12.4.1987; Rel. 7: Malta, Salina Bay, 16.4.1987; Rel. 8: Malta, Fiddien, 16.4.1987; Rel. 9: Malta, Hagar Qim (Qrendi), 1.5.2007; Rel. 10: Malta, Pembroke (St. Andrews), 29.4.2007; Rel. 11–13: Malta, Hagar Qim (Qrendi), 21.5.2017

Table 15.4 *Chamaeleo gummiferis-Brachypodietum retusi*

Number of relevé	1	2	3	4	5	6	7	8	9	10	11	12
Plot size (m^2)	50	50	30	50	20	30	30	20	50	30	10	5
Plant cover (%)	100	100	100	100	100	100	100	90	90	80	80	100
Char. Association												
Chamaeleon gummifer	2	2	1	3	1	2	3	2	1	2	2	2
Bellis sylvestris	1	2	1	2	2	+	2	.	.	.	+	.
Char. All. (*Thero-Brachypodion ramosi*)												
Brachypodium retusum	5	4	5	4	4	4	3	4	5	4	5	5
Char. Cl. (*Lygeo sparti-Stipetetea tenacissimae*)												
Dactylis hispanica	1	1	+	1	1	1	2	1	2	1	1	1
Asphodelus ramosus	2	2	1	+	3	2	3	+	1	2	2	1
Charybdis pancration	1	+	1	1	2	1	1	1	+	2	2	1
Carlina corymbosa	.	+	1	1	2	1	2	2	+	1	1	1
Reichardia picroides var. *picroides*	+	+	+	1	+	.	+	1	+	1	.	+
Convolvulus elegantissimus	1	+	+	+	.	+	+	1	.	+	.	1
Bituminaria bituminosa	.	1	+	+	.	1	2	+	2	.	.	1
Foeniculum vulgare	.	.	.	.	.	.	.	+	1	1	1	+
Leontodon tuberosus	.	+	.	.	+	+	+	.	.	.	.	.
Pallenis spinosa	.	.	.	.	.	.	.	+	+	+	.	+
Hyparrhenia hirta	.	.	.	.	.	.	.	2	1	1	1	.
Kundmannia sicula	.	+	+	1	+	.	.	.	.	.	.	.
Sedum sediforme	.	.	.	.	.	.	.	1	.	+	1	.
Hyoseris radiata	.	.	.	.	.	.	.	+	.	+	.	.
Lathyrus articulatus	.	.	+	.	+	.	.	.	.	.	.	.
Anthyllis vulneraria subsp. *maura*	.	.	.	.	.	.	.	+	.	+	.	.
Convolvulus althaeoides	.	.	.	.	.	.	.	1	1	.	.	.
Centaurea sicula	.	.	.	.	.	.	.	.	+	.	.	+
Euphorbia pinea	.	.	.	.	.	.	.	.	+	.	.	2
Ferula melitensis	.	.	.	.	.	.	.	1	.	1	.	.
Sanguisorba minor	.	.	.	.	.	.	.	+	.	.	.	.
Other species												
Lotus ornithopodioides	+	.	+	+	+	.	+	+	+	1	.	.
Thymbra capitata	1	2	2	1	2	2	2	.	.	.	.	.
Teucrium fruticans	1	1	+	.	.	1	1	1	.	.	.	.
Fedia graciliflora	+	+	.	+	.	+	.	.	.	.	.	.
Scorpiurus muricatus	.	+	.	.	+	+	+	.	.	.	.	.
Phagnalon rupestre	+	+	.	+	+	.	.	.	.	.	.	.
Teucrium flavum subsp. *flavum*	.	.	+	1	+	.	+	.	.	.	.	.
Asparagus aphyllus	.	.	.	.	.	.	.	1	+	+	.	.
Sonchus oleraceus	.	+	+	.	.	+	.	.	.	.	.	.
Plantago serraria	.	.	+	1	1	.	.	.	.	.	.	.
Asperula aristata	.	.	.	.	1	.	.	+	.	+	.	.

(continued)

Table 15.4 (continued)

Number of relevé	1	2	3	4	5	6	7	8	9	10	11	12
Plot size (m^2)	50	50	30	50	20	30	30	20	50	30	10	5
Plant cover (%)	100	100	100	100	100	100	100	90	90	80	80	100
Daucus carota	.	.	.	.	.	.	.	1	+	1	.	.
Phagnalon graecum	.	.	.	.	.	.	.	1	.	2	1	.
Allium subhirsutum	.	.	.	.	.	.	.	+	.	+	.	.
Anacamptis pyramidalis	.	.	.	.	.	.	.	+	.	+	.	.
Galactites elegans	.	.	.	.	.	.	.	+	+	.	.	.
Lotus cytisoides	.	.	.	.	.	.	.	.	+	+	.	.
Moraea sisyrinchium	.	.	.	.	.	.	.	.	.	+	.	.
Loncomelos narbonensis	.	.	.	.	.	.	.	+	.	.	.	.
Anacamptis urvilleana	.	.	.	.	.	.	.	+	.	.	.	.
Lotus edulis	.	.	.	.	.	.	+	.	.	.	.	.

Rel. 1–7: Malta, Bahtija Valley, 11.4.1987; Rel. 8: Malta, Wadi Babu, near Blue Grotto (Zurrieq), 1.5.2007; Rel. 9: Malta, Hagar Qim (Qrendi), 1.5.2007; Rel. 10: Malta, Dingli, 1.5.2007; Rel. 11: Gozo, Triq Kalati Punic, 19.5.2017; Rel. 12: Malta: Hagar Qim (Qrendi), 21.5.2017

grassland linked to less exposed rocky slopes of the inland hills, which is characterized by several hemycriptophytes and geophytes, such as *Chamaeleon gummifer* (=*Carlina gummifera*), *Dactylis hispanica*, *Asphodelus ramosus*, *Bellis sylvestris*, *Charybdis pancration*, *Carlina corymbosa*, *Reichardia picroides*, *Bituminaria bituminosa*, *Convolvulus elegantissimum*, etc. From the phytosociological point of view, this vegetation, belonging to *Thero-Brachypodion ramosi* Br.-Bl. 1925, alliance of the *Lygeo sparti-Stipetea tenacissimae*, was described by Brullo et al. (2006) from this Archipelago and named *Chamaeleo gummiferis-Brachypodietum retusi*, having as characteristic species *Chamaeleon gummifer* and *Bellis sylvestris*, which grow mixed to the stolons of *Brachypodium retusum*. Dynamically, this association seems closely correlated to the degradation processes of the maquis of *Periploco angustifoliae-Euphorbietum dendroidis*.

15.5 *Diplotaxio tenuifoliae-Oryzopsietum miliaceae* Brullo 1984 (Table 15.5)

In road environments, such as the edges of asphalt roads and country paths, a perennial herbaceous vegetation, in which *Oryzopsis miliacea* and *Dittrichia viscosa* have a significant physiognomic role, is frequent. It is a typical hemicryptophytical vegetation, confined to more or less anthropic habitats, mainly of extra-urban areas. For its ecological and floristical features, it belongs to the *Bromo-Oryzopsion miliaceae* O. Bolòs 1970, alliance of *Lygeo sparti-Stipetea tenacissimae*, which groups associations linked to synantropic stands (de Bolós 1975a; Brullo 1984b;

Table 15.5 *Diplotaxio tenuifoliae-Oryzopsietum miliaceae*

Number of relevé	1	2	3	4	5	6	7	8	9	10	11	12	13	14	15
Plot size (m^2)	10	20	20	15	20	40	10	10	25	20	15	10	20	20	20
Plant cover (%)	90	100	100	100	80	100	100	100	100	100	100	90	100	100	80
Char. Association															
Diplotaxis tenuifolia	3	4	3	2	2	3	2	2	2	+	1	1	2	2	1
Diff. Subassociation															
Antirrhinum tortuosum	.	.	.	.	.	.	.	.	.	2	2	4	1	3	2
Char. All. (*Bromo-Oryzopsion miliaceae*)															
Oryzopsis miliacea	1	2	3	2	1	2	3	2	3	3	3	2	4	3	2
Dittrichia viscosa	1	2	2	2	1	2	3	2	3	2	3	1	2	1	3
Plumbago europaea	+	.	+	1	.	.	.	1	.	.	+	1	+	.	.
Lepidium graminifolium	.	.	+	.	.	.	+	.	1	.	.	.	.	+	+
Char. Ord. (*Cymbopogono hirti-Brachypodietalia ramosi*) and Cl. (*Lygeo sparti-Stipetetea tenacissimae*)															
Foeniculum piperitum	3	2	3	4	3	3	3	3	3	2	1	.	1	2	1
Hyparrhenia hirta	.	1	2	.	2	1	1	2	1	1	2	.	1	2	+
Convolvulus elegantissimus	.	+	+	1	.	+	+	+	1	+	+	1	+	.	1
Lobularia maritima	+	1	1	.	+	1	+	2	.	.	+	+	1	.	+
Euphorbia pinea	+	+	.	.	1	+	+	1	.	1	+	1	+	.	+
Dactylis hispanica	.	+	+	.	1	+	.	+	1	2	1	+	.	1	1
Asphodelus ramosus	.	+	1	1	.	2	+	1	.	.	1	.	.	1	1
Reichardia picroides var. *picroides*	.	+	+	+	.	+	+	.	.	2	+	.	+	+	.
Pallenis spinosa	.	+	+	1	.	+	.	.	.	1	+	1	.	+	1
Carlina corymbosa	2	1	1	.	.	1	+	.	1	1	.	+	+	.	.
Charybdis pancration	+	+	.	+	1	1	.	+	.	.	1	.	.	1	1
Brachypodium retusum	+	1	1	.	1	1	.	.	.	.	2	1	+	1	.
Bituminaria bituminosa	+	+	.	1	+	.	.	.	.	+	.	+	+	+	.
Lathyrus articulatus	+	+	.	+	.	.	.	.	+	.	.	+	+	1	.

(continued)

Table 15.5 (continued)

Number of relevé	1	2	3	4	5	6	7	8	9	10	11	12	13	14	15
Plot size (m^2)	10	20	20	15	20	40	10	10	25	20	15	10	20	20	20
Plant cover (%)	90	100	100	100	80	100	100	100	100	100	100	90	100	100	80
Hyoseris radiata	.	.	+	+	.	.	.	.	.	+	1	.	+	+	+
Anthyllis vulneraria subsp. *maura*	.	.	+	+	.	.	.	.	.	1	+	.	1	+	.
Verbascum sinuatum	.	.	+	.	+	+	.	.	.	.	.	+	1	+	.
Phagnalon graecum	.	+	.	.	+	+	1	.	+	.	.	.	.	.	.
Clinopodium nepeta	.	.	+	+	.	+	.	+	1	.	.	.	.	.	.
Salvia verbenaca	.	.	.	.	.	.	+	.	.	+	.	+	+	.	.
Other species															
Jacobaea maritima subsp. *bicolor*	1	2	2	2	1	1	.	+	.	2	1	.	2	+	.
Daucus carota	+	1	1	1	+	+	.	+	.	+	.	+	+	.	.
Asparagus acutifolius	1	.	1	+	.	+	.	.	1	1	.	+	+	+	.
Sonchus oleraceus	.	.	+	.	.	.	+	.	.	+	+	.	+	+	.
Carlina lanata	+	.	1	+	+	.	.	.	.	+	.	.	+	.	.
Galagtites elegans	.	+	.	.	.	.	.	+	.	+	+	+	.	+	.
Hedypnois cretica	.	+	.	.	.	.	+	.	.	.	1	1	.	.	+
Symphyotrichum squamatum	.	.	+	+	+	.	+	.	+	.	.	.	.	.	.
Oxalis pes-caprae	.	.	+	.	.	.	.	.	.	1	2	+	.	.	+
Parietaria judaica	.	.	.	.	+	+	.	.	+	.	1	.	1	.	.
Echium parviflorum	.	.	.	+	.	+	+	.	.	.	.	.	+	.	.
Avena barbata	+	.	.	.	.	.	.	+	.	.	+	+	.	.	.
Hypericum triquetrifolium	.	+	.	.	1	.	.	.	.	.	.	+	.	+	.
Anisantha diandra	.	.	.	.	+	.	.	.	.	.	.	+	1	.	1
Asparagus aphyllus	.	.	.	.	.	.	.	.	.	.	.	.	.	+	.

Rel. 1–3: Malta, Mellieha Ridge, 23.9.1985; Rel. 4: Malta, Mtahleb, 24.9.1985; Rel. 5–6: Malta, Mistra Valley, 10.4.1984; Rel. 7: Gozo, Nadur, 27.11.1985; Rel. 8: Malta, Luqa, 28.9.1985; Rel. 9–10: Malta, Roman Baths, 26.9.1985; Rel. 11–13: Malta, Selmun Palace, 10.4.1985; Rel. 14: Malta, Ta Saliba, 26.9.1985; Rel. 15: Gozo, Nadur, 27.11.1985

Brullo et al. 2001c, 2010). Well-represented and frequent are the species of this alliance and order, among them *Plumbago europea*, *Lepidium graminifolium*, *Foeniculum piperitum*, *Hyparrhenia hirta*, *Convolvulus elegantissimus*, *Lobularia maritima*, *Dactylis hispanica*, *Asphodelus ramosus*, *Reichardia picroides*, *Pallenis spinosa*, *Bituminaria bituminosa*, *Verbascum sinuatum*, etc. For the occurrence of *Diplotaxis tenuifolia*, often with high coverage values, this vegetation is attributable to *Diplotaxio tenuifoliae-Oryzopsietum miliaceae*, association described by Brullo (1984b) and distributed in the Central Mediterranean territories (Brullo et al. 2010). It is a xeric association growing on various substrata (limestones, marls, calcarenites, volcanites etc.). In the Maltese territory within the association two different variants can be recognized, represented by subass. *diplotaxietosum tenuifoliae*, corresponding to the typical aspect, and by subass. *antirrhenetosum tortuosi* subass. nov. hoc loco (Holotypus: 14, hoc loco), floristically characterized by *Antirrhinum tortuosum*, which is circumscribed to the cooler and shady road stands. This anthropogenic vegetation showing a pioneer character and can be considered a permanent coenosis of constantly disturbed habitats.

15.6 *Feruletum melitensis* ass. nov. hoc loco (Table 15.6)

Holotypus: rel. 1, hoc loco

In the uncultivated fields and more frequently in the old abandoned land, it is frequent a very peculiar subnitrophilous vegetation, physiognomically characterized by *Ferula melitensis* (Fig. 15.1c), which can be considered a Maltese vicariant of *F. communis*. It is a flashy plant, higher than 2 m, endemic to Maltese islands, described by Brullo et al. (2018). Basing on its floristic set and ecological requirements, this vegetation can be attributed to the *Bromo-Oryzopsion miliaceae*. In particular, it differs from the other communities of this alliance for the occurrence and dominance of *Ferula melitensis*, species proposed as characteristic of a new association named *Feruletum melitensis*. Physiognomically and ecologically, it shows close relations with *Thapsio-Feruletum communis* Brullo 1984a, b, another association of the *Bromo-Oryzopsion miliaceae*, described from Sicily, where it grows in very similar habitats (Brullo et al. 2010). On the whole, these two phytocenosis are well distinct from the structural and floristic aspects and therefore represent two geographic vicariant syntaxa. Recently, the nitrophilous communities rich in species of *Ferula*, surveyed in the Apennins, has been included by Biondi (see Biondi et al. 2016, 2017) in an independent class and order represented by the *Charybdido pancratii-Asphodeletea ramosi* and *Asphodeletalia ramosi,* gathering several alliances. This syntaxonomical arrangement seems quite acceptable and justified in the case of communities spread in territories affected by temperate-mesic bioclimate, as those ones of the central Italy, while in more dry conditions, as in habitats with Mediterranean markedly arid climate, this type of vegetation is enriched in steppe elements, especially cespitose grasses of *Lygeo sparti-Stipetea tenacissimae*.

Table 15.6 *Feruletum melitensis*

Number of relevé	1	2	3	4	5	6	7	8	9	10	11	12	13	14	15	16	17
Plot size (m^2)	50	100	100	80	50	40	50	100	100	100	50	50	100	100	100	100	100
Plant cover (%)	100	90	100	100	100	100	100	100	90	100	100	100	100	100	90	100	100
Char. Association																	
Ferula melitensis	3	2	2	1	2	2	1	1	+	1	1	1	1	4	3	3	3
Char. All. (*Bromo-Oryzopsion miliaceae*)																	
Oryzopsis miliacea	3	2	2	1	2	2	1	1	+	1	1	.	1	1	1	.	1
Dittrichia viscosa	2	2	2	1	2	+	2	+	.	1	+	1	+	1	+	.	+
Char. Ord. (*Cymbopogono hirti-Brachypodietalia ramosi*) and Cl. (*Lygeo sparti-Stipetetea tenacissimae*)																	
Charybdis pancration	1	1	1	2	2	1	1	2	2	3	2	2	2	1	2	1	2
Asphodelus ramosus	1	+	2	2	3	2	3	1	2	2	1	1	+	2	3	4	1
Foeniculum piperitum	1	2	2	2	+	1	2	2	2	2	1	1	1	3	2	1	2
Dactylis hispanica	1	2	2	1	2	1	1	2	1	2	3	2	1	.	.	1	+
Carlina corymbosa	1	1	1	+	1	1	2	3	1	3	3	3	2	.	.	1	2
Reichardia picroides var. *picroides*	+	1	1	+	+	+	+	.	+	+	1	+	+	1	+	.	1
Bituminaria bituminosa	.	+	+	.	+	2	+	.	2	3	2	3	2	2	2	+	1
Convolvulus althaeoides	1	1	2	+	1	+	1	1	1	2	1	1	+	.	.	.	.
Hyparrhenia hirta	.	1	.	2	+	1	1	.	1	2	1	1	1	1	2	.	2
Lobularia maritima	+	+	.	1	1	.	.	1	.	1	1	+	1	1	+	1	1
Euphorbia pinea	1	.	1	+	.	+	+	1	+	1	+	.	1	.	+	.	.
Narcissus tazetta	.	+	+	1	.	.	1	.	1	+	+	2	1	2	.	.	+
Hyoseris radiata	1	2	2	2	1	1	2	1	+	+	.	+	.	.	.	.	.
Anthyllis vulneraria subsp. *maura*	+	.	.	+	+	+	+	.	.	+	+	1	+	1	+	.	.
Lathyrus articulatus	+	.	1	+	+	.	+	.	1	2	2	.	.	.	.	1	+
Kundmannia sicula	1	+	1	.	+	+	1	.	.	1	1	+	.	.	.	.	.

Convolvulus elegantissimus	.	+	.	1	1	2	+	1	+	1	.	+	.	.	.	.	.
Verbascum sinuatum	1	+	1	+	.	.	+	+	+	.	.	+	.	.	.	.	.
Sanguisorba minor	.	+	1	1	+	.	+	.	+	.	.	+	+	.	.	.	.
Pallenis spinosa	.	1	1	+	.	.	1	1	.	+	1	+	.	.	.	.	.
Phagnalon graecum	1	+	1	+	.	+	.	.	+	+	.	.	.	.	.	.	.
Salvia verbenaca	.	.	1	+	.	.	.	+	.	+	.	1	1	.	.	.	.
Sedum sediforme	.	.	.	+	+	1	.	+	.	.	+	.	.	.	.	.	+
Hyparrhenia sinaica	.	.	.	.	.	+	1	+	.	1	+	.	.	.	.	.	.
Clinopodium nepeta	+	+	.	.	+	.	.	+	.	+	.	.	.	.	.	.	.
Brachypodium retusum	+	.	.	.	1	.	.	+	.	.	+	.	+	.	.	.	.
Andropogon distachyos	.	.	.	.	.	.	.	.	.	.	.	.	.	.	.	.	1
Other species																	
Oxalis pes-caprae	2	1	2	2	1	1	+	+	+	1	1	+	+	2	1	1	1
Sonchus oleraceus	+	+	+	+	1	+	+	.	+	+	+	+	+	+	+	.	+
Chamaeleon gummifer	.	1	1	+	1	2	1	+	2	2	2	1	+	.	+	+	+
Asparagus acutifolius	1	2	2	+	1	1	+	.	+	.	+	1	.	.	+	2	1
Galagtites elegans	1	+	.	+	+	.	.	+	+	+	.	+	.	1	+	+	1
Daucus carota	1	1	1	+	1	2	.	+	+	.	.	.	+	.	.	1	.
Jacobaea maritima subsp. *bicolor*	1	.	+	1	1	+	.	.	1	1	.	.	.	.	.	1	.
Teucrium fruticans	.	.	+	.	.	.	+	1	.	1	1	1	+	.	.	.	.
Smyrnium olusatrum	1	+	+	.	+	.	.	1	.	1	+	.	.	.	.	.	.
Reseda alba	+	+	.	+	.	.	+	.	+	+	.	.	.	.	.	.	.
Parietaria judaica	.	+	.	.	1	.	+	+	.	+	.	.	+	.	.	.	.
Thymbra capitata	.	.	+	.	1	.	+	+	.	.	.	1	1	.	.	.	.
Gladiolus italicus	.	+	1	+	.	.	.	.	.	1	.	.	.	.	.	.	.
Leopoldia comosa	.	.	+	.	+	.	.	.	+	.	.	.	+	.	.	.	.

(continued)

Table 15.6 (continued)

Number of relevé	1	2	3	4	5	6	7	8	9	10	11	12	13	14	15	16	17
Plot size (m^2)	50	100	100	80	50	40	50	100	100	100	50	50	100	100	100	100	100
Plant cover (%)	100	90	100	100	100	100	100	100	90	100	100	100	100	100	90	100	100
Scandix pecten-veneris	.	.	+	+	.	.	+	.	.	.	+	.	.	.	.	.	.
Ruta chalepensis	.	.	.	.	.	1	.	+	.	1	2	.	.	.	.	.	.
Antirrhinum tortuosum	.	.	.	.	.	.	.	+	2	+	.	.	.	.	.	.	.
Allium melitense	.	.	.	.	.	.	1	.	.	.	.	+	.	.	.	.	.
Arisarum vulgare	.	.	.	.	.	.	.	.	.	.	.	.	.	1	1	.	.
Glebionis coronaria	.	.	.	.	.	.	.	.	.	.	.	.	.	+	+	.	.
Borago officinalis	.	.	.	.	.	.	.	.	.	.	.	.	.	+	+	.	.
Anisantha sterilis	.	.	.	.	.	.	.	.	.	.	.	.	.	+	1	.	.
Hippocrepis multisiliquosa	.	.	.	.	.	.	.	.	.	.	.	.	.	+	.	.	.
Erodium malacoides	.	.	.	.	.	.	.	.	.	.	.	.	.	+	.	.	.
Trifolium nigrescens subsp. *nigrescens*	.	.	.	.	.	.	.	.	.	.	.	.	.	+	.	.	.
Sulla coronaria	.	.	.	.	.	.	.	.	.	.	.	.	.	.	.	+	.
Mercurialis annua	.	.	.	.	.	.	.	.	.	.	.	.	.	+	.	.	.
Ajuga iva	.	.	.	.	.	.	.	.	.	+	.	.	.	.	.	.	.
Lotus cytisoides	.	.	.	.	.	.	.	.	.	.	.	.	.	.	.	+	.
Vulpia ciliata	.	.	.	.	.	.	.	.	.	.	.	+	.	.	.	.	.

Rel. 1–3: Malta, St. Andrews, 11.4.1987; Rel. 4–6: Malta, Naxxar, 12.4.1987; Rel. 7–8: Malta, Mellieha Ridge, 11.4.1987; Rel. 9–11: Malta, Gigenti Valley, 13.4.1987; Rel. 12–13: Comino, 15.4.1987; Rel 14–15: Malta, White Rocks, St. Julias, 8.3.2014; Rel. 16: Gozo, Xlendi, 9.3.2014; Rel. 17: Gozo, Hala, 9.3.2014

15.7 *Romuleo melitensis-Ranunculetum bullati* ass. nov. hoc loco (Table 15.7)

Holotypus: ril. 3, hoc loco.

During the autumnal period in small flat surfaces placed within the maquis, garigues and grasslands, very peculiar plant communities characterized by geophytes and hemicryptophytes having their flowering at the early beginning of the rainy season (September–November) occur. In fact, this vegetation appears already well-structured in autumn when several cryptophytes have their optimum, as *Ranunculus bullatus* subsp. *bullatus, Romulea melitensis, Prospero autumnalis, Narcissus serotinus, Allium chamaemoly* subsp. *chamaemoly, Leontodon tuberosus, Colchicum cupanii, Arisarum vulgare,* etc. It is the first in the sinfenosuccession of thermo-xerophilous series, which is replaces in springtimes by calcicolous ephemeral grasslands of *Stipo-Trachynetea distachyae.* According to literature (Biondi et al. 2001, 2016, 2017; Farris et al. 2007; Perez Latorre et al. 2008), several associations quite similar to that one surveyed in Malta were descrived from Spain, Sardinia and Italy, although their occurrence in other Mediterranean countries is most likely. In particular for the dominance of *Ranunculus bullatus* subsp. *bullatus,* the Maltese vegetation is related to *Scillo autumnalis-Ranunculetum bullati* Pérez Latorre & Cabezudo in Pérez Latorre et al. 2008, described from southern Spain on siliceous substrata. This association is included by the authors in the *Ranunculion bullati,* alliance referred to the *Tuberarietea guttatae.* It shows also significant floristic differences than to Maltese vegetation, mainly for the occurrence of *Lobularia maritima, Leucojum autumnale, Charybdis maritima* and *Narcissus papyraceus.* Other associations chacterized by *Ranunculus bullatus* subsp. *bullatus* are the *Cyclamino hederifolii-Ranunculetum bullati* and *Allio chamaemoly-Ranunculetum bullati* described from Italian Peninsula by Biondi et al. (2017) and referred to a new alliance *Cyclamino hederifolii-Ranunculion bullati.* However, the releves of Maltese community differ from two aforesaid association and their related alliance both for more xerophilous requirements and floristic set too. Therefore, the phytocoenosis surveyed in Malta seems that can be better arranged in the *Leontodonto tuberosi-Bellidion sylvestris* Biondi, Filigheddu & Farris 2001. Initially, this alliance was attributed by Biondi et al. (2001) to the order *Brachypodio ramosi-Dactyletalia hispanicae*, included in the *Artemisietea vulgaris* Lohmeyer, Preising & Tüxen ex von Rochow 1951, while later Biondi et al. (2017) transferred it within the new order of the class *Charybdido pancratii-Asphodeletea ramosi* Biondi in Biondi et al. 2016, named it *Bellido sylvestris-Arisaretalia vulgaris*, which includes also the *Cyclamino hederifolii-Ranunculion bullati* and *Ranunculion bullati.* In fact, the *Leontodonto tuberosi-Bellidion sylvestris* is a syntaxon linked to dry thermo-Mediterranean bioclimate and gathers plant communities distributed in the central-Mediterranean territories. The Maltese vegetation differs from the other known syntaxa of this alliance for the dominance of *Ranunculus bullatus* subsp. *bullatus* and *Romulea melitensis*, which are proposed as characteristics of a new association, named *Romuleo melitensis-Ranunculetum bullati.* Basing on personal unpublished relevés,

Table 15.7 *Romuleo melitensis-Ranunculetum bullati*

Number of relevé	1	2	3	4	5	6	7
Plot size (m^2)	2	2	2	2	3	2	2
Plant cover (%)	70	60	80	70	50	60	40
Char. Association							
Ranunculus bullatus subsp. *bullatus*	3	3	4	3	2	3	2
Romulea melitensis	+	1	1	2	2	+	+
Char. All. (*Leontodonto tuberosi-Bellidion sylvestris*) and Ord. (*Bellido sylvestris-Arisaretalia vulgaris*)							
Prospero autumnalis	3	2	2	2	3	3	2
Allium chamaemoly subsp. *chamaemoly*	2	2	3	+	1	1	1
Arisarum vulgare	1	2	2	2	.	1	+
Leontodon tuberosus	.	+	+	.	1	1	1
Narcissus serotinus	.	+	+	.	1	1	+
Bellis sylvestris	+	1	+	.	1	1	.
Colchicum cupanii	.	.	1	2	1	.	.
Char. Cl. (*Lygeo sparti-Stipetetea tenacissimae)*							
Asphodelus ramosus	1	1	+	+	1	.	+
Charybdis pancration	1	+	+	.	.	1	1
Carlina corymbosa	.	.	1	1	.	1	1
Transgr. Cl. (*Stipo-Trachynietea distachyae)*							
Moraea sisyrinchium	1	1	2	1	+	+	1
Medicago littoralis	1	2	2	1	+	+	.
Catapodium balearicum	+	1	1	+	.	+	+
Linum strictum	+	1	+	1	+	.	+
Trifolium scabrum	+	+	.	+	+	+	.
Stipella capensis	.	+	1	1	.	+	+
Evax pygmaea	.	+	+	.	+	+	.
Euphorbia exigua	+	+	.	+	+	.	.
Hypochoeris achyrophorus	.	+	+	1	.	+	.
Lagurus ovatus subsp. *ovatus*	.	.	+	.	+	+	1
Echium parviflorum	.	+	.	+	+	.	+
Valantia muralis	.	+	+	.	.	+	.
Trigonella monspeliaca	.	.	+	+	+	.	.
Ononis sieberi	+	.	.	+	.	+	.
Sedum rubens	.	.	+	1	.	+	.
Anagallis arvensis	.	.	+	+	.	+	.
Plantago coronopus	.	.	+	+	.	.	+
Other species							
Bellis annua	.	+	1	2	2	+	1
Catapodium rigidum	+	+	.	+	.	+	.

Rel. 1–7: Malta, Dingli, 29.11.1987

an allied plant community occurs also in southern Sicily, where it seems quite frequent in the habitats very similar to those ones of Maltese Archipelago. According to Mucina et al. (2016) as concerns its arrangement at level of class, the *Leontodonto tuberosi-Bellidion sylvestris* basing on ecology and mainly on floristic characteristics must be included in the *Lygeo sparti-Stipetea tenacissimae*.

Chapter 16
Chasmo-nitrophilous Vegetation

The synanthropic communities colonizing the walls of old buildings and road stoneworks are very frequent in the Maltese Islands. This casmophilous vegetation, showing nitrophilous exigences, is well represented in urban and suburban habitat and is characterized by quite specialized chamaephytes or hemicryptophytes. It is a perennial vegetation, belonging to *Parietarietea judaicae* class, which is observable throughout the year.

16.1 *Oxalido-Parietarietum judaicae* (Br.-Bl. 1952) Segal 1969 (Table 16.1)

The urban and suburban walls, represented by masonries and rocky faces more or less disturbed by human activities, are colonized by perennial plant communities belonging to *Pariarietea judaicae* Oberd. 1977 class. According to Brullo and Guarino (1998, 2002), this class, widespread in the Euro-Mediterranean territories, gathers numerous associations characterized by quite specialized chamaephytes or hemicryptophytes, such as *Parietaria judaica*, *Capparis spinosa*, *Antirrhinum tortuosum*, *Hyoseris radiata*, *Sonchus tenerrimus*, *Ficus carica*, *Erysimum cheirii*, *Cymbalaria muralis*, etc. This type of vegetation is very spread, apart from the anthropized rocky surfaces, also in the walls of old buildings or road stoneworks. In the Maltese Islands one of the most common association of the *Parietarietea judaicae* is the *Oxalido-Parietarietum judaicae*, which shows quite mesic requirements, localizing in the more fresh and shaded stands. Floristically, it is dominated by *Parietaria judaica* and can be considered the most nitrophilous community of this class, since an increase of the nitrogen component in the environment effects a greater vegetative luxuriance. Releves referable to this vegetation was previously published from Maltese walls by Brandes (1991). The association is quite common in Mediterranean regions and is quoted mainly from France, Iberian Peninsula,

S. Brullo et al., *The Vegetation of the Maltese Islands*, Geobotany Studies,
https://doi.org/10.1007/978-3-030-34525-9_16

Table 16.1 *Oxalido-Parietarietum judaicae*

Number of relevé	1	2	3	4	5	6	7	8	9	10
Plot size (m^2)	50	40	20	10	40	20	20	30	20	20
Plant cover (%)	30	50	40	40	20	60	40	60	40	40
Char. Association										
Parietaria judaica	2	3	3	3	2	4	3	4	3	3
Char. All. (*Parietarion judaicae*) and Cl. (*Parietarietea judaicae*)										
Capparis spinosa	1	+	.	+	+	.	.	+	+	+
Sonchus tenerrimus	+	.	+	+	.	1	+	.	.	+
Antirrhinum tortuosum	.	2	.	.	+	.	.	1	1	+
Hyoseris radiata	.	+	.	+	+	+	.	.	.	.
Ficus carica	.	+	.	.	.	+	.	.	1	.
Hyoscyamus albus	.	.	+	.	+	.	.	+	.	.
Cheiranthus cheiri	.	.	.	.	.	+	.	.	.	.
Other species										
Mercurialis annua	.	+	.	+	1	+	+	.	+	+
Jacobaea maritima subsp. *bicolor*	+	1	.	1	+	.	.	.	.	.
Antirrhinum siculum	.	.	+	+	.	+	+	.	.	.
Oxalis pes-caprae	1	.	.	.	.	2	+	.	.	.
Beta maritima	+	+	.	.	.	.	.	.	+	.
Reichardia picroides var. *picroides*	+	+	.	.	.	.	.	.	,	+
Sonchus oleraceus	.	.	.	.	.	.	+	.	.	.
Nicotiana glauca	.	.	.	.	.	.	+	.	.	.
Daucus carota	.	+	.	.	.	.	.	.	.	.
Lobularia maritima	.	+	.	.	.	.	.	.	,	.
Carlina corymbosa	.	+	.	.	.	.	.	.	.	.
Limbarda crithmoides subsp. *longifolia*	.	1	.	.	.	.	.	.	.	.

Rel. 1–2: Malta, La Valletta, 17.4.1987; Rel. 3–5: Gozo. Mgarr, 25.9.1985; Rel. 6: Malta, Msida, 30.11.1985; Rel. 7–8: Malta, Rabat, 12.4.1987; Rel. 9–10. Malta, Sliema, 30-11.1985

Yugoslavia, Italy, etc. (Braun-Blanquet et al. 1952; de Bolós 1967; Rivas-Martinez 1960, 1969; Horvatić 1963; Bartolo and Brullo 1986; Oberdorfer 1977; Brullo and Guarino 1998, 2002).

16.2 *Parietario judaicae-Matthioletum incanae* Vigo & Terrada 1969 (Table 16.2)

On the walls of old masonries as La Valletta fortresses is frequent a community belonging to *Pariarietea judaicae*, characterized by the occurrence and usually dominance of *Matthiola incana* subsp. *incana*. It prefers mainly the surfaces more protected, as those ones north facing with cemented walls. It can be referred to *Parietario judaicae-Matthioletum incanae*, association described by Vigo and

Table 16.2 *Parietario judaicae-Matthioletum incanae*

Number of relevé	1	2	3	4	5	6
Plot size (m^2)	40	20	40	25	30	25
Plant cover (%)	50	70	70	70	50	50
Char. Association						
Matthiola incana subsp. *incana*	2	4	3	3	2	2
Char. All. (*Parietarion judaicae*) and Cl. (*Parietarietea judaicae*)						
Parietaria judaica	3	3	2	3	2	3
Sonchus tenerrimus	2	1	+	1	+	+
Antirrhinum tortuosum	.	+	+	.	+	1
Ficus carica	+	.	1	+	.	+
Capparis spinosa	1	.	.	.	.	.
Hyoseris radiata	+	.	.	.	.	.
Other species						
Limbarda crithmoides subsp. *longifolia*	.	2	3	3	3	2
Antirrhinum siculum	+	+	.	1	1	.
Diplotaxis tenuifolia	.	.	+	.	+	.
Diplotaxis tenuifolia	.	.	+	.	+	.
Mercurialis annua	+	.	.	.	.	.
Jacobaea maritima subsp. *bicolor*	+	.	.	.	.	.
Oxalis pes-caprae	1	.	.	.	.	.
Nicotiana glauca	.	.	1	.	.	.
Dittrichia viscosa	.	.	.	.	.	+

Rel. 1–6: Malta, La Valletta, 17.4.1987

Terradas (1969) from Catalonia (Spain) and after recorded also from Baleares Islands by de Bolós (1996). In the Maltese Islands, *Matthiola incana* subsp. *incana* grows together with *Parietaria judaica*, *Sonchus tenerrimus*, *Antirrhinum tortuosum*, *Ficus carica*, *Limbarda crithmoides* subsp. *longifolia*, etc. Previously, relevés referable to this vegetation were published from La Valletta (Malta) by Brandes and Brandes (1999).

16.3 *Capparidetum rupestris* O. Bolòs & Molinier 1958 (Table 16.3)

In the sunniest stands the *Oxalido-Parietarietum judaicae* is usually replaced by a markedly thermophilous community dominated by *Capparis spinosa* (*C. rupestris*). This is a synanthropic vegetation localized mainly in the upper part of the walls of old buildings and also on the urban rocky places, preferably with southern exposure. It must be attributed to *Capparidetum rupestris*, association widespread in the Mediterranean area described by de Bolós and Molinier (1958). Its thermophilous character is attested by the usually low coverage of the species more mesophilous of

Table 16.3 *Capparidetum rupestris*

Number of relevé	1	2	3	4	5	6	7	8	9	10	11	12
Plot size (m^2)	50	50	80	50	30	20	20	50	20	50	50	50
Plant cover (%)	30	60	30	40	20	30	35	20	40	40	40	40
Char. Association												
Capparis spinosa	3	4	3	3	2	2	2	2	3	3	3	3
Char. All. (*Parietarion judaicae*) and Cl. (*Parietarietea judaicae*)												
Parietaria judaica	+	1	1	1	+	.	+	1	1	1	1	2
Sonchus tenerrimus	+	+	+	+	+	.	.	+	+	1	1	+
Antirrhinum tortuosum	.	+	+	.	.	1	1	.	+	.	1	.
Ficus carica	.	.	+	1	.	.	.	+	.	+	.	1
Hyoseris radiata	.	+	.	+	.	.	+	+	.	.	.	.
Other species												
Jacobaea maritima subsp. *bicolor*	1	1	1	.	+	+	1	+	.	.	.	.
Limbarda crithmoides subsp. *longifolia*	1	+	1	.	.	1	1	.	.	.	.	.
Antirrhinum siculum	+	.	+	.	+	.	.	+	.	1	.	.
Hyoscyamus albus	.	+	.	.	.	.	+	+	.	.	.	1
Oxalis pes-caprae	.	.	.	+	.	.	.	.	.	+	+	.
Beta maritima	.	.	.	.	.	+	+	.	.	.	.	.
Valantia muralis	.	.	.	.	.	+	+	.	.	.	.	.
Phagnalon graecum	.	.	.	.	.	1	+	.	.	.	.	.
Mercurialis annua	.	+	.	.	.	.	.	.	1	.	.	.
Anisantha madritensis	.	.	.	.	.	+	.	.	.	.	.	.
Daucus carota	.	.	.	.	.	+	.	.	.	.	.	.
Reseda alba	.	.	.	.	.	.	.	.	.	+	.	.
Nicotiana glauca	.	.	.	.	.	.	.	.	.	.	.	2
Dittrichia viscosa	.	.	.	.	.	.	.	.	.	.	.	1
Lobularia maritima	.	.	.	.	.	.	.	.	.	.	.	1

Rel. 1–3: Gozo, Mgarr. 25.9.1985; Rel. 4: Malta, Msida, 30.11.1985; Rel. 5–6: Malta, Dragunara, 24.6.1973; Rel. 7: Malta, Sliema, 24.6.1973; Rel. 8–9: Malta, Rabat, 12.4.1987; Rel. 10: Gozo, Xlendi, 13.4.1987; Rel. 11: Malta, Girgenti, 16.4.1987; Rel. 12: Malta, La Valletta, 24.6.1973

Parietarietea judaicae class, such as *Parietaria judaica*, *Hyoseris radiata* and *Sonchus tenerrimus*. The *Capparidetum rupestris* is well represented in the Maltese archipelago, as well as in many Mediterranean territories (de Bolós 1967; de Bolós and Molinier 1958; Hruska 1987; Bartolo and Brullo 1986; Oberdorfer 1977; Brullo and Guarino 1998, 2002; Rivas-Martínez et al. 2001, 2011).

Table 16.4 *Antirrhinetum siculi*

Number of relevé	1	2	3	4	5	6	7
Plot size (m^2)	20	30	20	20	50	20	50
Plant cover (%)	70	50	40	50	40	60	50
Char. Association							
Antirrhinum siculum	3	3	3	3	3	3	3
Char. All. (*Parietarion judaicae*) and Cl. (*Parietarietea judaicae*)							
Parietaria judaica	2	1	1	1	+	3	2
Sonchus tenerrimus	1	1	1	2	1	+	.
Capparis spinosa	+	.	1	.	1	1	1
Antirrhinum tortuosum	+	+	.	.	+	.	.
Hyoseris radiata	.	+	+	+	.	.	.
Ficus carica	.	1	.	.	.	.	.
Cheiranthus cheiri	.	+	.	.	.	.	.
Other species							
Oxalis pes-caprae	2	1	+	+	.	.	.
Hyoscyamus albus	.	.	.	.	+	+	+
Sonchus oleraceus	.	.	.	+	+	.	.
Nicotiana glauca	.	.	.	.	.	1	1
Mercurialis annua	.	+	.	+	.	.	.
Jacobaea maritima subsp. *bicolor*	.	+	.	.	.	.	.
Daucus carota .	+	.	.	.	.	.	.
Phagnalon graecum	.	+	.	.	.	.	.
Lobularia maritima	.	.	+	.	.	.	.

Rel. 1–3: Malta, Msida, 30.11.1985; Rel. 4–5: Malta, Rabat, 12.4.1987; Rel. 6–7: Malta, La Valletta, 17.4.1987

16.4 *Antirrhinetum siculi* Bartolo & Brullo 1986 (Table 16.4)

In conditions of marked xericity and edaphic sterility the *Capparidetum rupestris* is replaced by the *Antirrhinetum siculi*, an association characterized by the occurrence and dominance of *Antirrhinum siculum*, a central-Mediterranean endemism. The vegetation at issue is linked to cemented walls, where the plants colonize the narrow and slightly humified crevices. This synanthropic habitat privilege clearly the typical chasmophytes, such as *Anthirrinum siculum*, that in the natural habitat is a member of a rupestrian association of the *Dianthion rupicolae*. Apart from Maltese islands, where was previously recorded by Brandes (1991) and Brandes and Brandes (1999), the *Antirrhinetum siculi* is until known only from Sicily and Southern Italy (Bartolo and Brullo 1986; Brullo et al. 2001a, b, c; Brullo and Guarino 1998, 2002).

Table 16.5 *Hyoscyamo albi-Parietarietum judaici*

Number of relevé	1	2	3
Plot size (m^2)	30	20	20
Plant cover (%)	50	50	40
Char. Association			
Hyoscyamus albus	3	3	2
Char. All. (*Parietarion judaicae*) and Cl. (*Parietarietea judaicae*)			
Parietaria judaica	3	3	3
Antirrhinum tortuosum	1	+	1
Ficus carica	.	.	+
Matthiola incana subsp. *incana*	.	+	.
Other species			
Antirrhinum siculum	+	+	1
Dittrichia viscosa	1	.	+
Limbarda crithmoides subsp. *longifolia*	.	+	.

Rel. 1–3: Malta, La Valletta, 17.4.1987

16.5 *Hyoscyamo albi-Parietarietum judaicae* Segal 1969 (Table 16.5)

The walls of old buildings of urban or suburban stands near the coast are colonized by sunny-thermophilous plant community of halo-tolerant type characterized by *Hyoscyamus albus*, usually growing together with *Parietaria judaica*. For its floristic characteristics and ecological requirements, this vegetation can be referred to *Hyoscyamo albi-Parietarietum judaicae*, chasmo-nitrophilous association described by Segal (1969) from Mediterranean area. In the Maltese Island it is not very common, but it seems linked to masonries dunget by seabirds, as well as in the rest of the other Medierranean sites (Segal 1969; Bartolo and Brullo 1986; Ruiz-Téllez 1991; Caneva et al. 1995; Brullo and Guarino 1998, 2002; Brullo et al. 2001c).

Chapter 17
Annual Halophilous Vegetation

During the winter-spring period in stands characterized by more or less salt soils, such as in the rocky reefs, a microphytic vegetation differentiated by rare precocious endemic species occurs. Other communities are localized also in the salt marshes, mainly in the clearings mixed to perennial vegetation of *Sarcocornietea fruticosae*. From the phytosociological point of view, they belong to *Frankenietalia pulverulentae*, order of *Saginetea maritimae*.

17.1 *Desmazerio pignattii-Senecionetum pygmaei* Brullo & Scelsi 1998 (Table 17.1)

Along the rocky coast of Maltese Islands in the stands closest to the shore, a subhalophitic ephemeral vegetation, dominate by several microphytes is frequent. Usually, it is mixed with the perennial associations of the *Crucianellion rupestris* characterized by *Limonium* sp.pl., where it colonizes the clearings. Very significant it is the localization in this community of two rare Maltese-Hyblaean endemisms, represented by *Desmazeria pignattii* and *Senecio pygmaeus* (see Brullo and Pavone 1985). These two species are a member of a very peculiar micro-association rich in therophytes of the *Frankenion pulverulentae* Rivas-Martínez ex Rivas-Martinez and Costa 1976, Mediterranean alliance of the *Saginetea maritimae* Westhoff, Leeuwen & Adriani 1962. This class gathering ephemeral subhalophilous associations distributed in Euro-Mediterranean territories, usually localized in coastal stands (Rivas-Martinez and Costa 1976; Brullo 1988b; Brullo et al. 1994; Brullo and Siracusa 1996; Brullo and Scelsi 1998; Brullo and Giusso 2003). Among the species of these syntaxa occur mainly *Parapholis incurva, Plantago coronopus, Catapodium balearicum, Sagina maritima, Spergularia marina, Bellis annua, Frankenia pulverulenta, Hymenolobus revelierei* subsp. *sommierianum, Senecio leucanthemifolius, Polygonum subspathaceus*, etc.

S. Brullo et al., *The Vegetation of the Maltese Islands*, Geobotany Studies,
https://doi.org/10.1007/978-3-030-34525-9_17

Table 17.1 *Desmazerio pignattii-Senecionetum pygmaei*

Number of relevé	1	2	3	4	5	6	7	8	9	10	11	12	13	14	15	16	17	18	19	20	21	22	23	24	25	26
Plot size (m²)	0.5	1	1	0.6	0.5	1	2	1	1.5	0.5	0.5	0.8	0.6	1.5	1	2	1	0.6	0.5	1	1	0.5	1	1	2	1
Plant cover (%)	20	40	50	30	50	70	50	50	70	70	80	50	50	70	40	70	30	50	50	50	70	50	60	50	50	50
Char. Association																										
Desmazeria pignattii	+	2	2	2	1	+	2	3	3	2	1	2	1	2	2	2	3	2	2	1	+	+	+	1	1	2
Senecio pygmaeus	+	1	2	+	.	1	.	.	.	+	+	2	3	+	2	+	1	2	+	1	.	.	.	.	.	.
Diff. Subassociation																										
Anthemis urvilleana	.	.	.	.	.	.	.	.	2	3	1	2	2	3	2	2	1	2	3	2	4	3	4	3	3	2
Romulea melitensis	.	.	.	.	.	.	.	.	.	+	.	1	2	2	1	2	+	.	+	+	2	1	1	2	+	2
Plantago crypsoides	.	.	.	.	.	.	.	.	.	.	+	+	.	.	.	.	.	.	+	2	2	2	1	2	1	.
Char. All. (*Frankenion pulverulentae*) and Cl. (*Saginetea maritimae*)																										
Parapholis incurva	.	2	1	1	1	+	+	.	.	+	+	.	+	.	1	+	+	1	1	.	1	+	1	1	1	1
Catapodium balearicum	.	1	+	1	2	.	.	.	2	1	2	+	1	+	+	.	+	+	1	+	+	+	.	+	2	2
Plantago coronopus	.	2	2	1	2	+	2	.	2	2	3	+	2	3	1	2	.	+	+	.	.	.	.	.	.	.
Sagina maritima	.	.	.	.	+	.	.	+	1	+	1	+	+	1	+	1	+	.	+	+	.	.	1	.	+	+
Spergularia marina	.	.	+	+	.	.	+	.	.	1	.	1	+	1	.	.	+	.	+	1	.	.	2	1	+	1
Frankenia pulverulenta	.	.	.	.	.	.	.	.	.	.	.	.	1	.	.	.	+	+	+	.	+	.	1	1	+	+
Bellis annua	.	.	.	+	+	.	.	.	.	+	.	1	+	+	.	.	.	.	+	.	.	.	.	.	.	.
Polypogon subspathaceus	.	.	.	.	.	.	.	.	.	.	.	.	.	.	.	.	.	.	.	.	+	.	+	+	.	+
Hymenolobus revelierei subsp. *sommierianum*	.	.	.	.	.	.	.	.	.	.	.	.	.	+	.	.	+	.	+	.	.	.	.	.	.	.
Senecio leucanthemifolius	.	.	.	.	.	.	.	.	+	.	+	.	+	.	.	.	.	.	.	.	.	.	.	.	.	.
Juncus hybridus	.	.	.	.	.	.	.	.	.	.	.	.	.	.	.	+	.	.	.	.	.	.	.	.	.	.
Bupleurum semicompositum	.	.	.	.	.	1	.	.	.	.	.	.	.	.	.	.	.	.	.	.	.	.	.	.	.	.

Other species																										
Moraea sisyrinchium	.	2	1	.	2	.	+	1	1	1	+	.	+	+	.	.	.	+	.	1	1	+	+	1	1	1
Silene sedoides	1	+	1	.	1	2	1	2	2	.	.	+	.	.	.	.	1	2	1	2	.	.	.	.	.	.
Valantia muralis	+	1	.	+	.	+	+	.	+	.	+	1	.	.	.	.	+	.	.	.	+	+	.	.	+	+
Mesembryanthemum nodiflorum	.	+	+	.	+	.	.	.	.	.	.	+	.	+	.	.	.	.	+	.	.	.	1	2	1	+
Euphorbia sommieriana	.	.	.	+	.	+	1	1	+	.	+	1	.	.	.	.	+	+	.	.	.	.	.	.	.	.
Asteriscus aquaticus	.	.	+	.	1	+	.	1	+	.	.	+	.	.	.	.	.	.	+	.	.	.	.	.	.	.
Medicago littoralis	.	.	+	+	.	.	.	.	2	1	+	1	.	.	.	.	.	.	.	.	.	.	.	.	.	.
Trifolium scabrum	.	.	.	.	.	.	+	.	1	.	.	+	.	.	.	+	1	.	.	+	.	.	.	.	.	.
Anisantha fasciculata	.	.	.	.	.	.	.	.	.	.	.	.	.	.	.	.	.	.	.	.	1	.	+	+	.	+
Frankenia hirsuta	.	.	.	.	+	.	.	.	.	+	+	.	1	.	.	.	.	.	.	.	.	.	.	.	.	.
Evax pygmaea	.	+	.	+	.	.	.	.	.	+	.	.	.	+	.	.	.	.	.	.	.	.	.	.	.	.
Lotus cytisoides	+	+	.	+	.	.	.	.	.	.	.	.	.	.	.	.	.	.	.	.	.	.	.	.	.	.
Prospero autumnalis	.	.	.	.	.	.	.	.	.	.	.	.	.	.	.	.	.	.	.	.	.	.	.	+	1	1
Limonium zeraphae	1	.	.	+	.	+	.	.	.	.	.	.	.	.	.	.	.	.	.	.	.	.	.	.	.	.
Echium parviflorum	.	+	.	+	.	.	.	.	.	.	.	.	.	.	.	+	.	.	.	.	.	.	.	.	.	.

Rel. 1–3: Malta, Mtahleb, 12.4.1987; Rel. 4–5: Gozo, Dwejra, 11.4.1984; Rel. 6: Malta, Mtahleb, 12.4.1987; Rel. 7–8: Gozo, Ta Cenc, 13.4.1987; Rel. 9–11: Gozo, Dwejra, 13.4.1987; Rel. 12: Malta, Dragunara, 8.4.1984; Rel. 13: Malta, St, Dimitri, 8.4.1984; Rel. 14–20: Comino, 15.4.1987; Rel. 21–26: Gozo, Dwejra Point, 30.04.2007

This association, described by Brullo and Scelsi (1998) as *Desmazerio pignattii-Senecionetum pygmaei* from southern Sicily, is widespread in the all Maltese Archipelago, where it is frequent on the calcareous outcrops near the sea (Fig. 14.1b). Here, it is represented by two subassociations, one corresponding to the subass. *typicum* (rel. 1–8), observed in the more open and exposed surfaces, while the second one is proposed as *anthemidetosum urvilleanae* subass. nov. hoc loco (rel. 9–20, holotypus: rel. 12, hoc loco), circumscribed to the most protected stands. The latter is differentiated by *Anthemis urvilleana*, *Romulea melitensis* and *Plantago crypsoides*, showing a less xerophilous exigences than the typical aspect.

17.2 *Anthemido urvilleanae-Frankenietum pulverulentae* ass. nov. hoc loco (Table 17.2)

Holotypus: rel. 1, hoc loco.

On coastal stands, often trampled, characterized by silty-clay soils, a halophilous-ephemeral community characterized by microphytes with prostrate habit occurs. It is localized in small surfaces among the calcareous rocks, which are damp or slightly flooded during the winter period. Among the most significant species frequent in these habitats are *Frankenia pulverulenta, Parapholis incurva, Plantago coronopus, Sagina maritima, Spergularia marina,* etc. This vegetation for its ecology and floristic set falls in the *Saginetea maritimae* class, which in the Mediterranean territories is represented by the *Frankenion pulverulentae*, alliance of *Franketalia pulverulentae*.

Table 17.2 *Anthemideto urvilleanae-Frankenietum pulverulentae*

Number of relevé	1	2	3	4	5
Plot size (m^2)	2	5	2	3	5
Plant cover (%)	70	70	80	70	80
Char. Association					
Anthemis urvilleana	1	1	+	1	+
Char. All. (*Frankenion pulverulentae*)					
Frankenia pulverulenta	2	2	3	1	1
Parapholis incurva	1	1	+	2	3
Juncus hybridus	+	2	1	1	+
Spergularia marina	2	1	2	2	1
Char. Cl. (*Saginetea maritimae*)					
Plantago coronopus	3	3	2	2	3
Sagina maritima	1	1	2	+	+
Other species					
Medicago littoralis	1	+	1	1	+
Plantago lagopus	.	+	+	2	+
Beta maritima	.	+	.	1	.
Moraea sisyrinchium	.	.	+	+	.

Rel. 1–5: Marfa Ridge.White Tower, 11.4.1987

Several associations of this syntaxa are distributed in various Mediterranean countries (Rivas-Martinez and Costa 1976; Brullo 1988b; Brullo and Giusso 2003). In particular, the Maltese vegetation differs from the other known associations for the occurrence of the endemic *Anthemis urvilleana*, which is proposed as characteristic of a new association named *Anthemido urvilleanae-Frankenietum pulverulentae.*

This association shows some relations with *Parapholido-Frankenietum pulverulentae* Rivas-Martínez ex Castroviejo & Porta 1976 from western Mediterranean, and also with *Frankenio pulverulentae-Anthemidetum secundirameae* Brullo & Scelsi 1998 from Sicily, the latter can be considered as a geographical vicariant.

17.3 *Parapholidetum filiformis* Brullo, Scelsi & Siracusa 1994 (Table 17.3)

In the salt marshes mixed to the perennial communities of *Sarcocornietea fruticosae*, an ephemeral meadow physiologically differentiated by the dominance of *Parapholis filiformis* is frequent. It is linked to halomorphic soils rich in sandy

Table 17.3 *Parapholidetum filiformis*

Number of relevé	1	2	3	4	5
Plot size (m^2)	30	20	20	10	20
Plant cover (%)	90	80	80	90	90
Char. Association					
Parapholis filiformis	5	4	4	5	5
Char. All. (*Frankenion pulverulentae*)					
Frankenia pulverulenta	2	1	+	+	1
Parapholis incurva	1	+	.	1	+
Hordeum marinum	+	.	+	+	1
Juncus hybridus	+	.	1	2	2
Polypogon monspeliensis	.	1	+	+	+
Spergularia marina	.	+	+	+	.
Char. Cl. (*Saginetea maritimae*)					
Plantago coronopus	1	1	+	+	+
Sagina maritima	.	1	+	.	.
Other species					
Melilotus massanensis	1	+	+	1	1
Cressa cretica	.	1	+	+	1
Suaeda spicata	1	1	.	+	+
Bellis annua	.	+	+	.	+
Rumex bucephalophorus subsp. *bucephalophorus*	+	+	.	.	+
Soda inermis	.	.	+	+	.
Symphyotrichum squamatum	.	+	+	.	.

Rel.1–5: Malta, Mellieha Bay, 26.6.1973

component, wet during the the winter period. This vegetation is usually favored by anthropic disturbance which determines a certain ruderalization of these retrodunal habitats. It is a halo-subnitrophilous association referable to the *Parapholidetum filiformis* Brullo, Scelsi & Siracusa 1994, belonging to the *Frankenion pulverulentae*. This alliance and higher syntaxa are here represented by a significant set of characteristic species. Previously, this association was recorded from Sicily, Sardinia and Italy (Brullo 1988b; Brullo and Scelsi 1998; Brullo et al. 1994; Brullo and Giusso 2003; Tomaselli et al. 2010; Minissale and Sciandrello 2017).

Chapter 18
Ruderal Vegetation

The Maltese islands, as well as all the other Mediterranean insular environments, due to the anthropic disturbance dating back to some millennia ago, show a rich nitrophilous flora, which is spread in many secondary habitats. Several and diversified plant communities are frequent in these ruderal stands, represented by footpaths, roadsides, ruble accumulations, organic matter deposits, etc. They are mainly represented by annual or more rarely perennial associations having a winter-spring or sometimes an autumnal vegetative cycle. The phytosociological class involved in these types of vegetation are *Polygono arenastri-Poetea annuae, Artemisietea vulgaris*, *Chenopodietea*, *Epilobietea angustifolii* and *Nicotiano glaucae-Ricinetea communis*.

18.1 *Polycarpo tetraphylli-Spergularietum rubrae* Brullo & Marcenò 1976 (Table 18.1)

On trampled surfaces of footpaths, dirt roads and suburban areas, a highly specialized synanthropic vegetation of the *Polygono arenastri-Poetea annuae* Rivas-Martínez 1975 class occurs (Rivas-Martinez 1975; Brullo 1980a). In the Maltese Islands, the most common aspect of this class is the *Polycarpo tetraphylli-Spergularietum rubrae*, an association described by Brullo and Marcenò (1976) from Sicily, belonging to *Polycarpion tetraphylli* Rivas-Martínez 1975. Floristically, it is differentiated by *Spergularia rubra*, *Polycarpon tetraphyllum* and *Arenaria leptoclados*, which grow together with other nitrophilous microphytes, such as *Sagina apetala, Ochlopoa annua, Plantago coronopus, Polygonum arenastrum, Coronopus didymus,* etc. This association, characterized by therophytes with prostrate habit, is frequent on ruderal stands covered by a very thin layer of soil, often mixed with cobblestones.

S. Brullo et al., *The Vegetation of the Maltese Islands*, Geobotany Studies,
https://doi.org/10.1007/978-3-030-34525-9_18

Table 18.1 *Polycarpo tetraphylli-Spergularietum rubrae*

Number of relevé	1	2	3	4	5	6	7	8	9	10	11	12	13
Plot size (m^2)	5	3	4	3	3	2	4	10	5	8	10	10	5
Plant cover (%)	70	70	80	70	70	50	60	30	30	50	60	50	70
Char. Association and All. (*Polycarpion tetraphylli*)													
Spergularia rubra	2	3	3	3	3	2	3	2	2	2	2	3	3
Polycarpon tetraphyllum	2	3	3	2	1	+	3	1	1	2	1	1	+
Arenaria leptoclados	.	+	.	.	+	.	.	.	.	.	.	.	.
Char. Cl. (*Polygono-Poetea annuae*)													
Sagina apetala	3	1	2	2	3	3	1	2	1	2	3	1	3
Ochlopoa annua	1	2	2	2	1	2	1	1	1	2	1	1	1
Plantago coronopus	2	1	.	1	+	.	1	.	+	.	2	2	2
Polygonum arenastrum	.	1	.	.	1	.	.	.	1	+	.	.	.
Coronopus didymus	.	.	.	.	.	.	.	1	2	1	+	.	.
Matricaria chamomilla	.	.	2	.	+	.	1	.	.	.	.	.	.
Other species													
Catapodium rigidum	1	.	.	+	.	+	+	.	+	+	+	+	.
Catapodium balearicum	+	.	.	.	+	+	.	.	.	.	+	+	.
Plantago lagopus	+	.	+	+	+	.	.	.	.	.	.	.	.
Oxalis pes-caprae	+	+	.	+	.	.	+	.	.	.	.	.	.
Trifolium tomentosum	.	.	.	.	.	.	.	+	1	1	1	.	.
Trifolium nigrescens subsp. *nigrescens*	+	.	.	+	.	.	+	.	.	.	.	.	.
Erodium cicutarium	+	+	.	.	.	.	.	+	.	.	.	.	.
Conyza bonariensis	.	.	.	.	.	.	.	+	+	+	.	.	.
Hordeum leporinum	.	.	.	.	.	.	.	+	.	.	.	+	.
Sagina maritima	.	.	.	.	.	.	.	1	.	+	.	.	.
Polypogon subspathaceus	.	.	.	.	.	.	.	.	+	.	+	.	.
Galium murale	+	.	.	.	.	+	.	.	.	.	.	.	.
Trisetaria aurea	.	.	.	+	.	.	+	.	.	.	.	.	.
Geranium molle	.	.	.	.	.	.	+	.	.	.	.	.	+
Lobularia maritima	+	.	.	.	.	.	.	.	.	.	.	.	.
Bromus scoparius	.	.	.	.	+	.	.	.	.	.	.	.	.
Capsella rubella	.	.	.	.	+	.	.	.	.	.	.	.	.
Rostraria cristata	.	.	.	.	+	.	.	.	.	.	.	.	.

Rel. 1: Malta, Tal-Blata, 9.4.1984; Rel. 2: Malta, Sliema, 8.4.1984; Rel. 3: Gozo, Munxar, 13.4.1987; Rel. 4: Malta, Wied Incita, 16.4.1987; Rel. 5–7: Gozo, Victoria, 11.4.1984; Rel. 8–11: Malta, La Valletta, 17.4.1987; Rel. 12–13: Gozo, Wied il Mielah, 13.4.1987

18.2 *Trisetario aureae-Crepidetum bursifoliae* Brullo 1980 (Table 18.2)

In the less nitrophilous and more xeric surfaces, such as characterized by rather superficial soils very rich in minute crushed stone, the *Polycarpo tetraphylli-Spergularietum rubrae* is replaced by the *Trisetario aureae-Crepidetum bursifoliae*. This association belonging always to the *Polycarpion tetraphyllii* was described by

Table 18.2 *Trisetario aureae-Crepidetum bursifoliae*

Number of relevé	1	2	3	4	5	6	7
Plot size (m^2)	3	2	5	5	5	4	6
Plant cover (%)	60	50	50	50	50	50	70
Char. Association							
Trisetaria aurea	3	2	3	3	3	2	3
Crepis bursifolia	.	1	+	.	.	.	.
Char. All. (*Polycarpion tetraphylli*)							
Spergularia rubra	+	2	+	+	1	1	1
Polycarpon tetraphyllum	.	+	1	+	1	3	1
Char. Cl. (*Polygono-Poetea annuae*)							
Ochlopoa annua	1	2	2	1	1	2	2
Matricaria chamomilla	.	+	2	1	+	1	+
Sagina apetala	3	2	.	2	2	1	3
Plantago coronopus	+	+	.	2	2	.	+
Polygonum arenastrum	.	.	.	1	.	.	.
Other species							
Plantago lagopus	.	+	.	+	+	.	.
Trifolium nigrescens subsp. *nigrescens*	.	.	.	+	1	.	+
Catapodium balearicum	+	1	.	+	.	.	.
Bromus scoparius	+	.	.	+	+	.	.
Catapodium rigidum	.	.	.	.	+	1	.
Erodium cicutarium	.	+	.	.	.	+	.
Trifolium tomentosum	1	.	.	.	+	.	.
Galium murale	.	+	1	.	.	.	.
Rostraria cristata	.	.	.	.	.	+	+
Oxalis pes-caprae	.	.	.	.	+	+	.
Lobularia maritima	.	.	.	+	.	.	.

Rel. 1: Gozo, Dahlet Qorrot Beach, 13.4.1987; Rel. 2–3: Malta, Sliema, 8.4.1984; Rel. 4–6: Malta, Addolorata, 17.4.1987; Rel. 7: Malta, Buggiba, 9.4.1984

Brullo (1980a) and later recorded also from South Italy (Brullo et al. 2001c). It is differentiated by *Trisetaria aurea* and *Crepis bursifolia*, which grow together with other ruderal therophytes, such as *Spergularia rubra*, *Polycarpon tetraphyllum, Ochlopoa annua, Sagina apetala*, and so on. In the Maltese Islands this association is quite rare and only occasionally was observed.

18.3 *Galio muralis-Trifolietum suffocati* ass. nov. hoc loco (Table 18.3).

Holotypus: ril. 1, hoc loco.

In the more fresh and shaded stands the alliance *Polycarpion tetraphylli* is represented by a mesic microvegetation, characterized by the dominance of

Table 18.3 *Galio muralis-Trifolietum suffocati*

Number of relevé	1	2	3	4	5	6	7
Plot size (m²)	3	4	4	3	5	2	10
Plant cover (%)	50	40	50	50	40	60	50
Char. Association							
Galium murale	1	2	2	1	2	3	2
Trifolium suffocatum	3	2	2	3	2	3	1
Char. All. (*Polycarpion tetraphylli*)							
Polycarpon tetraphyllum	+	+	1	+	.	1	2
Filago congesta	1	+	1	+	.	.	+
Spergularia rubra	.	.	.	+	+	.	2
Char. Cl. (*Polygono-Poetea annuae*)							
Plantago coronopus	2	+	1	.	+	+	.
Sagina apetala	+	.	.	+	1	1	1
Ochlopoa annua	1	+	.	+	+	.	+
Polygonum arenastrum	.	+	+	.	.	.	.
Matricaria chamomilla	.	.	.	1	.	.	.
Other species							
Catapodium balearicum	1	+	.	.	+	+	+
Evax pygmaea	1	+	2	+	.	.	.
Trisetaria aurea	+	1	+	.	.	.	1
Trifolium nigrescens subsp. *nigrescens*	+	.	+	+	.	.	.
Plantago lagopus	.	.	+	.	+	.	1
Catapodium rigidum	.	+	.	.	.	1	.
Medicago littoralis	.	.	.	.	.	.	1
Malva parviflora	.	.	.	.	.	.	+
Ajuga iva	.	.	.	.	.	.	+
Erodium cicutarium	.	.	.	+	.	.	.

Rel. 1–3: Malta, Wied Incita, 16.4.1987; Rel. 4–6: Malta, Girgenti, 16.4.1987; Rel. 7: Malta, Hagar Qim (Qrendi), 1.5.2007

Galium murale and *Trifolium suffocatum*. This species are proposed as characteristics of a new association, named *Galio muralis-Trifolietum suffocati*. Usually, it is localized on the cobbled of gardens and archaeological areas, where it prefers the surfaces less subject to tread. In this synanthropic plant community several ruderal species are frequent, such as *Spergularia rubra*, *Polycarpon tetraphyllum, Plantago coronopus, Ochlopoa annua, Sagina apetala*, *Polygonum arenastrum*, etc.

As concerns its distribution, this association was found, apart from in the Maltese Islands, also in southern Sicily, especially in coastal places (unpublished data).

Table 18.4 *Spergulario rubrae-Filaginetum congestae*

Number of relevé	1	2	3	4
Plot size (m^2)	5	5	10	5
Plant cover (%)	30	70	70	50
Char. Association				
Filago congesta	2	4	3	3
Char. All. (*Polycarpion tetraphylli*)				
Polycarpon tetraphyllum	2	1	2	2
Spergularia rubra	2	2	3	3
Char. Cl. (*Polygono-Poetea annuae*)				
Sagina apetala	+	1	1	.
Polygonum arenastrum	+	+	1	.
Ochlopoa annua	.	.	+	+
Other species				
Trisetaria aurea	2	2	2	+
Plantago lagopus	1	1	2	1
Malva parviflora	.	+	+	1
Catapodium balearicum	+	.	.	+
Euphorbia helioscopia	.	+	.	.
Ajuga iva	.	.	1	.

Rel.1–3: Malta, Hagar Qim (Qrendi), 01.05.2007; Rel. 4: Malta, Manjdra (Qrendi), 01.05.2007

18.4 *Spergulario rubrae-Filaginetum congestae* ass., nov. hoc loco (Table 18.4)

Holotypus: rel. 2, hoc loco.

In the paths of countryside, softly subject to footsteps with soil rich in sandy component, a vegetation dominated by *Filago congesta* occurs. This compact and prostrate therophyte grows with other ruderal microphytes, such as *Spergularia rubra* and *Polycarpon tetraphyllum.* For its ecology and floristic peculiarity, this plant community shows some relations with other associations of *Polycarpion tetraphylli*, among them *Schismo barbati-Filaginetum congestae* O. de Bolós 1975 from Spain and *Evaco-Filaginetum congestae* Bartolo, Brullo, Minissale et Spampinato 1990 from Lampedusa, but it differs from these ones for the floristic set. Therefore, the Maltese vegetation is proposed as *Spergulario rubrae-Filaginetum congestae* ass. nov., having *Filago congesta* as characteristic species.

18.5 *Carthamo lanati-Onopordetum argolici* ass. nov. hoc loco (Table 18.5)

Holotypus: rel. 2 hoc loco.

In the ruderal stands at the road edges where there are deposits of waste or carryng materials, often a herbaceous vegetation dominated by big spiny

Table 18.5 *Carthamo lanati-Onopordetum argolici*

Number of relevé	1	2	3	4	5
Plot size (m^2)	10	8	15	10	10
Plant cover (%)	80	70	60	70	70
Char. Association					
Onopordon argolicum	2	2	1	3	2
Char. All. (*Onopordion illyrici*)					
Scolymus grandiflorus	2	1	2	1	1
Cynara cardunculus	3	2	2	.	1
Char. Ord. (*Carthametalia lanati)* and Cl. (*Artemisetea vulgaris*)					
Carthamus lanatus	1	3	2	2	1
Silybum marianum	1	+	.	1	2
Notobasis syriaca	.	+	1	1	.
Scolymus hispanicus	.	1	+	.	2
Scolymus maculatus	1	.	.	1	2
Centaurea calcitrapa	1	.	.	1	2
Cichorium intybus	.	1	+	.	.
Lactuca serriola	+	.	+	.	.
Other species					
Oxalis pes-caprae	2	1	1	2	+
Verbascum sinuatum	1	.	1	+	+
Beta maritima	1	1	+	.	.
Melilotus sulcata	+	.	.	+	.
Limbarda crithmoides subsp. *longifolia*	1	+	.	.	.
Scorpiurus muricatus	+	.	+	.	.
Carduus pycnocephalus subsp. *pycnocephalus*	.	+	.	.	+

Rel. 1–2: Malta, Wied iz Zurrieq, 27.6.1973; Rel. 3: Malta, Rabat, 30.11.1985; Rel. 4: Comino, 27.9.1985; Rel. 5: Gozo, Ramla Bay, 25.9.1985

Carduaceae can be observed. In these stands quite significant it is the occurrence of several nitrophilous species of the *Onopordion illyrici* Oberd. 1954, alliance of *Carhametalia lanati* Brullo in Brullo & Marceno 1985b belonging to *Artemisietea vulgaris* Lohmeyer et al. in Tx. ex von Rochow 1951. Among the characteristics of these syntaxa are frequent *Carthamus lanatus, Silybum marianum, Scolymus grandiflorus, Cynara cardunculus, Notobasis syriaca, Scolymus hispanicus,* etc.

In the Maltese Island the vegetation at issue differs from the already known associations of *Onopordion illyrici* (see Brullo and Marcenò 1985a, b; Barbagallo et al. 1987; Chronououlos et al. 2005) for the occurrence of *Onopordum argolicum*, East Mediterranean species, which is proposed as characteristic of a new association named *Carthamo lanati-Onopordetum argolici*.

18.6 *Chrysanthemo coronarii-Malvetum parviflorae* Ferro 1980 (Table 18.6)

This association, which is widespread in South-eastern and West Sicily (see Ferro 1980; Brullo 1983b), is localized mainly along the roadsides and edge of the paths. This is a ruderal vegetation linked to road habitats, which belongs to the *Hordeion leporini* Br.-Bl. in Br.-Bl. et al. 1936 corr. O. Bolós 1962, an alliance of *Chenopodietea* Br.-Bl. in Br.-Bl. et al. 1952 with a wide Mediterranean distribution (see Oberdorfer 1977; Rivas-Martinez 1978c; Rivas-Martinez and Izco 1954; Brullo 1983b; Cano-Ortiz et al. 2014). In the Maltese community there is a rich floristic set of this alliance, in which the most significant species for their physiognomic role are *Glebionis coronaria, Hordeum leporinum, Malva parviflora* and *Reseda alba.*

The *Chrysanthemo coronariae-Malvetum parviflorae*, frequent on calcareous and clay substrates, is a strictly thermo-xerophilic association and therefore is well adapted to the markedly arid climate of the Maltese islands. Even in Sicily, it is confined to the more xeric and arid territories, especially in the coastal places of the Hyblaean area and Gela plain, while in the inland is localized in the climatic belt of the *Oleo-Ceratonion.*

18.7 *Carduetum marmorati* Brullo 1983 corr. hoc loco (Table 18.7)

Syn.: *Carduetum australis* Brullo 1983.

In the stands with clayey or marly substrate, affected by a more fresh microclimate, the *Chrysanthemo coronarii-Malvetum parviflorae* is replaced by a vegetation dominated by *Carduus australis* subsp. *marmoratus*. It was descrived from Sicily by Brullo (1983a, b) as *Carduetum australis*, but this name is uncorrect, since according to current literature *C. australis* is represented in Sicily and Maltese Islands by the subsp. *marmoratus*. Therefore, this name must correct in *Carduetum marmorati* (art. 43 and 48c, I.C.N.P.). Similar to the previous one, it is a ruderal association linked to roadsides, very rich in nitrophytes of the *Hordeion leporini* and higher syntaxa.

As concerns its distribution, apart from Maltese Islands, it was also surveyed in Linosa and Lampedusa (Pelagian Archipelago) by Brullo (1983a, b), Bartolo et al. (1990) and Brullo and Siracusa (1996). In these countries, the association is always localized in more or less shaded places, close to dry walls which delimit the paths or more rarely on ruble accumulations, with soils rich in clay.

From a dynamic point of view, it is widespread in the belt of *Periploco angustifoliae-Euphorbietum dendroidis*, edaphoclimatic association of the Maltese islands and Pelagian Archipelago.

Table 18.6 *Chrysanthemo coronarii-Malvetum parviflorae*

Number of relevé	1	2	3	4	5	6	7	8
Plot size (m^2)	5	5	10	8	15	10	5	5
Plant cover (%)	100	100	100	100	100	100	100	100
Char. Association								
Malva parviflora	2	2	1	2	2	+	1	3
Char. All. (*Hordeion leporini*)								
Hordeum leporinum	3	3	2	3	2	4	3	3
Glebionis coronaria	3	2	3	1	3	2	3	1
Reseda alba	2	1	1	2	1	.	+	1
Rostraria cristata	.	+	+	.	+	1	1	+
Erodium ciconium	+	1	+	.	.	+	.	1
Sisymbrium officinale	+	.	.	.	+	+	.	+
Bromus scoparius	.	1	+	+	.	+	.	.
Plantago lagopus	.	1	+	.	.	+	.	+
Char. Ord. (*Brometalia rubenti-tectorum*)								
Medicago polymorpha	1	+	1	2	1	2	+	+
Galagtites elegans	2	3	2	2	2	1	2	.
Avena barbata	2	1	2	2	+	1	2	.
Lolium rigidum subsp. *rigidum*	2	2	2	1	+	1	.	.
Echium plantagineum	1	1	2	1	.	.	1	+
Stipella capensis	.	2	.	1	2	+	1	1
Anisantha madritensis	.	2	1	2	2	+	.	.
Lotus edulis	.	+	1	1	.	.	+	.
Catapodium rigidum	.	+	.	1	.	.	1	+
Hedypnois cretica	.	.	1	+	+	.	1	.
Sulla coronaria	.	.	1	+	.	1	1	.
Hirschfeldia incana	.	1	1	.	.	+	.	.
Char. Cl. (*Chenopodietea*)								
Sonchus oleraceus	2	1	2	1	1	+	1	+
Oxalis pes-caprae	1	+	1	1	2	1	.	.
Diplotaxis erucoides	+	1	.	+	+	1	.	+
Erodium malacoides	2	1	1	+	1	.	1	.
Borago officinalis	1	+	.	+	1	+	2	.
Geranium molle	.	+	.	+	+	+	+	.
Mercurialis annua	+	+	.	+	+	.	+	.
Anisantha diandra	.	1	1	+	.	1	1	.
Capsella rubella	.	.	+	+	.	.	+	1
Fumaria densiflora	+	+	.	.	.	+	.	.
Solanum nigrum	1	.	.	+	.	.	+	.
Senecio vulgaris	.	.	+	+	.	.	.	1
Malva nicaeensis	.	.	+	.	.	1	.	2
Urtica membranacea	.	+	.	.	.	+	.	.
Chenopodium murale	1	.	.	.	+	.	.	.

(continued)

Table 18.6 (continued)

Number of relevé	1	2	3	4	5	6	7	8
Plot size (m^2)	5	5	10	8	15	10	5	5
Plant cover (%)	100	100	100	100	100	100	100	100
Malva sylvestris	.	+	.	.	+	.	.	.
Ecballium elaterium	.	.	+	.	+	.	.	.
Other species								
Foeniculum piperitum	1	.	+	+	1	1	1	.
Papaver rhoeas	+	+	+	.	+	+	.	.
Beta maritima	1	+	.	.	.	+	+	.
Melilotus sulcata	.	.	+	1	.	+	.	1
Ochlopoa annua	+	.	.	+	.	.	+	.
Trisetaria aurea	.	+	1	.	1	.	.	.
Reichardia picroides var. *picroides*	.	+	+	.	.	.	+	.
Trifolium tomentosum	.	.	.	.	.	.	.	2
Hypochoeris achyrophorus	.	.	.	.	.	.	.	+

Rel. 1–4: Malta, St. Andrews, 11.4.1987; Rel. 5: Malta, Dingli, 12.4.1984; Rel. 6–7: Malta, Zurrieq, 10.4.1984; Rel. 8: Malta, La Valletta, 17.4.1987

18.8 *Chrysanthemo coronarii-Silybetum mariani* Brullo 1983 (Table 18.8)

In conditions of marked ruderalization and a certain edaphic humidity, the *Hordeion leporini* is represented by *Chrysanthemo coronariae-Silybetum mariani*. This association, described from Sicily by Brullo (1983b), in the Maltese islands is localized along the paths where there are ruble accumulations, mainly in stands on the bottom of the wadis. It is physiognomically differentiated by *Silybum marianum*, a big spiny Carduaceae usually occurring in plant communities belonging to *Carthametalia lanati*, assuming in this ruderal vegetation the role of differential, since indicator of a peculiar ecological situation. Besides, in this association the nitrophilous species of the *Hordeion leporini* are quite frequent, such as *Glebionis coronaria*, *Hordeum leporinum*, *Plantago lagopus* and *Carduus pycnocephalus* subsp. *pycnocephalus*.

It should be emphasized, however, that for its ecological exigences, this association is quite rare in the Maltese Islands.

18.9 *Hordeo leporini-Sisymbrietum orientalis* Oberdorfer 1977 (Table 18.9)

In sunny and arid roadside stands of Malta, a vegetation of the *Hordeion leporini* characterized by *Sisymbrium orientale* occurs. For its ecology and floristic set it must be attributed to the *Hordeo leporini-Sisymbrietum orientalis*, an association described from the Balkans by Oberdorfer (1977) and recorded in Sicily by Brullo

Table 18.7 *Carduetum marmorati*

Number of relevé	1	2	3	4	5	6	7	8	9	10	11
Plot size (m^2)	5	3	10	10	5	5	10	15	10	10	6
Plant cover (%)	100	100	100	80	90	100	100	80	90	100	100
Char. Association											
Carduus australis subsp. *marmoratus*	2	3	3	1	3	3	2	2	3	2	3
Char. All. (*Hordeion leporini*)											
Glebionis coronaria	2	1	3	3	1	+	+	.	2	1	2
Rostraria cristata	2	+	+	+	+	+	+	.	+	.	.
Carduus pycnocephalus subsp. *pycnocephalus*	.	2	1	1	.	2	1	.	1	1	+
Reseda alba	+	.	1	1	.	.	+	+	.	+	.
Plantago lagopus	.	.	.	+	+	.	.	1	.	+	1
Erodium ciconium	.	+	.	.	+	.	.	.	1	.	.
Anacyclus clavatus	.	+	+	.	.	.	.	.	.	.	.
Char. Ord. (*Brometalia rubenti-tectorum*)											
Galagtites elegans	2	2	1	2	2	3	1	2	2	2	1
Anisantha madritensis	.	2	2	1	1	2	2	1	+	+	1
Avena barbata	+	1	1	+	+	.	.	1	2	1	2
Medicago polymorpha	.	1	2	+	+	+	1	1	+	1	2
Stipella capensis	1	+	1	.	.	1	+	2	.	.	.
Catapodium rigidum	.	1	.	+	+	1	1	.	.	.	.
Lotus edulis	.	.	+	.	+	+	+	.	.	+	.
Avena sterilis	.	1	.	+	+	1	1	.	.	.	.
Lolium rigidum subsp. *rigidum*	.	.	1	+	+	+	+	.	.	.	.
Hedypnois cretica	.	+	.	+	.	.	+	.	.	+	.
Trifolium stellatum	.	.	.	.	+	.	+	+	.	.	.
Sulla coronaria	,	.	.	.	+	.	+	+	.	.	.
Anisantha rubens	.	.	.	.	+	.	,	+	.	.	+
Hirschfeldia incana	.	.	.	.	.	.	+	+	.	+	.
Vulpia ciliata	+	.	.	.	.	.	.	+	.	.	.
Astragalus hamosus	.	.	.	.	.	.	.	+	.	.	.
Char. Cl. (*Chenopodietea*)											
Anisantha diandra	.	+	+	.	1	+	2	1	2	2	2
Sonchus oleraceus	+	+	.	.	2	+	1	1	.	1	1
Borago officinalis	+	.	.	.	.	+	+	.	+	.	1
Stellaria media	.	.	+	.	.	+	.	+	+	.	.
Erodium malacoides	.	.	.	.	+	+	.	+	.	.	.
Sisymbrium irio	.	.	.	.	.	+	+	.	+	.	.
Mercurialis annua	.	.	.	.	.	.	+	+	.	+	.
Malva parviflora	+	+	.	.	.	.	.	.	.	.	.
Urtica membranacea	.	.	.	.	.	.	.	+	.	.	+
Oxalis pes-caprae	.	.	.	.	.	.	.	.	.	.	+
Diplotaxis erucoides	.	.	.	.	.	.	.	.	.	2	2
Galium aparine	.	.	.	.	.	.	.	.	.	.	1

(continued)

Table 18.7 (continued)

Number of relevé	1	2	3	4	5	6	7	8	9	10	11
Plot size (m^2)	5	3	10	10	5	5	10	15	10	10	6
Plant cover (%)	100	100	100	80	90	100	100	80	90	100	100
Solanum nigrum	.	.	.	.	.	.	.	.	.	+	.
Fumaria wirtgenii	.	.	1	.	.	.	.	.	.	.	.
Other species											
Lobularia maritima	1	.	+	+	.	.	.	1	+	1	.
Trifolium tomentosum	.	.	.	+	+	.	+	.	.	.	.
Echium parviflorum	.	.	+	.	.	.	.	.	1	.	.
Rhodalsine geniculata	+	+	.	.	.	.	.	.	.	.	.
Rumex bucephalophorus subsp. *bucephalophorus*	.	+	.	.	.	.	.	.	.	.	.
Anthemis urvilleana	.	.	.	.	.	.	+	.	.	.	.
Papaver rhoeas	.	.	.	.	.	.	.	.		.	+
Foeniculum piperitum	.	.	.	.	.	.	.	.	.	1	.

Rel. 1–3: Gozo, Dwerja, 11.4.1984; Rel. 4: Malta, Mellieha, 11.4.1987; Rel. 5–6: Gozo, Ta Cenc, 13.4.1987; Rel. 7–8: Gozo, Dwerja, 13.4.1987; Rel. 9–11: Gozo, Xlendi, 13.4.1987

(1983b). It is an association occurring in the urban and suburban stands of coastal places, characterized by the dominance of *Hordeum leporinum* and several species of the alliance and higher rank. In the Maltese archipelago, it is fairly rare, probably due to the remarkable aridity of coastal environments, that is not favorable to the settlement of the *Hordeion leporini* coenosis.

18.10 *Acantho mollis-Smyrnietum olusatri* Brullo & Marcenò 1985 (Table 18.10)

This association is localized in fresh and humid ruderal stands, always shaded, such as those ones near rural or suburban areas, valleys, abandoned gardens, etc. The *Acantho mollis-Smyrnietum olusatri* represents an association with clearly thermophilic characters, which is linked to micro-climatic conditions of the sciaphilo-nitrophilous type. Characteristic species of this vegetation are *Acanthus mollis* and *Smyrnium olusatrum*, which play also a significant physiognomic role.

From the syntaxonomical viewpoint, this association was included by Brullo and Marcenò (1985b) in the *Allion triquetri* O. Bolós 1967, an alliance belonging to *Urtico-Scrophularietalia peregrinae*, which in Malta are represented by several species, such as *Arum italicum, Parietaria judaica, Urtica membranacea, Galium aparine, Scrophularia peregrina, Allium triquetrum*, etc.

The *Acantho mollis-Smyrnietum olusatri*, apart from Malta, where was previously recorded by Brandes (1991), is widespread in Sicily and some neighboring

Table 18.8 *Chrysanthemo coronarii-Silybetum mariani*

Number of relevé	1	2	3	4	5
Plot size (m^2)	10	20	20	10	10
Plant cover (%)	100	100	100	100	100
Char. Association					
Silybum marianum	4	4	3	5	4
Char. All. (*Hordeion leporini*)					
Glebionis coronaria	2	3	3	1	2
Hordeum leporinum	2	1	2	2	2
Plantago lagopus	+	1	+	+	.
Carduus pycnocephalus subsp. *pycnocephalus*	2	1	1	.	+
Bromus scoparius	.	.	+	1	.
Rostraria cristata	.	+	.	+	.
Char. Ord. (*Brometalia rubenti-tectorum*)					
Galagtites elegans	1	1	.	+	1
Avena barbata	+	.	.	1	+
Medicago polymorpha	.	+	+	1	.
Lolium rigidum subsp. *rigidum*	.	.	+	+	1
Anisantha madritensis	1	.	+	.	.
Medicago truncatula	.	.	+	.	.
Astragalus hamosus	.	.	+	.	.
Char. Cl. (*Chenopodietea*)					
Oxalis pes-caprae	.	1	1	2	1
Anisantha diandra	.	+	2	1	1
Urtica membranacea	1	.	1	+	+
Sonchus oleraceus	+	1	.	+	2
Malva sylvestris	+	+	.	1	.
Mercurialis annua	.	1	.	+	+
Galium aparine	1	.	2	2	.
Erodium malacoides	+	.	.	+	.
Capsella rubella	+	.	.	.	+
Ecballium elaterium	+	.	.	.	.
Other species					
Beta maritima	1	+	+	.	.
Raphanus raphanistrum	+	.	1	.	.
Foeniculum piperitum	.	.	+	.	1
Vicia sativa	.	.	+	+	.
Parietaria judaica	.	.	.	+	+
Papaver hybridum	.	.	+	.	.
Carthamus lanatus	.	.	.	+	.
Plantago afra subsp. *afra*	.	.	.	+	.

Rel. 1–2: Gozo, Xlendi, 11.4.1984; Rel. 3–4: Gozo, Xlendi, 13.4.1987; Rel. 5: Gozo, Mgarr ix-Xini, 11.4.1984

Table 18.9 *Hordeo leporini-Sisymbrietum orientalis*

Number of relevé	1	2	3	4
Plot size (m^2)	10	10	8	5
Plant cover (%)	90	80	60	60
Char. Association				
Sisymbrium orientale	+	1	1	1
Char. All. (*Hordeion leporini*)				
Hordeum leporinum	4	4	3	3
Sisymbrium officinale .	1	+	1	.
Rostraria cristata	1	+	.	.
Glebionis coronaria	.	.	+	1
Char. Ord. (*Brometalia rubenti-tectorum*)				
Hirschfeldia incana	1	1	+	+
Lolium rigidum subsp. *rigidum*	+	+	1	1
Avena barbata	+	1	1	+
Urospermum picroides	+	.	.	+
Anisantha madritensis	.	.	+	1
Galagtites elegans	.	.	+	.
Char. Cl. (*Chenopodietea*)				
Merculialis annua	+	.	+	+
Malva sylvestris	+	.	+	+
Emex spinosa	+	.	+	+
Anisantha diandra	+	+	1	.
Geranium molle	1	+	.	.
Sonchus oleraceus	+	.	+	.
Capsella rubella	.	+	.	.
Other species				
Trifolium resupinatum	1	.	.	.
Trisetaria aurea	+	.	.	.

Rel. 1–4: Malta, Sliema, 12.4.1987

islands, as well as in various territories of southern Tirreno as Sardinia and Italian Peninsula (Bartolo et al. 1990, Biondi et al. 1989; Ceschin et al. 2003). Other associations with *Smyrnium olusatrum* were described from several countries of West Mediterranean by Rivas Goday (1964), de Bolós and Molinier (1958), de Bolós (1967), Izco and Géhu (1978), Biondi et al. (1989, 1990), Julve (1993), Amigo and Romero (1997), Vázquez and Buján (1997) and Costa et al. (2012).

18.11 *Lavatero creticae-Malvetum nicaeensis* ass. nov. hoc loco (Table 18.11)

Holotypus: rel. 5 hoc loco.

During the spring season, in urban and suburban environments, limitedly tu ruderal stands with deep and well nitrified soils, it is frequent an annual vegetation

Table 18.10 *Acantho mollis-Smyrnietum olusatri*

Number of relevé	1	2	3	4	5	6	7	8	9	10	11	12	13
Plot size (m^2)	50	10	50	50	20	20	10	10	10	6	10	5	5
Plant cover (%)	100	100	100	100	100	100	100	100	100	100	100	100	100
Char. Association and All. (*Allion triquetri*)													
Acanthus mollis	3	1	3	2	2	3	3	2	1	3	4	3	1
Smyrnium olusatrum	3	4	3	3	4	3	2	4	4	3	2	3	4
Cynoglossum creticum	.	.	.	.	.	.	.	.	.	.	.	+	.
Char. Ord. (*Urtico-Scrophularietalia peregrinae*)													
Arum italicum	2	2	2	1	1	2	3	2	3	2	2	2	3
Parietaria judaica	1	2	2	3	2	3	3	2	1	1	1	1	2
Urtica membranacea	1	2	2	1	2	3	2	1	.	1	+	2	+
Galium aparine	3	1	2	2	1	2	2	1	1	1	1	2	1
Scrophularia peregrina	+	.	1	+	.	+	.	.	.	+	+	2	.
Fumaria capreolata	+	.	+	1	.	+	.	.	.	.	.	.	.
Arisarum vulgare	.	.	.	.	.	.	.	.	+	1	.	+	.
Char. Cl. (*Chenopodietea*)													
Oxalis pes-caprae	3	2	3	2	2	3	3	3	2	2	3	3	2
Sonchus oleraceus	+	1	+	+	+	+	.	2	+	+	2	1	+
Mercurialis annua	2	2	2	1	1	+	.	1	+	.	+	.	+
Borago officinalis	+	1	1	1	.	+	.	1	+	.	.	.	+
Anisantha diandra	.	1	1	+	1	+	.	.	+	+	1	2	.
Stellaria media	+	.	1	.	+	+	.	.	.	+	+	1	.
Geranium molle	+	+	.	+	.	.	.	.	.	.	.	.	+
Glebionis coronaria	.	1	.	.	+	.	+	+	.	.	.	.	.
Galagtites elegans	1	.	.	.	.	.	.	.	.	.	1	+	.

Urtica pilulifera	.	1	.	.	.	.	.	.	.	.	.	.	.
Ecballium elaterium	1	.	.	.	.	.	.	.	.	.	.	.	.
Solanum nigrum	.	.	.	.	.	.	.	.	.	.	.	.	+
Other species													
Phragmites australis	2	.	1	2	.	1	.	.	.	.	.	.	.
Rumex conglomeratus	.	.	+	.	.	1	1	.	.	.	.	.	.
Bromus racemosus	.	.	+	+	.	+	.	.	.	.	.	.	.
Allium subhirsutum	.	.	.	.	.	.	.	.	+	.	1	+	.
Foeniculum piperitum	1	.	.	.	.	+	.	.	.	.	.	.	.
Potentilla reptans	.	.	1	.	.	+	.	.	.	.	.	.	
Ranunculus sardous	.	.	+	.	.	+	.	.	.	.	.	.	.
Ranunculus macrophyllus	.	.	.	.	.	.	.	.	.	1	2	.	.
Theligonum cynocrambe	.	.	.	.	.	.	.	.	.	.	1	+	.
Lotus ornithopodioides	.	.	.	.	.	.	.	.	.	.	+	.	.

Rel. 1: Malta, Salina Bay, 9.4.1984; Rel. 2: Malta, Mellieha, 10.4.1984; Rel. 3–4: Gozo, Xlendi Valley, 11.4.1984; Rel. 5–7: Gozo, Xlendi Valley, 15.11.1986; Rel. 8: Malta, Mgarr, 30.11.1985; Rel. 9: Malta, St. Andrews, 11.4.1986; Rel. 10: Malta, Buskett, 12.4.1987; Rel. 11–12: Malta, Salina Bay, 16.4.1987; Rel. 13: Malta, Addolorata, 17.4.1987

Table 18.11 *Lavatero creticae-Malvetum nicaeensis*

Number of relevé	1	2	3	4	5
Plot size (m^2)	5	5	8	10	10
Plant cover (%)	40	60	70	70	60
Char. Association					
Malva nicaeensis	1	2	2	2	2
Lavatera cretica	.	.	.	.	+
Char. All. (*Malvion parviflorae*) and Ord. (*Chenopodietalia muralis*)					
Malva parviflora	+	2	3	1	2
Chenopodium murale	2	+	2	3	2
Conyza bonariensis	.	+	+	.	.
Sisymbrium irio	+	.	.	.	+
Char. Cl. (*Chenopodietea*)					
Sonchus oleraceus	1	+	2	1	1
Stellaria media	.	+	+	.	+
Glebionis coronaria	.	.	.	3	+
Mercurialis annua	2	+	.	.	.
Hordeum leporinum	.	.	.	1	+
Solanum nigrum	+	.	1	.	.
Senecio vulgaris	.	+	+	.	.
Urtica membranacea	.	2	.	.	.
Euphorbia helioscopia	+	.	.	.	.
Helminthotheca echioides	.	.	+	.	.
Borago officinalis	.	.	+	.	.
Galactites elegans	.	.	.	+	.
Other species					
Parietaria judaica	1	2	1	.	1
Oryzopsis miliacea	+	1	.	.	.
Beta maritima	.	.	.	2	.
Spergularia rubra	.	.	.	+	.
Silene colorata	.	.	.	+	.
Daucus carota	.	.	+	.	.

Rel. 1–2: Malta, Girgenti Valley, 16.4.1987; Rel. 3: Malta, Buskett, 16.4.1987; Rel. 4: Gozo, Xlendi Tower, 30.4.2007; Rel. 5: Malta, Sliema, 12.4.1987

belonging to the *Chenopodietalia muralis* Br.-Bl. in Br.-Bl. et al. 1936. The occurrence in this plant community of *Malva nicaeensis* and *M. parviflora*, markedly nitrophilous species, allows to attribute it to *Malvetum parvifloro-nicaensis,* association described by Braun-Blanquet & Maire (1924) and succesively recorded by Braun-Blanquet et al. (1936) as provisional association for the Ibero-Mauritanian territories. Unfortunately, as already emphasized by de Foucault (2012), this syntaxon must be considered a nomen dubium (art. 16, 37), since in any of the three releves published by Braun Branquet and Maire (1924) are present contemporally the two species that give the name to the association. Besides, de

Foucault (2012) points out that the releves carried out in Sicily by Brullo and Marcenò (1985b) and attributed by them to *Malvetum parvifloro-nicaeensis*, are very different from Morocco those ones. As concerns the releves carried out in the Maltese islands, they are very similar to the Sicilian ones and therefore can be attributed to a new association, named *Lavatero creticae-Malvetum nicaeensis*. The characteristic species of this syntaxon are *Lavatera cretica* and *Malva nicaeensis*, which usually grow together with *Malva parviflora*. For its ecology and floristic set, this syntaxon can be included in the *Malvion parviflorae* (Rivas-Martinez 1978) Brullo in Brullo & Marceno 1985, alliance of *Chenopodietalia muralis* Br.-Bl. in Br.-Bl. et al. 1936, gathering communities with a strictly spring cycle.

18.12 *Spergulario marinae-Taraxacetum minimi* ass. nov. hoc loco (Table 18.12)

Holotypus: rel. 6, hoc loco.

In the ruderal stands of urban stands characterized by a low edaphic salinity, due to sea aerosol, a very peculiar nitrophilous vegetation belonging for its ecology and floristic set to *Malvion parviflorae* was surveyed. It is differentiated by the dominance of the *Taraxacum minimum* and *Spergularia marina*, which are proposed as characteristic of a new association named *Spergulario marinae-Taraxacetum minimi*. It is linked to road environments, especially at the edge of tree lined avenues and flowerbeds of promenade of some coastal urban centers, where it has its optimum in the winter period until early spring.

The *Spergulario marinae-Taraxacetum minimi* can be considered as a subalophilous vicariant of *Lavatero creticae-Malvetum nicaeensis*, which usually likes the urban environments but with not salt soils.

18.13 *Lavateretum arboreae* Br.-Bl. & Molinier 1935 (Table 18.13)

On ruble accumulations or deposits rich in organic material, within urban or rural areas mainly near the coast, there is a strictly ruderal vegetation in which *Lavatera arborea* plays a physiognomic relevant role. This woody nitrophyte, covering often very large surfaces, features a peculiar association validly described by Braun-Blanquet and Molinier (1935) as "associàtion à *Lavatera arborea*", while from the later authors it is wrongly quoted as *Lavateretum ruderale*. For this reason, Rivas-Martinez (1978b) believed to propose a new name *Lavateretum arboreo-cretici* in place of *Lavateretum ruderale,* but for obvious reasons it is a *nomen superfluum* (art.29c). Unfortunately, this name is reported in literature by many authors (cf. Brullo and Marcenò 1985b; de Foucault 2012) For its floristic set and ecological

Table 18.12 *Spergulario marinae-Taraxacetum minimi*

Number of relevé	1	2	3	4	5	6	7	8
Plot size (m^2)	5	5	10	6	5	8	6	10
Plant cover (%)	50	50	50	50	40	60	60	50
Char. Association								
Taraxacum minimum	3	3	2	3	2	3	2	1
Spergularia marina	.	+	+	.	+	1	2	1
Char. All. (*Malvion parviflorae*) and Ord. (*Chenopodietalia*)								
Sisymbrium irio	+	1	+	1	2	+	2	1
Chenopodium murale	.	+	+	.	+	1	2	1
Malva parviflora	.	.	+	+	.	1	1	+
Conyza bonariensis	.	+	+	.	.	+	.	+
Hyoscyamus albus	.	+	+	.	.	+	.	.
Lavatera arborea	.	.	.	.	+	1	.	.
Lavatera cretica	.	.	.	.	.	+	.	+
Char. Cl. (*Chenopodietea*)								
Sonchus oleraceus	+	+	+	.	+	+	+	.
Chenopodium album	+	.	+	+	.	1	1	2
Symphyotrichum squamatum	.	+	+	1	+	.	+	+
Oxalis pes-caprae	+	+	+	.	+	+	+	.
Stellaria media	+	+	.	+	1	+	.	+
Urtica membranacea	.	.	+	+	1	.	+	+
Hordeum leporinum	.	+	+	.	.	+	1	.
Erodium malacoides	.	.	+	+	.	+	.	.
Glebionis coronaria	.	.	+	.	.	.	+	.
Capsella rubella	.	.	.	.	.	.	+	+
Other species								
Beta maritima	1	1	2	1	+	1	1	2
Convolvulus arvensis	2	1	2	2	2	1	2	.
Cynodon dactylon	.	2	1	.	+	1	.	.
Polygonum aviculare	.	.	.	+	+	.	+	.
Polycarpon tetraphyllum	.	.	.	.	+	+	.	+

Rel. 1–3: Malta, La Valletta, 12.4.1987; Rel. 4–8: Malta, Sliema, 12.4.1987

requirements, this synantropic association was usually included in the *Chenopodietalia muralis* Br.-Bl. in Br.-Bl. et al. 1936 and in particolar at first within the *Chenopodion muralis* Br.-Bl. in Br.-Bl. et al 1936, while later Rivas-Martinez (1978b) attributed it to the suballiance *Malvenion parviflorae*, sintaxon elevated to rank of alliance (*Malvion parviflorae*) by Brullo and Marcenò (1985b). Among the species of this alliance and order are quite frequent *Lavatera cretica, Malva parviflora, Chenopodium muralis, Hyosciamus albus, Ecballium elaterium,* etc.

Nevertheless, this syntaxonomical arrangement is in contrast with that one more recently proposed by Rivas-Martínez et al. (2001, 2002, 2011) and Mucina et al. (2016). In fact, these authors feel most appropriate to attribute the associations

Table 18.13 *Lavateretum arboreae*

Number of relevé	1	2	3	4	5	6
Plot size (m^2)	20	50	30	50	50	20
Plant cover (%)	100	100	100	100	100	100
Char. Association						
Lavatera arborea	1	3	4	4	4	4
Char. All. (*Malvion parviflorae*) and Ord. (*Chenopodietalia*)						
Lavatera cretica	5	4	1	1	2	1
Chenopodium murale	1	+	+	1	1	3
Ecballium elaterium	1	2	+	+	.	.
Malva parviflora	+	+	.	1	2	.
Hyoscyamus albus	.	1	+	2	.	.
Char. Cl. (*Chenopodietea*)						
Oxalis pes-caprae	2	2	2	2	2	3
Glebionis coronaria	1	2	1	+	+	2
Sonchus oleraceus	1	+	1	1	1	2
Hordeum leporinum	1	+	1	+	+	1
Erodium malacoides	+	1	+	+	1	.
Mercurialis annua	1	+	+	+	+	.
Malva sylvestris	+	.	.	1	+	.
Borago officinalis	.	.	1	.	+	.
Anisantha diandra	.	.	1	2	.	.
Stellaria media	.	.	.	1	.	.
Convolvulus arvensis	.	.	.	+	.	.
Erodium cicutarium	.	.	.	+	.	.
Reseda alba	.	.	.	.	.	+
Sulla coronaria	.	.	.	.	.	1
Galactites elegans	.	.	.	.	.	1
Other species						
Beta maritima	1	1	1	2	1	2
Carthamus lanatus	+	1	+	.	.	.
Reichardia picroides var. *picroides*	.	.	+	.	+	+
Melilotus messanensis	.	.	+	.	+	.
Avena sterilis	.	.	+	.	+	.
Plantago lagopus	.	.	.	+	.	.
Daucus carota	.	.	.	+	.	.
Anagallis arvensis	.	.	.	+	.	.

Rel. 1–3: Malta, Sliema, 8.4.1984; Rel. 4: Malta, Sliema, 12.4.1987; Rel. 5: Malta, Ghallis, 8.4.1984; Ril. 6: Malta, St. Thomas, Marsaskala, 18.5.2017

dominated by *Lavatera arborea* to a special alliance of *Pegano harmalae-Salsoletea vermiculatae* Br.-Bl. & O. Bolós 1958, represented by *Medicagini citrinae-Lavaterion arboreae* O. Bolós, Folch and Vigo in O. Bolós and Vigo 1984. Even de Foucault (2015) is of the same opinion, but he includes these plant communities in an another alliance of the *Pegano harmalae-Salsoletea vermiculatae*, proposed as

Beto maritimae-Malvion arboreae nom. illeg. However, these hypothesis does not seem supported both in floristic and in ecological terms too, since in the releves regarding these associations are usually absent the shrubby species of the aforesaid class, while the nitrophilous therophytes of *Malvion parviflorae* and related higher syntaxa are always very frequent (see holotype of *Lavateretum arboreae* pubblished by Braun Blanquet and Molinier 1935). Besides, the vegetation at issue is linked to strictly ruderal stands, with markedly nitrified soil, therefore well diversified from natural or semi-natural ones, with arid and weakly nitrified soils, typical of the *Pegano harmalae-Salsoletea vermiculatae* associations.

This association, widespread in Maltese Archipelago where was surveyed also by Brandes (1991), occurs in several coastal places of France (Braun-Blanquet and Molinier 1935; Braun-Blanquet et al. 1936, 1952; Géhu and Géhu-Franck 1961, 1969; Bensettiti et al. 2004; de Foucault 2012, 2015), Spain (de Bolós 1967; Rivas-Martinez 1978b; Ladero et al. 1981; Rivas-Martínez et al. 2001, 2011), Balears (de Bolós and Molinier 1958; Brandes 1988; de Bolós 1996), Croatia (Horvatić 1963; Hećimović 1984; Pandža et al. 2005; Stancić et al. 2008) Italy (Cristofolini et al. 1967), Sardinia (Biondi et al. 1990, 2001), Sicily (Brullo and Marcenò 1985b; Bartolo et al. 1990; Gianguzzi 1999).

18.14 *Mesembryanthemetum crystallino-nodiflori* O. Bolos 1957 (Table 18.14)

This subnitrophilous association in the Maltese Islands is linked to coastal places limitedly to halomorphic clayey or marly soils. Physiognomically, it is dominated by *Mesembryanthemum crystallinum* and *M. nodiflorum*, succulent therophytes with creeping habit, which are mainly frequent within perennial communities characterized by *Lygeum spartum* or *Darniella melitensis*. Previously, this vegetation was recorded from several countries of the western Mediterranean, such as Iberian Peninsula (de Bolós 1957, 1967; Rivas-Martínez et al. 2001), Baleares (de Bolós and Molinier 1958; de Bolós 1996), France (Bensettiti et al. 2004), Tunisia (Poittier Alapetite 1954), Corsica (Géhu et al. 1988), Sicily (Bartolo et al. 1982, 1990; Brullo and Siracusa 1996). Based on the literature data, the *Mesembryanthemetum cristallino-nodiflori* was previously included in various phytocociological classes, such as *Pegano harmalae-Salsoletea vermiculatae*, *Frankenietea pulverulentae* and *Stellarietea mediae* (see de Bolós 1957, 1962; Peinado et al. 1989; Bartolo et al. 1982, 1990; Brullo and Siracusa 1996; Rivas-Martínez et al. 1993, 2001). In fact, due to its ecology, since it is an ephemeral ruderal vegetation growing in synanthropic stands near villages or sites visited by sea birds, the *Mesembryanthemetum cristallino-nodiflori* shows a floristic set rich in elements of the *Chenopodietalia muralis* and particularly of the *Mesembryanthemion cristallini* Rivas- Martinez et al. 1993, which differs from *Chenopodion muralis* for its subhalophilous character.

Table 18.14 *Mesembrryanthemetum crystallino-nodiflori*

Number of relevé	1	2	3	4	5
Plot size (m^2)	20	10	10	15	5
Plant cover (%)	40	60	60	60	40
Char. Association and All. (*Mesembryanthemion cristallini*)					
Mesembryanthemum nodiflorum	3	4	4	4	3
Mesembryanthemum crystallinum	1	2	1	.	1
Char. Ord. (*Chenopodietalia*)					
Hyoscyamus albus	1	+	.	+	.
Lavatera arborea	+	1	.	+	1
Chenopodium murale	.	+	1	.	1
Malva parviflora	.	1	+	.	.
Char. Cl. (*Chenopodietea*)					
Glebionis coronaria	+	+	+	.	1
Sonchus oleraceus	.	+	+	+	1
Chenopodium album	.	+	.	.	.
Other species					
Beta maritima	1	+	1	.	1
Parapholis incurva	+	+	.	+	1
Soda inermis	2	.	.	+	.
Plantago coronopus	+	1	.	.	.
Spergularia marina	.	.	+	.	1
Glaucium flavum	+	.	.	.	.
Limbarda crithmoides subsp. *longifolia*	+	.	.	.	.

Rel. 1: Malta, Sliema, 24.6.1973; Rel. 2–3: Malta, Dragunara, 8.4.1984; Rel. 4: Malta, Delimara, 10.4.1984; Ril. 5: Malta, St. Thomas, Marsaskala, 18.5.2017

The *Mesembryanthemetum crystallino-nodiflori* can be considered a vicariant of the *Lavateretum arboreae* in markedly halo-xeric conditions and lower soil nitrification, which favor the succulent annual species, having also a large size.

18.15 *Amarantho deflexi-Dysphanietum ambrosiodis* ass. nov. hoc loco (Table 18.15)

Holotypus: rel. 1, hoc loco.

During the summer-autumnal period in urban and suburban habitat, it was observed a ruderal vegetation physiognomically dominated by *Dysphania ambrosioides* (=*Chenopodium ambrosioides*). This species grows together with some species of the *Chenopodion muralis* and *Chenopodietalia muralis*, such as *Conyza bonariensis, Ecballium elaterium, Chenopodium murale, Amaranthus deflexus, Chenopodium vulvaria,* etc. On the whole, this plant community shows close relations with *Amarantho blitoidis-Chenopodietum ambrosioidis* O. Bolós

Table 18.15 *Amarantho deflexi-Dysphanietum ambrosiodis*

Number of relevé	1	2	3	4	5
Plot size (m^2)	10	10	6	15	10
Plant cover (%)	90	40	50	70	70
Char. Association					
Dysphania ambrosioides	3	2	2	2	3
Amaranthus deflexus	1	+	+	+	.
Char. All. (*Chenopodion muralis*) and Ord. (*Chenopodietalia*)					
Conyza bonariensis	2	1	1	3	2
Ecballium elaterium	+	+	+	2	.
Chenopodium murale	+	.	+	1	2
Chenopodium vulvaria	+	+	.	.	.
Char. Cl. (*Chenopodietea*)					
Chenopodium album	2	1	1	1	+
Symphyotrichum squamatum	2	2	3	+	1
Anisantha diandra	2	1	1	2	+
Sonchus oleraceus	1	+	1	1	+
Malva sylvestris	+	1	1	.	+
Urtica membranacea	+	+	+	.	.
Mercurialis annua	+	+	1	.	.
Solanum nigrum	+	.	+	.	1
Setaria verticillata	.	.	+	.	+
Amaranthus graecizans subsp. *sylvestris*	.	+	.	.	+
Other species					
Parietaria judaica	2	1	1	.	1
Cynodon dactylon	.	1	+	.	.
Reichardia picroides var. *picroides*	.	.	.	+	.

Rel.1–3: Malta, Delimara, 26.9.1985; Rel. 4–5: Malta, Zurrieq, 24.9.1985

(1962) 1967 (=*Chenopodietum muralis* B.-Bl. 1936 *chenopodietosum ambrosioidis* O. Bolós 1962), association with marked xero-heliophilous and acidophilous features, descrived from Spain and recorded also in France (de Bolós 1962, 1967; Julve 1993; Curcò and Masip 2000; Rivas-Martínez et al. 2001; de Foucault 2012). However, some significant floristic and ecological differences allows to distinguish quite well the Maltese vegetation from the aforesaid association. These differences regard the absence in the Maltese releves of *Amaranthus blitoides*, *A. muricatus*, *Chenopodium botrys* and *Mirabilis jalapa*, as well as their localization in fresh and shady calcicolous stands, often characterized by weak edaphic umidity. Therefore, it is described as a new association, named *Amarantho deflexi-Dysphanietum ambrosiodis*, having as charateristic species *Amaranthus deflexus*. In the Maltese Islands, the association is quite rare, with tendency to locate in the cooler and shaded stations or in any case where there is some editable moisture. This association occurs also in Sicily, where it was surveyed by Brullo and Marcenò (1985b), but uncorrectly ascribed to *Amarantho blitoidis-Chenopodietum ambrosioidis*.

18.16 *Parietario lusitanicae-Veronicetum cymbalariae* Brullo & Marcenò 1985 (Table 18.16)

In shady and weakly nitrified habitas represented by non-cemented walls and rocky surfaces a very peculiar microfitic vegetation occurs, which have its optimum in the springtime. It is an ephemeral aspect belonging to the *Valantio-Galion muralis* Brullo in Brullo & Marcenò 1985, alliance of the *Geranio purpurei-Cardamietalia hirsutae* Brullo in Brullo & Marcenò 1985, which for the occurrence of *Veronica cymbalaria* and *Parietaria lusitanica* can be attributed to *Parietario lusitanicae-Veronicetum cymbalariae*. In this association, described by Brullo and Marcenò (1985b) from Sicily, *Valantia muralis, Galium muralis, Geranium purpureum, Campanula erimus* and *Cardamine hirsuta* are very frequent too. This association with a stricty basiphilous character is linked to semi-natural and a little anthropized places.

Table 18.16 *Parietario lusitanicae-Veronicetum cymbalariae*

Number of relevé	1	2	3	4	5
Plot size (m^2)	1	1	0.6	0.5	0.5
Plant cover (%)	70	70	50	60	40
Char. Association					
Veronica cymbalaria	2	1	1	2	+
Char. All. (*Valantio muralis-Galion muralis*)					
Valantia muralis	3	3	1	2	+
Galium murale	1	+	+	1	2
Campanula erinus	+	2	.	+	.
Char. Ord. (*Geranio purpurei-Cardaminetalia hirsutae*)					
Parietaria lusitanica	2	2	3	1	2
Geranium purpureum	2	1	1	1	.
Centranthus calcitrapa	.	2	+	.	+
Theligonum cynocrambe .	.	+	1	2	.
Cardamine hirsuta	.	.	1	1	+
Char. Cl. (*Chenopodietea*)					
Oxalis pes-caprae	1	1	+	1	.
Mercurialis annua	.	1	+	1	+
Sonchus oleraceus	1	1	.	.	1
Galagtites elegans	+	.	.	.	+
Stellaria media	.	.	.	+	.
Other species					
Allium subhirsutum	1	1	.	.	+
Umbilicus horizontalis	+	+	.	1	.
Vulpia ciliata	.	+	.	+	.

Rel. 1–2: Gozo, Dahlet Qorrot (Nadur), 11.4.1987; Rel. 3: Malta, Wied Incita, 16.4.1987; Rel. 4–5: Malta, Naxxar, 12.4.1987

Table 18.17 *Galio aparinae-Conietum maculati*

Number of relevé	1	2	3
Plot size (m^2)	5	5	8
Plant cover (%)	100	100	100
Char. Association			
Conium maculatum	5	5	4
Char. All. (*Balloto-Conion maculati*) and Cl. (Artemisietea vulgaris)			
Galium aparine	1	2	1
Ballota nigra subsp. *uncinata*	1	.	1
Other species			
Arum italicum	2	2	1
Smyrnium olusatrum	1	1	2
Oxalis pes-caprae	1	+	+
Parietaria judaica	.	1	2
Urtica membranacea	.	+	+
Notobasis syriaca	+	.	.

Rel. 1–3: Malta, Luqa, 12.4.1987

18.17 *Galio aparine-Conietum maculati* Rivas-Martinez ex Lopez 1978 (Table 18.17)

This markedly nitrophilous association is localized in wet places with fresh soils rich in organic matter and constant edaphic moisture. It occurs along the edges or streams with hypertrophic wathers, due to discharges of slurries or sewages. The differential species of this association is *Conium maculatum*, which for its big size and high coverage has a significant physiognomic role.

The *Galio aparine-Conietum maculati* described from Spain by Lopez (1978) is included by Brullo (cfr. Brullo and Marcenò 1985b) in the *Balloto-Conion maculati*, alliance belonging to *Galio aparines-Alliarietalia petiolatae* Görs & Müller 1969 of the class *Epilobietea angustifolii* R. Tx. & Preising ex von Rochow 1951 (=*Galio aparines-Urticetea dioicae* Passarge ex Kopecky 1969). Among the species of these syntaxa are frequent *Galium aparine* and *Ballota nigra* subsp. *uncinata*.

This association, which is quite rare in the Maltese Islands, was previously recorded from the Iberian Peninsula (Lopez 1978; Diaz Gonzalez et al. 1988; Rivas-Martínez et al. 2001), Central Italy (Baldoni and Biondi 1993; Pedrotti 2008) and Sicily (Brullo and Marcenò 1985b).

18.18 *Calystegio sylvaticae-Arundinetum donacis* Brullo, Scelsi & Spampinato 2001 (Table 18.18)

At the bottom of the canals or at the edges of streams with more or less permanent fresh water and weakly nitrified slimy soils, there is often a megaforb vegetation dominated by *Arundo donax*. These are densely reeds, usually monophytic, located

Table 18.18 *Calystegio sylvaticae-Arundinetum donacis*

Number of relevé	1	2	3
Plot size (m^2)	50	50	50
Plant cover (%)	100	100	100
Char. Association			
Calystegia sylvatica	1	2	1
Char. All. (*Cynancho acuti—Convolvulion sepium*) and Cl. (*Epilobietea angustifolii*)			
Arundo donax	5	5	5
Galium aparine	2	2	+
Other species			
Symphyotrichum squamatum	1	1	+
Rubus ulmifolius	2	1	2
Oxalis pes-caprae	2	2	1
Parietaria judaica	1	.	+
Stellaria media	+	.	+
Conyza canadensis	+	.	.

Rel. 1–2: Malta, Buskett, 12.4.1987; Rel. 3 Malta, Girgenti Valley, 16.4.1987

in open and sunny habitats, especially within agricultural areas localized in large depressions. In this vegetation it is possible to observed some species climbing and twisting on the high and robust stems of *Arundo donax*, as in particolar *Calystegia sylvatica*. From the syntaxonomical point of view, it can be ascribed to *Calystegio sylvaticae-Arundinetum donacis*, association described by Brullo et al. (2001c) from S. Italy, but occurring also in Sicily. According to Mucina et al. (2016), this coenosis must included in the *Cynancho acuti-Convolvulion sepium* Rivas Goday & Rivas-Martinez ex Rivas-Martinez 1977, alliance of *Convolvuletalia sepium* R. Tx. ex Moor 1958, while as concerns the class it goes more correctly referred to *Epilobietea angustifolii* R. Tx. & Preising ex von Rochow 1951. For the dominance of *Arundo donax*, this association shows close relation with the *Arundini donacis-Convolvuletum sepium* R. Tx. & Oberd. ex O. Bolós 1962, hygro-subnitrophilous vegetation of eutrophic type widespread in many Mediterranean countries (de Bolós 1967; Brullo and Marcenò 1985b; Biurrun 1999; Curcò and Masip 2000; Costa et al. 2000; Rivas-Martínez et al. 2001; de Foucault 2011; Jasprica et al. 2016).

18.19 *Nicotiano glaucae-Ricinetum communis* de Foucault 2013 (Table 18.19)

In the Maltese Islands, as well as in many Mediterranean and Macaronesian countries, a peculiar shrub vegetation rich in tropical neophytes is often well developped in arid and anthropized habitats on nitrified soils. It is usually characterized by *Ricinus communis, Nicotiana glauca, Ipomea indica*, etc., perennial xenophytes linked to deep soils on various substrata of ruderal stands. Communities dominated

Table 18.19 *Nicotiano glaucae-Ricinetum communis*

Number of relevé	1	2
Plot size (m^2)	20	50
Plant cover (%)	100	100
Char. All. (*Nicotiano glaucae-Ricinion communis*) and Ord. (*Nicotiano glaucae-Ricinetalia communis*)		
Nicotiana glauca	5	1
Car, Cl. (*Nicotiano glaucae-Ricinetea communis*)		
Ricinus communis	1	5
Ipomoea indica	1	+
Other species		
Foeniculum vulgare	2	2
Oxalis pes-caprae	2	1
Leucaena leucocephala	1	.
Glebionis coronaria	2	.
Agave americana	.	+
Medicago arborea	.	1
Pennisetum setaceum	.	+
Rubus ulmifolius	.	2

Ril. 1: Malta, Qrendi, 21.5.2017; Ril. 2: Malta, Mellieha Rigde, 22,5.2017

by these species was described for the first time by Braun-Blanquet and Maire (1924) from Morocco and named "broussaille à *Ricinus* et *Nicotiana*" but without support of phytosociological releves. Later, de Foucault (1993a) proposed it as *Nicotiano glaucae-Ricinetum communis* (Br.-Bl. & Maire 1924) ass. nov., but according to Theurillat and Moravec (1996) it is a nom. inval. (art. 2b, 3f). This association was valided by de Foucault (2013b) on a releve carried out in Tunisia. Other associations characterized by the two aforesaid species were recorded from Canary Islands and Madeira by Sunding (1972), Rivas-Martínez et al. (1993, 2001, 2011), Costa et al. (1997) and Espírito-Santo (2003), they are *Polycarpaeo-Nicotianetum glaucae* Sunding 1972 and *Tropaeolo majoris-Ricinetum communis* Rivas-Martínez et al. 1993. Floristically, the Maltese vegetation can be attributed to *Nicotiano glaucae-Ricinetum communis*.

The communities at issue was arranged by Rivas-Martínez et al. (1999) in a well distinct alliance and order, represented by *Nicotiano glaucae-Ricinion communis* and *Nicotiano glaucae-Ricinetalia communis*, syntaxa included in the *Pegano harmalae-Salsoletea vermiculatae*. Basing on its ecology and floristic peculiarity, the inclusion of these ruderal perennial phytocoenosis in this class must be rejected, since it is not supported by the occurrence of species of this syntaxon, but usually in the releves only xenophytes having an exotic origin are observable. In fact, the *Pegano harmalae-Salsoletea vermiculatae* is a syntaxon with a Mediterranean-Saharo-Arabian and Macaronesian distribution, differentiated by thermo-xerophilous species endemic of this area, represented mainly by succulent Chenopodiaceae. Besides, the associations belonging to this class are usually linked to natural habitats, even if they often require substrata with weak occurrence of salts and nitrates. Instead, the communities of the *Nicotiano glaucae-Ricinion communis* are strictly synanthropic, localized always in markedly ruderal stands; sometimes, as in the Canaries, they penetrating in more or less disturbed stands already occupied by communities rich in Canarian species of *Pegano harmalae-Salsoletea vermiculatae.* These anomalous situations have led the first authors to erroneous syntaxonomical interpretations. On the basis of these considerations, a new class is here described, which includes this type of nitrophilous vegetation; it is named *Nicotiano glaucae-Ricinetea communis* class nov. hoc loco (holotypus: *Nicotiano glaucae-Ricinetalia communis* Rivas-Martínez, Fernández-González & Loidi Rivas-Martínez et al. 1999, hoc loco), the floristic and ecological characteristic are the same as the order. Plant communities referable to *Nicotiano glaucae-Ricinetum communis* were recorded from several Mediterranean countries by de Foucault (1993b, 2015), La Mantia et al. (2009), Gaudillat and Louvel-Glaser (2015).

Chapter 19
Weedy Commensal Vegetation

Another type of nitrophilous vegetation regards the cultural habitats, where it is localized a very specialized and selected commensal flora. In the Maltese Islands there are many cultivated lands, represented by cereal, leguminous and horticultural crops, orchads and vineyards, which are characterized by different weedy communities. They are linked to agronomic activities carried out in these cultural environments, as fertilization, plowing, irrigation, weeding, etc. The cereal crops are characterized by a segetal weedy vegetation, belonging to *Papaveretea rhoeadis*, while in the other types of crops the commensal vegetation is referable to *Chenopodietea*. Besides, within this weedy communities can be distinguished associations having a winter-spring cycle and also a summer-autumnal cycle, which often occur in the same places, but in different period.

19.1 *Rapistro rugosi-Meliloletum infestae* Bartolo, Brullo, Fagotto & Grillo 1983 (Table 19.1)

One of the most widespread agricultural activities in the Maltese Islands is the cereal cultivation, which is clearly correlated with the climate dryness, which does not allow, except where it is possible a field irrigation, a land exploitation with more profitable crops. Traditionally, in fact, these winter-spring crops have always been used, which exploit the meteoric waters generally concentrated from October to March.

According to Ferro (1990) and Brullo et al. (2007), in the cereal fields there is a highly specialized and peculiar thermo-xerophilous weed flora. For the most part, it is represented by species of *Ridolfion segetum* Nègre ex Rivas-Martínez et al. 1997 and higher sintaxa (*Gladiolo italici-Ridolfietalia segetum* Mucina in Mucina et al. 2016 and *Papaveretea rhoeadis* Brullo et al. 2001c), such as *Papaver rhoeas*, *Raphanus raphanistrum*, *Rhagadiolus stellatus*, *Galium tricornutum*, *Papaver*

S. Brullo et al., *The Vegetation of the Maltese Islands*, Geobotany Studies,
https://doi.org/10.1007/978-3-030-34525-9_19

Table 19.1 *Rapistro rugosi-Meliloletum infestae*

Number of relevé	1	2	3	4	5	6	7	8	9	10	11	12
Plot size (m^2)	100	150	200	150	100	200	200	200	150	100	100	100
Plant cover (%)	100	100	100	100	100	100	100	100	100	100	100	100
Cultivated species												
Triticum durum	5	5	5	5	5	5	5	5	5	5	5	5
Char. Association												
Melilotus infesta	1	+	+	1	+	2	+	2	1	2	1	2
Rapistrum rugosum	.	+	+	.	.	+	1	+	+	+	.	.
Char. All. (*Ridolfion segetum*)												
Nigella damascena	+	+	1	+	1	1	.	+	+	i	1	.
Bupleurum lancifolium	1	+	.	.	.	.	+	+	.	+	+	.
Linaria chalepensis	.	.	+	.	.	1	+	.	.	.	.	1
Misopates orontium	.	+	+	+	.	.	.	.	+	.	.	.
Allium trifoliatum	.	+	+	.	.	.	+	.	.	.	+	.
Ridolfia segetum	.	.	.	.	.	.	+	.	2	1	.	.
Char. Ord. (*Gladiolo italici-Ridolfietalia segetum*)												
Raphanus raphanistrum	1	2	+	1	2	2	+	1	1	2	+	1
Rhagadiolus stellatus .	1	+	+	1	1	1	1	+	.	+	1	+
Galium tricornutum	1	2	2	1	2	2	+	.	2	1	1	+
Phalaris canariensis	+	+	1	+	+	+	+	+	.	.	+	+
Leopoldia comosa	1	1	1	+	1	.	.	+	1	1	+	.
Gladiolus dubius	1	1	1	.	+	+	.	+	.	+	+	1
Adonis microcarpa	.	1	+	+	.	.	+	+	+	+	1	.
Papaver hybridum	1	1	1	.	.	+	1	.	+	+	+	.
Coronilla scorpioides	+	+	.	.	+	+	+	+	.	.	+	.
Gladiolus italicus	.	+	.	1	.	.	+	1	1	.	.	+
Neslia paniculata	.	+	+	.	.	.	1	+	+	.	.	+
Anchusa italica	.	+	.	+	.	+	.	.	+	+	+	.
Lathyrus cicera	+	.	.	+	+	.	.	.	.	+	.	1
Allium nigrum	.	.	.	1	2	.	.	+	1	1	.	.
Lathyrus ochrus	.	.	+	+	.	+	.	.	.	1	.	.
Bifora testiculata	.	.	.	1	+	.	.	.	.	+	.	.
Char. Cl. (*Papaveretea rhoeadis*)												
Papaver rhoeas	1	1	2	+	1	2	2	2	+	1	1	2
Euphorbia exigua	+	.	+	+	+	+	.	1	+	.	+	+
Scandix pecten-veneris	+	.	1	1	+	+	.	+	+	.	+	1
Buglossoides arvensis	1	+	.	.	+	+	+	+	.	+	.	.
Sinapis arvensis	.	1	1	.	.	.	+	+	+	+	.	+
Reseda lutea	.	+	+	.	.	1	+	+	.	.	.	1
Lolium rigidum subsp. *rigidum*	2	.	.	+	+	.	.	.	.	1	+	1
Papaver dubium	.	.	+	1	1	.	.	.	.	.	+	.
Torilis nodosa	.	.	.	+	.	+	+	.	.	.	.	.

(continued)

Table 19.1 (continued)

Number of relevé	1	2	3	4	5	6	7	8	9	10	11	12
Plot size (m^2)	100	150	200	150	100	200	200	200	150	100	100	100
Plant cover (%)	100	100	100	100	100	100	100	100	100	100	100	100
Other species												
Oxalis pes-caprae	1	1	1	2	+	1	1	.	+	+	.	+
Melilotus sulcata	1	+	1	1	1	+	+	+	.	+	+	.
Sherardia arvensis	1	+	.	+	+	+	+	+	+	.	+	+
Sonchus oleraceus	+	+	.	1	1	+	+	.	.	+	+	+
Avena barbata	.	+	1	1	2	+	2	.	.	1	2	1
Anagallis arvensis	.	.	+	+	.	+	1	+	+	+	.	+
Catapodium rigidum	.	+	+	+	.	+	+	+	.	+	+	.
Fedia graciliflora	.	.	+	+	1	+	.	+	.	.	+	1
Hypericum triquetrifolium	.	1	1	.	+	+	.	.	.	+	1	+
Lotus edulis	.	+	+	.	.	.	+	.	+	+	+	.
Glebionis coronaria	1	1	.	+	.	.	.	+	+	.	.	1
Galagtites elegans	1	+	.	.	+	.	.	.	+	+	.	.
Lotus ornithopodioides	.	.	1	+	.	.	+	+	+	.	.	.
Daucus carota	.	.	+	+	+	.	.	+	+	.	.	.
Convolvulus arvensis	.	+	+	+	.	.	1	.	.	+	.	.
Fumaria wirtgenii	.	+	+	.	.	.	+	+	+	.	.	.
Arisarum vulgare	.	+	+	+	.	.	.	+	1	.	.	.
Urospermim picroides	.	+	.	.	.	+	+	.	.	+	.	.
Medicago polymorpha	.	.	.	.	+	+	.	+	.	.	.	1
Vicia sativa	.	.	+	.	.	.	.	.	+	+	.	.
Medicago orbicularis	.	+	+	.	.	.	.	+	.	.	.	.
Hedypnois cretica	.	.	.	+	+	.	.	.	.	.	+	.
Medicago rugosa	.	.	+	+	.	.	.	.	.	+	.	.
Hippocrepis multisiliquosa	.	.	.	+	+	.	.	.	.	.	+	.
Fumaria densiflora	.	+	.	.	.	.	+	.	.	+	.	.
Theligonum cynocrambe	.	.	1	.	1	.	.	.	.	+	.	.
Diplotaxis erucoides	.	.	.	.	+	+	.	.	+	.	.	.
Aira cupaniana	.	+	+	.	.	.	.	.	.	.	.	.
Orobanche ramosa	.	.	.	+	.	.	.	+	.	.	.	.
Lathyrus aphaca	.	.	.	.	.	+	.	.	.	.	.	.
Calendula tripterocarpa	.	.	.	+	.	.	.	.	.	.	.	.

Rel. 1–2: Malta, Girgenti, 16.4.1987; Rel. 3: Malta, Mtahleb, !6.4.1987; Rel. 4–5: Malta, Naxxar, 12.4.1987; Rel. 6–8: Malta, Mellieha Ridge, 11.4.1987; Rel. 9–11: Gozo, Nadur, 14.4.1987; Rel. 9–11: Gozo, Ta Pinu, 14.4.1987

hybridum, Buplerum lancifolium, Neslia apiculata, Ridolfa segetum, Galium tricornutum, Adonis microcarpa, Allium nigrum, Anchusa italica, etc.

In particular on the deep and prevalently clayey soils, the occurrence of *Melilotus infesta* and *Rapistrum rugosum* allows to attributed this weedy vegetation to

Rapistro rugosi-Melilotetum infestae, an association described from southern Sicily by Bartolo et al. (1983), where it is linked to the same type of substrate.

19.2 *Calendulo tripterocarpae-Hypecoetum procumbentis* Bartolo, Brullo, Minissale & Spampinato 1990 (Table 19.2)

Cereal crops on the limestone substrates in the Maltese Islands are less widespread than those on clay or marl, since the soils on the carbonatic rocks are usually more arid and superficial. In these segetal environments, the weed vegetation is quite rich in characteristic species of the *Ridolfion segetum* and higher syntaxa (*Gladiolo italici-Ridolfietalia segetum* and *Papaveretea rhoeadis*), while the occurrence of *Hypecoum procumbens*, nitrophyte rather rare in the Maltese islands, and *Calendula tripterocarpa*, species markedly xerophilous with South Mediterranean distribution. For these floristical and ecological peculiarity, this commensal vegetation can be attributed to *Calendulo tripterocarpae-Hypecoetum procumbentis*, association described by Bartolo et al. (1990), for the nearby island of Lampedusa (Pelagian Archipelago) where it occurs always on red carbonatic soils.

This segetal community can be considered a less xeric vicariant of *Adonido cupanianae-Anthemidetum incrassatae*, association diffused in very similar habitats of the Hyblaean area in southern Sicily, where it was described by Bartolo et al. (1983).

19.3 *Fumario densiflorae-Veronietum hederifoliae* Brullo & Marcenò 1985a, b (Table 19.3)

This nitrophilous association of *Fumarion wirtigenii-agrariae* described from Sicily by Brullo & Marcenò (1985b) is not very frequent in the Maltese islands. It was surveyed during the winter-spring period in leguminous and horticultural crops, limited to those ones localized at the bottom of the valleys in rather fresh stands, where the environmental micro-conditions are clearly more mesic than the surrounding area. All this justifies the occurrence of this weed association in Malta, normally distributed in areas with a more humid and fresh climate than occurring in this archipelago.

The species characteristic of this association are *Veronica hederifolia* and *Fumaria densiflora*, which grow together with some other nitrophytes of *Fumarion wirtigenii-agrariae* Brullo in Brullo & Marcenò 1985 and *Chenopodietea* Br.-Bl. in Br.-Bl. et al. 1952, such as *Fumaria parviflora, Fumaria wirtgenii, Fumaria gaillardotii, Oxalis pes-caprae, Sonchus oleraceus, Diplotaxis erucoides, Stellaria media, Senecio vulgaris*, etc.

Table 19.2 *Calendulo tripterocarpae-Hypecoetum procumbentis*

Number of relevé	1	2	3	4	5	6	7
Plot size (m^2)	100	100	100	100	100	100	100
Plant cover (%)	100	100	100	100	100	100	100
Cultivated species							
Triticum durum	5	5	5	5	5	5	5
Char. Association							
Calendula tripterocarpa	1	1	1	2	+	1	+
Hypecoum procumbens	+	1	.	.	+	+	+
Char. All. (*Ridolfion segetum*)							
Nigella damascena	+	1	1	1	.	+	.
Ridolfia segetum	+	.	1	+	1	.	.
Allium trifoliatum	+	+	1	.	.	.	+
Bupleurum lancifolium	.	+	+	.	.	.	+
Misopates orontium	+	+	.	.	+	.	.
Char. Ord. (*Gladiolo italici-Ridolfietalia segetum*)							
Galium tricornutum	2	1	2	1	1	1	2
Adonis microcarpa	1	+	+	1	1	.	+
Leopoldia comosa	1	+	+	.	1	1	1
Raphanus raphanistrum	1	+	+	+	.	1	1
Gladiolus dubius	1	1	1	1	+	+	.
Papaver hybridum	.	1	+	+	+	+	+
Phalaris canariensis	+	1	+	.	+	+	+
Coronilla scorpioides	.	+	.	.	+	1	+
Gladiolus italicus	+	.	.	.	+	+	.
Anchusa italica	.	+	1	+	.	.	.
Rhagadiolus stellatus	+	+	.	.	.	.	.
Bifora testiculata	.	.	.	.	.	1	.
Char. Cl. (*Papaveretea rhoeadis*)							
Papaver rhoeas	1	1	.	1	1	2	2
Lolium rigidum subsp. *rigidum*	.	1	1	1	.	+	+
Scandix pecten-veneris	.	+	+	1	1	.	.
Buglossoides arvensis	.	.	+	.	.	+	+
Sinapis arvensis	.	+	+	.	.	+	.
Euphorbia exiigua	+	.	.	.	.	.	+
Other species							
Oxalis pes-caprae	1	1	1	1	2	1	1
Sonchus oleraceus	.	+	+	+	+	.	+
Avena barbata	1	2	1	1	1	.	.
Anagallis arvensis	+	.	.	+	+	1	1
Glebionis coronaria	.	1	1	.	+	1	1
Lotus ornithopodioides	.	+	+	1	.	+	.
Sherardia arvensis	+	+	+	.	.	+	.
Hypericum triquetrifolium	.	.	+	+	.	1	.

(continued)

Table 19.2 (continued)

Number of relevé	1	2	3	4	5	6	7
Plot size (m^2)	100	100	100	100	100	100	100
Plant cover (%)	100	100	100	100	100	100	100
Vicia sativa	.	.	.	.	+	.	+
Convolvulus arvensis	.	+	.	.	.	.	+
Catapodium rigidum	.	.	.	.	+	+	.
Diplotaxis erucoides	+	+	.	.	.	.	.
Fumaria densiflora	.	.	+	.	.	.	.
Lotus edulis	.	.	.	.	+	.	.
Daucus carota	.	.	+	.	.	.	.
Loncomelos narbonensis	+	.	.	.	.	.	.
Phalaris paradoxa	.	.	+	.	.	.	.

Rel. 1–2: Malta, St. Andrews, 11.4.1987; Rel. 3: Malta, Zebbug, 17.4.1987; Rel. 4–5: Malta, Ghallis, 17.4.1987; Rel. 6: Malta, Salina Bay, 11.4.1987; Rel. 7: Malta, Luqa, 16.4.1987

19.4 *Fumarietum bicoloris* ass. nov. hoc loco (Table 19.4)

Holotypus: rel. 5 hoc loco

In leguminous crops, represented mainly by broad beans, with open and sunny surfaces, the weed vegetation is characterized by the occurrence and often dominance of *Fumaria bicolor*, having a West Mediterranean distribution. This species allows to differentiate a markedly more xerophilous association than *Fumarietum densiflorae-Veronicetum hederifoliae*, which is proposed as *Fumarietum bicoloris*. It is mainly linked to limestones characterized by red soil and is rich in nitrophilous species of *Fumarion wirtgenii-agrariae* and related order *Chenopodietalia*, among which *Fumaria agraria, F. parviflora, F.wirtgenii, Rumex bucephalophorus, Veronica cymbalaria, Calendula arvensis, Lamium amplexicaulis, Euphorbia helioscopia*, etc.

This association is also observed in crop rotation with segetal cultures, in which the commensal vegetation is represented by the *Calendulo tripterocarpae-Hypecoetum procumbentis*.

Another association closely related to this one is the *Fumarietum parvifloro-bastardii* Bartolo, Brullo, Minissale & Spampinato 1990, described for the Island of Lampedusa, which occurs in ecologically very similar habitats.

19.5 *Diplotaxio erucoidis-Brassicetum sylvestris* ass. nov. hoc loco (Table 19.5)

Holotypus: ril. 5 hoc loco.

In herbaceous or also perennial crops, subject to weeding treatment, as leguminous or vineyards cultivations on heavy clayey soils, the winter-spring cycle weeds

Table 19.3 *Fumario densiflorae-Veronicetum hederifoliae*

Number of relevé	1	2	3	4	5
Plot size (m^2)	50	50	100	50	50
Plant cover (%)	100	100	100	100	100
Cultivated species					
Vicia faba	5	4	4	5	5
Char. Association					
Veronica hederifolia	2	2	3	1	2
Char. All. (*Fumarion wirtgenii-agrariae*)					
Fumaria parviflora	1	3	1	2	+
Fumaria densiflora	+	.	1	1	2
Fumaria wirtgenii	+	+	.	2	1
Fumaria gaillardotii	1	.	1	1	.
Calendula arvensis	+	+	1	.	.
Sonchus asper	.	1	+	+	.
Lamium amplexicaule	.	.	+	.	1
Char. Cl. (*Chenopodietea*)					
Oxalis pes-caprae	+	+	+	+	1
Sonchus oleraceus	1	1	+	1	1
Diplotaxis erucoides	1	1	1	2	1
Stellaria media	2	1	1	1	1
Convolvulus arvensis	1	1	.	1	+
Anisantha diandra	.	+	+	.	1
Glebionis coronaria	.	+	+	.	+
Senecio vulgaris	.	.	+	.	+
Urtica membranacea	.	.	+	1	.
Erodium malacoides	.	1	.	.	+
Borago officinalis	1	.	.	+	.
Raphanus raphanistrum	.	+	.	.	.
Brassica rapa subsp. *sylvestris*	.	+	.	.	.
Other species					
Catapodium rigidum	1	+	.	.	+
Sherardia arvensis	.	+	+	.	.
Anagallis arvensis	.	.	+	+	.
Papaver hybridum	.	.	+	.	+
Scandix pecten-veneris	+	.	.	+	.
Lolium rigidum subsp. *rigidum*	.	.	.	1	.

Rel. 1–4: Malta, Ghajn il-Kbira, 16.4.1987; Rel. 5: Gozo, Nadur, 14.4.1987

are mainly represented by a plant community physiognomically characterized by *Brassica rapa* subsp. *sylvestris*. This species grows usually together with *Diplotaxis erucoides* and a rich set of species belonging to *Fumarion wirtgenio-agrariae* and related higher syntaxa. For its peculiarity it is proposed as a new association,

Table 19.4 *Fumarietum bicoloris*

Number of relevé	1	2	3	4	5	6
Plot size (m^2)	100	100	100	100	100	100
Plant cover (%)	100	100	100	100	100	100
Cultivated species						
Vicia faba	5	5	5	5	5	5
Char. Association						
Fumaria bicolor	2	2	+	1	2	1
Char. All. (*Fumarion wirtgenii-agrariae*)						
Fumaria agraria	2	1	+	.	1	1
Fumaria parviflora	+	+	1	.	1	2
Rumex bucephalophorus subsp. *bucephalophorus*	+	.	1	+	2	1
Fumaria wirtgenii	.	1	.	2	+	+
Veronica cymbalaria	+	.	1	+	.	+
Fumaria gaillardotii	.	.	.	.	.	+
Fumaria densiflora	.	1	.	.	.	.
Char. Ord. (*Chenopodietalia*)						
Calendula arvensis	1	+	+	.	+	1
Lamium amplexicaulis	1	.	1	+	+	.
Euphorbia helioscopia	.	+	.	.	1	+
Euphorbia peplus	.	+	+	.	1	.
Sonchus asper	.	.	+	+	.	.
Char. Cl. (*Chenopodietea*)						
Sonchus oleraceus	+	1	+	+	1	1
Senecio vulgaris	+	+	+	+	.	1
Convolvulus arvensis	+	1	.	1	+	+
Diplotaxis erucoides	+	2	1	1	.	.
Medicago polymorpha	+	.	+	+	1	.
Galium verrucosum	+	.	.	+	+	.
Chenopodium album	.	+	.	.	+	+
Erodium malacoides	.	+	+	.	+	.
Glebionis coronaria	.	1	+	.	.	.
Urtica urens	.	+	.	.	.	+
Malva sylvestris	.	.	.	.	.	+
Other species						
Sherardia arvensis	1	.	+	+	+	1
Anagallis arvensis	+	+	+	+	.	1
Papaver hybridum	.	1	+	1	+	+
Galium tricornutum	+	+	.	+	.	1
Papaver dubium	2	1	.	+	+	.
Phalaris canariensis	+	+	+	.	+	.
Raphanus raphanistrum	+	.	.	1	1	.
Scandix pecten-veneris	+	.	+	+	.	.

(continued)

Table 19.4 (continued)

Number of relevé	1	2	3	4	5	6
Plot size (m^2)	100	100	100	100	100	100
Plant cover (%)	100	100	100	100	100	100
Anisantha madritensis	+	.	+	.	.	+
Vicia sativa	1	.	+	+	.	.
Catapodium rigidum	1	.	+	+	.	.
Sinapis arvensis	.	+	.	+	+	.
Lolium rigidum subsp. *rigidum*	.	+	+	+	.	.
Beta maritima	.	.	.	+	+	.
Silene colorata	.	.	.	.	+	.
Neslia paniculata	.	.	.	.	+	.
Lathyrus ochrus	.	+	.	.	.	.

Rel. 1–2: Malta, Mistra Valley, 9.4.1984; Rel. 3: Malta, Bingemma, 11.4.1984; Rel. 4: Malta, Mellieha Ridge,12.4.1987. Rel. 5: Malta, Ramla, 13.4.1987; Rel. 6: Malta, Marfa Ridge, 9.4.1984

named *Diplotaxio erucoidis-Brassicetum sylvestris*. This commensal vegetation is relatively frequent both in Malta and in Gozo, representing a geographical vicariant of the *Diplotaxietum viminio-erucoidis* Brullo & Marcenò 1985 occurring in southern Sicily.

19.6 *Chrozophoro tinctoriae-Kickxietum integrifoliae* Brullo & Marcenò 1980 (Table 19.6)

During the summer-autumn period, in the wheat stubble of cereal fields and vineyards there is a xero-nitrophilous vegetation represented by *Chrozophoro tinctoriae-Kickxietum integrifoliae* Brullo & Marcenò 1980. This association belonging to the *Diplotaxion erucoidis* Br.-Bl. in Br.-Bl. et al. 1936 em. Brullo & Marcenò 1980, is characterized by *Kickxia spuria* subsp. *integrifolia* and *Andrachne telephioides*, which grow with many other therophytes with late summer cycles, such as *Hypericum triquetifolium, Chrozophora tinctoria, Helotropium europaeum, Amaranthus graecizans* subsp *sylvestris, Portulaca oleracea, Setaria verticillata*, etc.

Previously, this association was recorded for the Maltese islands by Brullo and Marcenò (1980), who described for this territory a special sub-association proposed as *euphorbietosum aleppicae*, which to be considered as a geographical variant. Differential species of this syntaxon is *Euphorbia aleppica*, nitrophyte absent in Sicily, where the association at issue is widespread and represented by the subass. *typicum*.

Table 19.5 *Diplotaxio erucoidis-Brassicetum sylvestris*

Number of relevé	1	2	3	4	5	6	7	8
Plot size (m^2)	100	100	100	100	100	100	100	100
Plant cover (mq)	80	70	70	60	70	60	70	60
Cultivated species								
Vicia faba	5	5	.	.	.	5	5	.
Pisum sativum	.	.	.	.	5	.	.	5
Vitis vinifera	.	.	5	5	.	.	.	.
Char. Association								
Brassica rapa subsp. *sylvestris*	3	2	2	3	2	1	2	2
Char. All. (*Fumarion wirtgenii-agrariae*)								
Fumaria parviflora	2	1	1	2	2	1	2	2
Fumaria wirtgenii	1	+	2	1	1	+	.	1
Fumaria densiflora	1	.	+	+	1	.	+	+
Fumaria gaillardotii	.	+	+	1	+	+	+	.
Rumex bucephalophorus subsp. *bucephalophorus*	.	1	+	1	+	.	+	1
Fumaria agraria	+	1	.	.	+	.	+	1
Veronica hederifolia	.	+	1	.	+	+	.	.
Char. Ord. (*Chenopodietalia*)								
Calendula arvensis	2	2	1	1	2	1	+	1
Sonchus asper	+	+	+	1	1	+	+	.
Lamium amlpexicaule	1	1	+	1	1	1	.	.
Euphorbia peplus	.	.	+	.	1	+	+	.
Euphorbia helioscopia	.	.	+	+	.	.	.	+
Char. Cl. (*Chenopodietea*)								
Diplotaxis erucoides	3	1	2	1	1	2	3	2
Glebionis coronaria	1	2	1	1	2	2	2	1
Oxalis pes-caprae	2	2	2	1	1	2	1	2
Convolvulus arvensis	1	1	2	.	+	1	1	+
Sonchus oleraceus	2	+	1	2	1	.	+	+
Medicago polymorpha	2	+	1	1	.	1	+	1
Mercurialis annua	1	.	+	+	.	+	1	1
Senecio vulgaris	+	1	+	+	1	+	.	+
Malva sylvestris	1	+	1	2	2	+	.	.
Galagtites elegans	.	.	+	.	.	+	1	1
Galium verrucosum	.	.	.	.	.	1	+	.
Other species								
Avena barbata	1	+	+	1	2	1	.	+
Anagallis arvensis	1	+	+	.	1	+	.	1
Sinapis arvensis	.	.	+	+	+	.	.	+
Lolium rigidum subsp. *rigidum*	+	+	.	+	.	+	.	.
Scandix pecten-veneris	.	+	.	+	.	+	.	.
Arisarum vulgare	.	+	+	.	.	.	+	.
Papaver hybridum	.	.	+	.	+	.	+	.
Reseda lutea	.	.	.	.	+	.	.	.

Rel. 1–5: Malta, Mtahleb, 11.4.1987; Rel. 6: Malta, St. Dimitri, 11.4.1987; Rel. 7: Malta, Salina Bay, 11.4.1987; Rel. 8: Gozo, Victoria, 14.4.1987

Table 19.6 *Chrozophoro tinctoriae-Kickietum integrifoliae*

Number of relevé	1	2	3	4	5	6	7
Plot size (m^2)	20	50	50	50	50	100	100
Plant cover (%)	25	30	25	40	50	40	60
Char. Association							
Kickxia spuria subsp. *integrifolia*	+	2	1	2	2	2	1
Euphorbia aleppica	1	+	1	1	.	+	.
Andrachne telephioides	.	.	.	1	+	+	.
Char. All. (*Diplotaxion erucoidis*)							
Hypericum triquetrifolium	+	1	1	1	2	1	3
Chrozophora tictoria	.	+	.	+	1	+	1
Helminthotheca echioides	.	+	1	.	+	+	.
Chamaesyce peplis	.	+	.	.	+	.	.
Char. Ord. (*Solano-Polygonetalia convolvuli*)							
Heliotropium europaeum	+	1	.	+	1	1	+
Portulaca oleracea	+	.	+	+	.	+	1
Amaranthus graecizans subsp. *sylvestris*	.	+	+	.	+	+	.
Amaranthus retroflexus	.	+	.	.	+	.	+
Setaria verticillata	.	+	.	.	+	.	.
Chenopodium opulifolium	.	.	.	.	.	+	.
Char. Cl. (*Chenopodietea*)							
Convolvulus arvensis	2	2	1	2	2	2	3
Sonchus oleraceus	1	1	.	+	+	+	.
Chenopodium album	1	.	2	2	.	.	+
Symphyotrichum squamatum	+	+	1	+	1	.	.
Diplotaxis tenuifolia	.	+	.	1	.	+	1
Chenopodium murale	.	+	+	.	.	.	1
Conyza canadensis	.	.	.	+	+	.	+
Conyza bonariensis	.	.	.	.	+	+	.
Solanum nigrum	.	+	.	.	+	.	.
Diplotaxis erucoides	.	.	+	.	.	+	.
Ecballium elaterium	.	.	.	+	+	.	.
Chenopodium vulvaria	.	.	.	.	.	+	1
Sonchus asper	.	.	.	.	1	.	+
Dysphania ambrosioides	.	.	.	.	.	.	+
Glebionis coronaria	.	.	.	.	.	+	.
Lamium amplexicaule	.	.	+	.	.	.	.
Other species							
Cynodon dactylon	+	+	1	.	+	.	1
Anagallis arvensis	+	1	.	+	1	+	.
Beta maritima	+	.	+	+	.	+	.
Reseda alba	+	.	+	+	.	.	.
Reichardia picroides var. *picroides*	+	.	.	+	+	.	.
Polygonum aviculare	+	.	+	.	.	.	.

(continued)

Table 19.6 (continued)

Number of relevé	1	2	3	4	5	6	7
Plot size (m^2)	20	50	50	50	50	100	100
Plant cover (%)	25	30	25	40	50	40	60
Papaver rhoeas	+	.	.	+	.	.	.
Vicia sativa	.	.	+	.	.	.	.
Dittrichia graveolens	.	.	.	.	+	.	.
Misopates orontium	.	+	.	.	.	.	.

Rel. 1–2: Malta, St. Julians, 25.6.1973; Rel. 3: Malta, Bahrija Valley, 25.6.1973; Rel. 4: Malta, Gharghur, 25.6.1973; Rel. 5–6: Malta, Mellieha Ridge, 26.9.1985; Rel. 7: Gozo, Nadur, 29.11.1985

19.7 *Setario pumilae-Echinochloetum colonum* A. & O. Bolós ex O. Bolós 1956 (Table 19.7)

During the summertime, in the irrigated crops such as cornfields, vegetable gardens and orchards, the weed vegetation is represented by plant communities belonging to *Spergulo arvensis-Erodion cicutariae* J. Tüxen in Passarge 1964 (=*Panico-Setarion viridis* Sissingh in Westhoff et al. 1946 nom. ambig.) according to Lososová (2009). In particular, the species more common of this alliance and the related order *Solano nigri-Polygonetalia convolvuli* (Sissingh in Westhoff et al. 1946) O. Bolós 1962 are *Echinocloa crus-galli, Setaria viridis, Setaria pumila, Portulaca oleracea, Cyperus rotundus, Amaranthus graecizans* subsp. *sylvestris, Digitaria sanguinalis*, etc.

For the occurrence of *Echinochloa colonum*, this nitrophilous vegetation can be referred to the *Setario pumilae-Echinochloetum colonum*, association previously recorded for the Iberian Peninsula (de Bolós 1956, 1967; Tüxen and Oberdorfer 1958) and Sicily (Brullo and Marcenò 1985b).

19.8 *Setario ambigui-Cyperetum rotundi* Brullo, Scelsi & Spampinato 2001c (Table 19.8)

In the citrus cultivations during the summer-autumn period, a weed vegetation linked to more or less shaded, irrigated and well-fertilized crops occurs. This type of vegetation was investigated in Malta by Maugeri et al. (1979), and later in Sicily by Maugeri (1980), who attributed initially it to a peculiar new weed association, named *Amarantho-Cyperetum rotundi* nom. inval. (art.5, I.C.N.P., cfr. Weber et al. 2000). For this syntaxon the new name *Setario ambiguae-Cyperetum rotundi* was proposed by Brullo et al. (2001c), which has as characteristic species *Setaria verticilliformis* (= S. *decipiens* Schimp., *S. ambigua* Guss.) and *Cyperus rotundus*. This association, belonging to *Spergulo arvensis-Erodion cicutariae*, is widespread in Sicily and neighboring islets, as well as in South Italy. Basing on literature, it is replaced in Croatia by *Fumario-Cyperetum rotundi* Horvatić 1960 and Crete by *Digitario sanguinalis-Cyperetum rotundi* Bergmeier 2008.

Table 19.7 *Setario pumilae-Echinochloetum colonum*

Number of relevé	1	2	3	4	5	6
Plot size (m^2)	100	100	100	100	100	100
Plant cover (%)	100	100	100	100	100	100
Cultivated species						
Zea mais	4	.	.	5	5	.
Lycopersicon esculentum	.	4	4	.	.	.
Solanum melongena	.	.	.	.	.	4
Char. Association						
Echinochloa colonum	2	1	2	2	+	1
Char. All. (*Spergulo arvensis-Erodion cicutarii*)						
Echinochloa crus-galli	+	+	1	1	1	.
Setaria viridis	+	.	1	2	+	1
Char. Ord. (*Solano-Polygonetalia convolvuli*)						
Portulaca oleracea	1	+	2	2	1	2
Setaria pumila	2	3	2	2	2	1
Amaranthus graecizans subsp. *sylvestris*	2	+	2	1	1	2
Cyperus rotundus	1	2	2	1	1	2
Chenopodium opulifolium	.	.	+	+	+	+
Amaranthus retroflexus	+	+	.	.	+	+
Sorghum halepense	.	1	+	+	.	1
Digitaria sanguinalis	+	.	.	+	+	.
Tribulus terrestris	.	.	+	+	.	.
Heliotropium europaeum	.	.	.	+	.	+
Amaranthus albus	.	+	+	.	.	.
Char. Cl. (*Chenopodietea*)						
Sonchus oleraceus	+	+	+	+	1	+
Convolvulus arvensis	1	1	+	1	+	+
Chenopodium album	.	+	+	.	1	+
Malva parviflora	.	+	+	.	.	.
Diplotaxis erucoides	.	.	+	+	.	.
Malva sylvestris	.	.	.	+	.	+
Medicago arabica	.	.	.	.	+	.
Rumex pulcher	.	.	+	.	.	.
Other species						
Medicago polymorpha	+	+	.	.	+	.
Cynodon dactylon	.	.	.	+	1	+
Beta maritima	.	+	.	+	.	.
Parietaria judaica	.	+	.	.	.	.

Rel. 1–4: Malta, Pwales Valley, 26.9.1985; Rel. 5: Gozo, Nadur, 27.11.1985; Rel. 1–4: Malta, Ghajn Rihana Valley, 23.9.1985

Table 19.8 *Setario ambigua-Cyperetum rotundi*

Number of relevé	1	2	3
Plot size (m^2)	50	50	30
Plant cover (%)	100	100	100
Cultivated species			
Citrus limon	5	5	.
Citrus sinensis	.	.	5
Char. Association			
Setaria verticilliformis	3	3	2
Cyperus rotundus	2	1	2
Char. All. (*Spergulo arvensis-Erodion cicutarii*)			
Setaria viridis	1	1	.
Echinochloa crus-galli	.	+	.
Char. Ord. (*Solano-Polygonetalia convolvuli*)			
Amaranthus graecizans subsp. *sylvestris*	2	1	2
Portulaca oleracea	1	1	2
Amaranthus retroflexus	+	2	1
Heliotropium europaeum	.	1	2
Digitaria sanguinalis	+	.	1
Sorghum halepense	.	.	+
Setaria pumila	.	.	+
Char. Clas. (*Chenopodietea*)			
Chenopodium album	2	2	+
Rumex pulcher	+	.	1
Conyza bonariensis	.	1	+
Solanum nigrum	+	1	.
Convolvulus arvensis	+	2	.
Sonchus oleraceus	.	1	+
Other species			
Cynodon dactylon	1	2	1
Parietaria judaica	2	1	.
Anagallis arvensis	+	.	.
Oryzopsis miliacea	.	.	1

Rel. 1–2: Gozo, San Blas, 27.11.1985; Rel. 3: Malta, St. Paul, 26.9.1985

19.9 *Bromo diandri-Brassicetum sylvestris* Brullo & Marcenò 1985 (Table 19.9)

In the spring period, the citrus groves of the Maltese islands show a very specialized weed vegetation, which is linked to remarkable sciaphilous conditions determinated by the tree canopy, as well as by a certain moisture and richness in nitrates of the soil.

In this commensal plant community the constant occurrence of *Brassica rapa* subsp. *sylvestris* is quite significant, since it allows to attribute it to *Bromo diandri-Brassicetum sylvestris*, basophilous association described by Brullo and Marcenò

Table 19.9 *Bromo diandri-Brassicetum sylvestris*

Number of relevé	1	2
Plot size (m^2)	50	50
Plant cover (%)	100	100
Cultivated species		
Citrus limon	5	4
Citrus sinensis	.	3
Char. Association		
Brassica rapa subsp. *sylvestris*	2	1
Anisantha diandra	2	1
Char. All. (*Veronico-Urticion urentis*) and Ord. (*Urtico-Scrophularietalia peregrinae*)		
Arum italicum	2	2
Galium aparine	1	2
Parietaria judaica	1	3
Fumaria capreolata	1	+
Urtica membranacea	+	+
Arisarum vulgare	.	+
Urtica urens	.	1
Char. Cl. (*Chenopodietea*)		
Oxalis pes-caprae	4	3
Mercurialis annua	1	1
Sonchus oleraceus	1	+
Convolvulus arvensis	+	1
Euphorbia peplus	+	1
Chenopodium murale	.	+
Borago officinalis	.	1
Other species		
Avena sterilis	1	+
Galium tricornutum	1	.
Daucus carota	+	.
Sherardia arvensis	+	.
Orobanche ramosa	.	1

Rel. 1–2: Gozo, San Blas, 13.4.1984

(1985b) from Sicily. It is included in the *Veronico persicae-Urticion urentis* Brullo in Brullo & Marcenò 1985, an alliance of *Urtico-Scrophularietalia peregrinae* Brullo in Brullo & Marcenò 1985, which gathers strictily sciaphilous and nitrophilous plant communities.

Among the species of alliance and order found in Malta there are: *Arum italicum, Parietaria judaica, Galium aparine, Fumaria capreolata, Urtica urens, Urtica membranacea, Arisarum vulgare*, etc.

Chapter 20
Subnitrophilous Vegetation of Uncultivated Lands

Another type of synantrophic vegetation frequent in the Maltese Islands is that one localized in the uncultivated lands or abandoned fields (Fig. 20.1), previously used for herbaceous crops. In these communities numerous therophytes well adapted to more or less nitrified soils are frequent and often dominant. Floristically, this vegetation is characterized by species belonging to *Echio plantaginei-Galactition tomentosae*, alliance of *Chenopodietea* widespread in the Mediterranean area.

20.1 *Chrysanthemo coronarii-Hippocrepidetum multisiliquosae* Brullo & Siracusa 1996 (Table 20.1)

The vegetation growing in the uncultivated areas as abandoned or resting fields in rotation with herbaceous crops is characterized by sub-nitrophilous therophytes. Many of these species, having thermophilous exigences, belongs to *Echio plantaginei-Galactition tomentosae* O. Bolòs & Molinier 1969, alliance of the *Brometalia rubenti-tectorum* (Rivas Goday and Rivas-Martinez 1973) Rivas-Martinez & Izco 1977, which has its highest expression in the Tyrrhenian territories, including Sicily where it was studied by Brullo (1983a) and Brullo and Marcenò (1985b).

Among the species of the above mentioned syntaxa in the Maltese Islands are frequent *Galactites elegans* (=*G. tomentosa*), *Glebionis coronaria*, *Lotus ornithopodioides, Trifolium nigrescens, Lolium rigidum, Avena barbata, Medicago polymorpha, Bromus madritensis, Hirdchilfeldia incana, Stipella capensis, Echium plantagineum, Hedypnois cretica, Urospermum picroides*, etc.

On the carbonate soils (limestones and marls), the *Echio plantaginei-Galactition tomentosae* is represented by a peculiar community, that for the occurrence and sometimes abundance of *Hyppocrepis multisiliquosa*, marked thermophilous species,

S. Brullo et al., *The Vegetation of the Maltese Islands*, Geobotany Studies,
https://doi.org/10.1007/978-3-030-34525-9_20

Fig. 20.1 Uncultivated lands with *Echio plantaginei-Galactition tomentosae* communities

can be attributed to *Chrysanthemo coronarii-Hyppocrepidetum multisiliquosae*, association described by Brullo and Siracusa (1996) for the island of Linosa (Pelagian Archipelago), where grows on andosoils of basaltic origin.

20.2 *Sullo coronariae-Ononidetum mitissimae* ass. nov. hoc loco (Table 20.2)

Holotypus: rel. 5, hoc loco

In the fields with clayey soils, the *Chrysanthemo coronarii-Hippocrepidetum multisilliquosae* is replaced by a different subnitrophilous vegetation, physiognomically differentiated by *Sulla coronaria* (=*Hedysarum coronarium*), which usually grows together with *Ononis mitissima*, both species closely linked to clayey substrata. Floristically, it shows some relations with *Hedysaro coronarii-Lavateretum trimestris*, association described from Sicily by Maugeri (1975), which has in common the dominance of *Sulla coronaria* and the ecology (cf. Brullo 1983a).

Table 20.1 *Chrysanthemo coronarii-Hippocrepidetum multisiliquosae*

Number of relevé	1	2	3	4	5	6	7	8
Plot size (m^2)	50	100	100	100	100	100	100	100
Plant cover	100	100	80	100	100	90	100	100
Char. Association								
Hippocrepis multisiliquosa	1	2	2	+	2	1	1	2
Char. All. (*Echio-Galctition tomentosae*)								
Lotus ornithopodioides	3	3	1	2	3	2	1	2
Trifolium nigrescens subsp. *nigrescens*	.	1	2	1	2	3	2	2
Urospermum picroides	+	1	+	+	1	.	+	.
Hypochoeris achyrophorus	+	+	+	.	1	.	.	.
Reichardia picroides var. *picroides*	.	+	.	1	+	+	.	.
Char. Ord. (*Brometalia rubenti-tectorum*)								
Galagtites elegans	2	2	3	2	1	3	2	2
Glebionis coronaria	3	1	+	1	+	1	2	+
Avena barbata	1	2	1	3	2	1	3	2
Lolium rigidum subsp. *rigidum*	2	1	2	1	2	2	1	2
Medicago polymorpha	+	1	2	3	2	1	2	2
Hedypnois cretica	+	1	+	+	+	2	+	+
Anisantha madritensis	1	2	1	1	2	2	2	1
Medicago truncatula	+	+	+	1	+	.	+	1
Hirschfeldia incana	+	1	+	+	2	1	+	.
Stipella capensis	.	1	2	2	.	+	1	2
Lotus edulis	+	1	.	+	+	+	.	+
Catapodium rigidum	+	1	1	.	+	1	+	.
Echium plantagineum	.	+	1	+	.	2	1	+
Avena sterilis	.	.	1	+	1	.	+	.
Vulpia ciliata	.	1	+	1	.	.	.	.
Char. Cl. (*Chenopodietea*)								
Erodium malacoides	2	1	2	+	1	1	+	+
Anisantha diandra	+	1	+	2	2	+	+	1
Oxalis pes-caprae	2	+	1	+	1	.	1	1
Sonchus oleraceus	+	1	+	1	+	.	1	+
Diplotaxis erucoides	+	.	1	1	1	+	1	1
Melilotus sulcata	+	1	+	+	1	+	+	.
Euphorbia helioscopia	.	+	1	.	+	+	+	1
Hordeum leporinum	+	1	+	+	.	+	.	+
Malva sylvestris	1	1	.	1	.	1	1	+
Mercurialis annua	+	+	.	+	+	.	+	.
Sonchus asper	.	+	.	.	+	+	.	.

(continued)

Table 20.1 (continued)

Number of relevé	1	2	3	4	5	6	7	8
Plot size (m^2)	50	100	100	100	100	100	100	100
Plant cover	100	100	80	100	100	90	100	100
Other species								
Beta maritima	+	+	+	.	+	1	+	+
Trachynia distachya	.	1	1	+	+	.	.	+
Vicia sativa	.	+	2	+	1	.	.	.
Anagallis arvensis	.	.	+	.	+	+	+	1
Scorpiurus muricatus	.	+	.	.	+	+	+	.
Lathyrus ochrus	.	+	.	.	+	.	.	+
Tripodium tetraphyllum	.	.	+	.	+	.	+	.
Linum bienne	.	+	+	.	+	.	.	.
Echium parviflorum	.	.	+	.	.	.	.	.
Convolvulus althaeoides	.	+	.	.	.	.	.	.

Rel. 1: Malta, Sliema, 8.4.1984; Rel. 2–3: Malta, Mistra Valley, 9.4.1984; Rel. 4–6: Malta, Mellieha, 12.4.1987; Rel. 7: Malta, Rabat, 16.4.1987; Rel. 8: Malta, Mosta, 16.4.1984

The Maltese vegetation, however, differs from the Sicilian one for the occurrence of *Ononis mitissima* and mainly because of the species characteristic of the last association are absent. This species is therefore proposed as characteristic of the new association *Sullo coronariae-Ononidetum mitissimae*, which from a syndinamic viewpoint represents a degradation stage within the edapho-climatophilous series of the *Convolvulo oleifolii-Lygeetum spartii*, steppic grassland colonizing the badlands.

20.3 *Meliloto messanensis-Hordeetum marini* Brullo 1983 (Table 20.3)

This association unlike the *Sullo coronariae-Ononidetum mitissimae* is localized on sandy-silty and rich in salt soils. The edaphic salinity is emphasized by the occurrence of *Hordeum marinum* and *Melilotus messanensis*, subhalophytes indicated as characteristics of this syntaxon. The *Meliloto messanensis-Hordeetum marini* was described by Brullo (1983a) for eastern Sicily, where it is localized near coastal saltmarshes, representing a halo-subnitrophilous vegetation linked to anthropic activity on stands with salty humid soils. In Malta it was observed in abandoned crops of brackish reclaimed areas or in disturbed stands adjacent to coastal marshes. Due to the peculiar edaphic conditions, this association is floristically rather impoverished, mainly as regards the species of *Echio plantaginei-Galactition tomentosae* and related order.

Table 20.2 *Sullo coronariae-Ononidetum mitissimae*

Number of relevé	1	2	3	4	5	6
Plot size (m^2)	50	50	50	100	50	50
Plant cover (%)	100	100	100	100	100	100
Char. Association						
Ononis mitissima	2	2	1	1	2	1
Char. All. (*Echio-Galctition tomentosae*)						
Medicago ciliaris	1	2	1	1	+	2
Urospermum picroides	1	.	+	1	1	1
Lotus ornithopodioides	+	+	.	+	+	.
Char. Ord. (*Brometalia rubenti-tectorum*)						
Sulla coronaria	4	4	5	4	5	5
Galagtites elegans	2	1	2	2	2	+
Avena barbata	2	2	1	2	+	+
Aegylops geniculata	1	+	+	1	1	+
Medicago polymorpha	2	1	2	2	1	1
Anisantha madritensis	1	+	1	1	.	+
Lolium rigidum subsp. *rigidum*	1	2	.	2	+	1
Medicago truncatula	+	.	+	+	.	+
Catapodium rigidum	+	+	+	+	.	.
Podospermum laciniatum	1	.	+	+	.	+
Astragalus hamosus	.	+	1	.	+	.
Char. Cl. (*Chenopodietea*)						
Sonchus oleraceus	+	+	.	+	1	1
Melilotus sulcata	+	+	.	+	+	+
Mercurialis annua	.	+	.	.	+	+
Borago officinalis	.	.	+	+	.	+
Senecio vulgaris	.	.	+	.	+	+
Malva sylvestris	+	+	.	+	.	.
Glebionis coronaria	.	.	.	+	.	
Other species						
Phalaris canariensis	1	+	1	+	1	1
Trachynia distachya	1	+	1	+	1	1
Foeniculum piperitum	1	1	1	.	+	+
Beta maritima	+	.	1	1	1	+
Melilotus messanensis	+	+	.	+	+	.
Cynara cardunculus	1	.	1	+	1	1
Hordeum marinum	1	+	+	.	+	.
Euphorbia exigua	+	.	+	+	.	.

Rel. 1–3: Malta, Gnien il-Kbir (Rabat), 13.4.1987; Rel. 4–5: Malta, Zebbug, 13.4.1987; Rel. 6: Malta, Gnejna Valley, 13.4.1987

Table 20.3 *Meliloto messanensis-Hordeetum marini*

Number of relevé	1	2	3	4	5
Plot size (m^2)	10	10	15	10	10
Plant cover (%)	60	60	80	90	50
Char. Association					
Hordeum marinum	3	3	4	5	3
Melilotus messanensis	1	2	1	.	+
Char. All. (*Echio-Galctition tomentosae*) and Ord. (*Brometalia rubenti-tectorum*)					
Podospermum laciniatum	2	2	2	1	+
Sulla coronaria	1	1	+	.	+
Hedypnois cretica	.	.	.	+	+
Aegylops geniculata	.	+	1	.	.
Galagtites elegans	.	.	.	+	+
Lolium rigidum subsp. *rigidum*	1	.	.	.	.
Medicago ciliaris	.	+	.	.	.
Char. Cl. (*Chenopodietea*)					
Sonchus oleraceus	+	+	+	+	.
Oxalis pes-caprae	1	.	+	.	1
Diplotaxis erucoides	.	.	+	.	.
Other species					
Parapholis incurva	2	1	2	1	+
Melilotus sulcata	+	1	1	+	.
Phalaris paradoxa	.	1	+	2	+
Trachynia distachya	+	+	.	+	.
Cynara cardunculus	.	1	+	.	+
Beta maritima	2	1	2	.	.
Sulla capitata	+	+	1	.	.
Centaurium tenuiflorum	.	.	.	3	1
Schenkia spicata	.	.	.	1	2

Rel. 1–3: Malta, Ghadira, 13.4.1987; Rel. 4–5: Malta, Ghajn Hadid Tower, 26.6.1973

Chapter 21
Syndinamic Considerations

In the last decades within the geobotanical investigations, regarding the ecological correlations between vegetation and environment, phytosociological studies have been carried out where the dynamic relationships of serial or catenal kind are taken into consideration. This methodological approach is part of the synphytosociology, which was developed initially by Rivas-Martinez (1976) and later revisited by Tüxen (1977, 1979), Gèhu (1979, 1991) and Rivas-Martinez (1983, 1985). According to Rivas-Martinez (2005), Rivas-Martinez and coaut. (2011), Pedrotti (2013) and Loidi (2017), these science is based on the concept of vegetation complexes, whose basic units are mainly represented by the sigmetum (= series), and geosigmetum (= geoseries), to which permasigmetum (= permaseries) and geopermasigmetum (= geopermaseries) are added too. In particular, it is called sigmetum the basic geobotanical unit of a local assemblage of all the communities that have the potential to develop toward a well definite single climax (see Pedrotti 2013). On the whole, it expresses the set of plant communities that can develop within an ecologically homogeneous space or tesela, including a single dynamic series of replacement and a single head series. Besides, it must be emphasized that into a sigmetum can be distinguished climatophilous and edaphophilous series, that in the latter are of two kinds (edaphoxerophilous or edaphohygrophilous). As concerns the climatophilous series, it is linked to the mesoclimate and only to rain water, with its head series usually represented by a forest as mature stage, since the highest expression of an evolved soil, while the edaphophilous series regard those ones in which the natural edaphic evolution is blocked by various environmental factors, determining more or less xerophytic soils (edaphoxerophilous series), as well as more or less wet soils (edaphohygrophilous series). In both cases the most mature stage usually coincides with a permanent woody vegetation, which is informative in defining the ecologic and phytogeographic role of the series. Instead, the geosigmetum is synonym with geoseries, which is the basic unit of the landscape phytosociology. It can be defined as a system of contiguous sigmeta replacing each other along an edaphic gradient inside of a bioclimatic belt of a given territory. Among these sigmeta, dynamic-catenal relations are established, distributing themselves according to the geological

S. Brullo et al., *The Vegetation of the Maltese Islands*, Geobotany Studies,
https://doi.org/10.1007/978-3-030-34525-9_21

and edaphic features of the surveyed area. Another kind of series is the permaseries o permasigmetum, regarding in particular perennial and stable plant communities occurring within a microtesela characterized by extremely specialized environmental conditions, such as coastal reefs, dunes, cliffs, rochy walls, salt marshes, etc. It is represented by a permanent and mature stage (edaphoclimax), usually mono-layered and pauci-specific, poor in perennial seral communities. It is a special kind of vegetation series from extreme environments, where a perennial climactic stage is present, without substitution perennial stages. Finally, the geopermaseries or geopermasigmetum is used in terms of catenal approach to the description of the adiacent sets of permasigmeta, clearly linked to different micro-topographic and edaphic conditions. In particular, the various permasigmeta belonging to a given geopermasigmetum are affected by a gradient of a well definited ecological situations, which are determinant in the realization of the catenal relationships. An azonal geopermaseries, like those localized in coastal habitat, can be recognized only within a bioclimatic belt and in quite uniform stands populated by permanent and perennial plant communities.

In the Maltese islands, although having a not very extended surface, several sigmeta, geosigmeta, permasigmeta and geopermasigmeta can be recognized. The most evolved potential vegetation is represented by holm oak woodlands corresponding to *Pistacio lentisci-Quercetum ilicis*. It can be considered the most mature stage of a climatophilous series corrisponding to *Pistacio lentisci-Querceto ilicis sigmetum,* strictly calcicolous, thermo-Medierranean dry, that currently is very rare and circumscribed to few stands characterized by mesoclimate condition (Fig. 21.1.1). Its first degradation stage is the *Erico multiflorae-Anthyllidetum melitensis,* followed in the succession by the *Hyparrhenietum hirto-sinaicae,* grasslands that for further soil erosion are replaced by ephemeral xeric micro-associations of *Stipo-Trachynetea distachyae*, which in order of degradation are *Ononido sieberi-Stipelletum capensis*, *Vulpio-Trisetarietum aureae* and lastly *Thero-Sedetum caerulei*. Other sigmeta occurring in these islands are represented by two edapho-xerophilous series, the first is that one of *Periploco angustifoliae-Euphorbieto dendroidis sigmetum* (Fig. 21.1.2), whose head series is *Periploco angustifoliae-Euphorbietum dendroidis*, maquis linked to thermo-Mediterranean dry climate localized on limestone outcrops with very superficial soil, widespread in all the islands. In conseguence of degradation processes this thermo-xerophilous shrub vegetation is replaced usually by garigues of the *Erico multiflorae-Coronilletum glaucae*, and subsequently by grassland of *Chamaeleo gummiferi-Brachypodietum retusi*, sometimes by the *Hyparrhenietum hirto-sinaicae,* and finally by ephemeral communiets of the *Plantagini coronopi-Catapodion balearici*, such as *Allietum lojaconoi*. The other edapho-xerophilous series is *Asparago aphylli-Tetraclineto articulatae sygmetum* (Fig. 21.1.3), high and dense maquis dominated by *Tetraclinis articulata*, colonizing mainly the more or less acclived rocky slopes, dynamically linked to *Erico multiflorae-Anthylledetum melitensis* in its variant with *Chiliadenus bocconei*. As concerns the edapho-hygrophilous series, they in the territory are represented by two sigmeta:

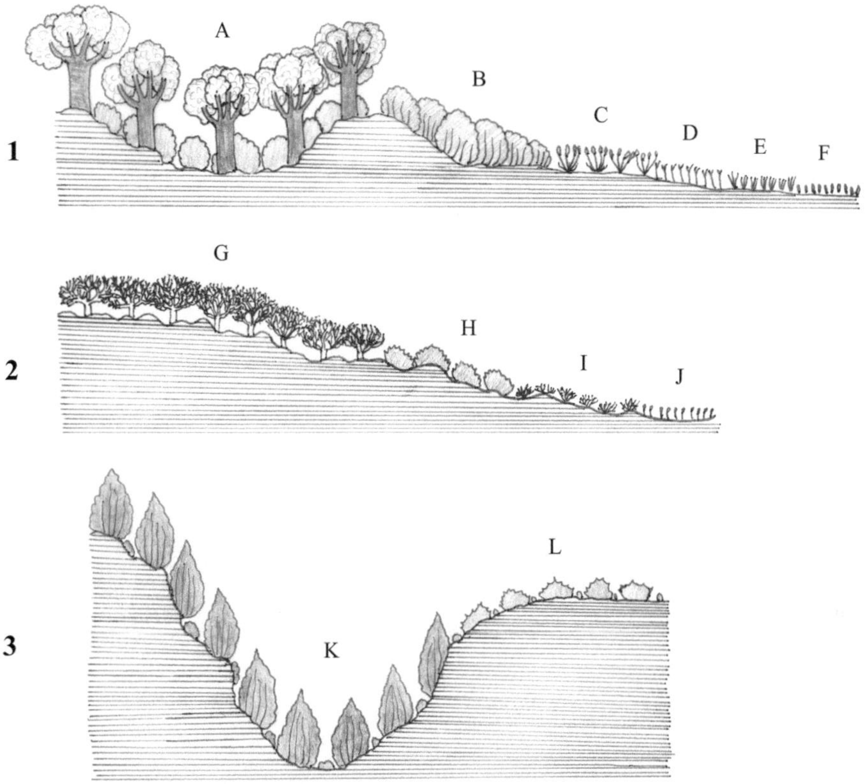

Fig. 21.1 Climatophilous and edapho-xerophilous vegetation series in the Maltese Islands: **1**—*Pistacio lentisci-Querceto ilicis sigmetum*, **2**—*Periploco angustifoliae-Euphorbieto dendroidis sigmetum*, **3**—*Asparago aphylli-Tetraclineto articulatae sigmetum*. (A) *Pistacio lentisci-Quercetum ilicis*, (B) *Erico multiflorae-Anthyllidetum melitensis*, (C) *Hyparrhenietum hirto-sinaicae*, (D) *Ononido sieberi-Stipelletum capensis*, (E) *Vulpio-Trisetarietum aureae*, (F) *Thero-Sedetum caerulei*, (G) *Periploco angustifoliae-Euphorbietum dendroidis*, (H) *Erico multiflorae-Coronilletum glaucae*, (I) *Chamaeleo gummiferi-Brachypodietum retusi*, (J) *Allietum lojaconoi*, (K) *Asparago aphylli-Tetraclinetum articulatae*, (L) *Erico multiflorae-Anthyllidetum melitensis* with *Chiliadenus bocconei*

1. *Ulmo canescentis-Saliceto pedicellatae sigmetum*, fluvial-alvear thermo-Mediterranean series currently relegated to short courses with freshwater in the inland of Malta, corresponding to basophilous riparian woodlands with *Ulmus canescens* and *Salix pedicellata* (Fig. 21.2.1). This wood vegetation is in catenal contact with permanent hygrophilous communities of *Magnocarici elatae-Phragmitetea australis* and *Potametea*, such as *Cyperetum longi, Polygono salicifolii-Phragmitetum australis, Helosciadetum nodiflori* and *Ranunculetum trichophylli*, constituting a geopermaseries.

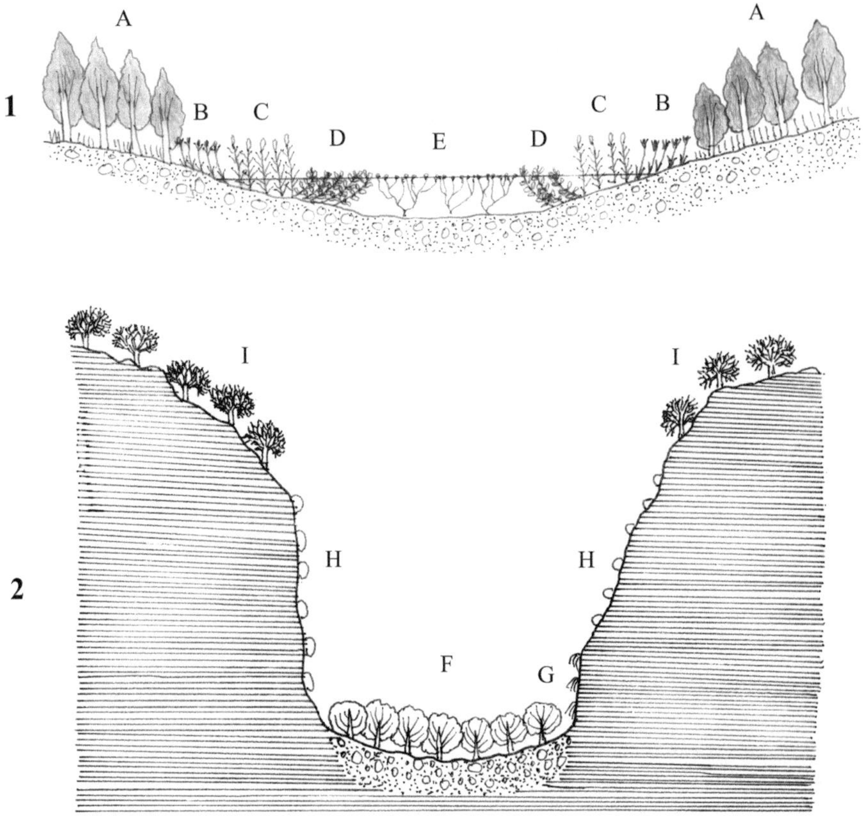

Fig. 21.2 Edapho-hygrophilous vegetation series in the Maltese Islands: **1**—*Ulmo canescentis-Saliceto pedicellatae sigmetum*, **2**—*Tamarici gallicae-Viticeto agni-casti sigmetum.* (A) *Ulmo canescenti-Salicetum pedicellatae*, (B) *Cyperetum longi*, (C) *Polygono salicifolii-Phragmitetum australis,* (D) *Helosciadetum nodiflori*, (E) *Ranunculetum trichophylli*, (F) *Tamarici gallicae-Viticetum agni-casti,* (G) *Eucladio verticillati-Adiantetum capilli-veneris,* (H) *Dianthion rupicolae*, (I) *Periploco angustifoliae-Euphorbietum dendroidis*

2. *Tamarici gallicae-Viticeto agni-casti sigmetum*, rivular-torrential thermo-Mediterraean dry series localized on the bottom of valleys along stream dried for most of the year (Fig. 21.2.2). These micro-woods have catenal contacts on the rocky slopes of the valleys with the edapho-xerophilous series of the *Periploco angustifoliae-Euphorbietum dendroidis* and some chasmophilous permaseries of *Dianthion rupicolae* and *Adiantion capilli-veneris,* caracterizing a geoseries.

Besides, several permasigmeta and geopermasigmeta can be identified within the natural landscape of these islands, with permanent plant communities structurally and ecologically well differentiated, physiognomically caracterized by dwarf shrubs or perennial herbaceous plants. Among the permasigmeta there are:

1. *Hyperico webbii-Chiliadeno bocconei permasigmetum*, chasmophilous thermo-Mediterranean dry permaseries localized mainly on the very high coastal calcareous cliffs, directly affected by sea aerosol and also on the vertical walls of wieds near their mouth (Fig. 21.3.1).

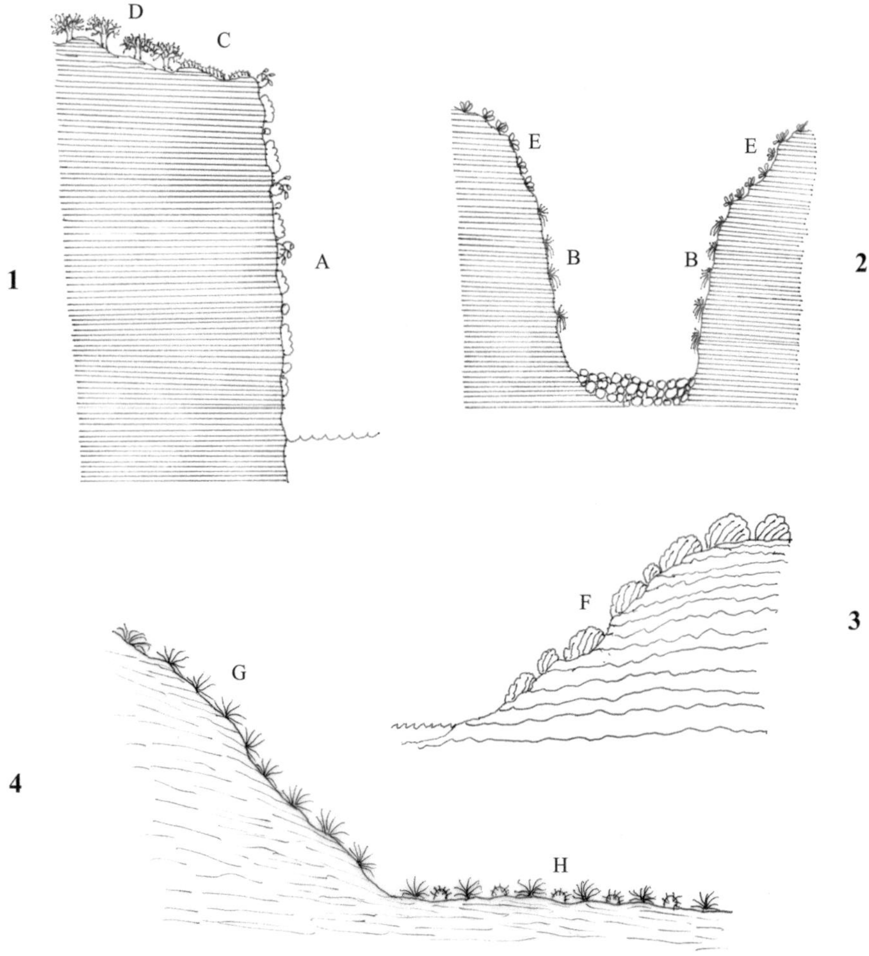

Fig. 21.3 Vegetation permaseries in the Maltese Islands: **1**—*Hyperico webbii-Chiliadeno bocconei permasigmetum*, **2**—*Putorio calabicae-Micromerieto microphyllae permasigmetum*, **3**—*Suaedo verae-Darnellio melitensis permasigmetum*, **4**—*Leontodonto tuberosi-Lygeo sparti permasigmetum* and *Convolvulo oleifolii-Lygeo sparti permasigmetum*. (A) *Hyperico webbii-Chiliadenetum bocconei*, (B) *Putorio calabricae-Micromerietum microphyllae*, (C) *Erico multiflorae-Coronilletum glaucae*, (D) *Periploco angustifoliae-Euphorbietum deindroidis*, (E) *Erico multiflorae-Anthyllidetum melitensis*, (F) *Suaedo verae-Darnielletum melitensis*, (G) *Leontodonto tuberosi-Lygeetum sparti*, (H) *Convolvulo oleifolii-Lygeetum sparti*

2. *Putorio calabricae-Micromerieto microphyllae permasigmetum*, chasmophilous thermo-Mediterranean mesic permaseries, linked to inland calcareous rocky walls of valleys (Fig. 21.3.2).
3. *Suaedo verae-Darnellio melitensis permasigmetum*, halo-nitrophilous thermo-Mediterranean dry permaseriies, linked to marly-clayey outcrops near the sea, dominated by succulent shrublands (Fig. 21.3.3).
4. *Leontodonto tuberosi-Lygeo sparti permasigmetum*, hemicryptophytic xerophilous thermo-Mediterranean dry permaseries, exclusive of eroded clayey budlands, which is represented by grasslands with *Lygeum spartum* and *Leontodon tuberosus* (Fig. 21.3.4G).
5. *Convolvulo oleifolii-Lygeo sparti permasigmetum*, hemicryptophytic xerophilous thermo-Mediterranean dry permasigmetum, localized on more or less flat eroded marly or clayey-marly substrata , which is represented by grasslands with *Lygeum spartum* and *Convolvulus oleifolius* (Fig. 21.3.4H).

Besides, in the Maltese Islands limitedly to particular coastal habitats, several geopermaseries formed by very differentiated permasigmeta showing among them close catenal contact, are frequent. The most representative are the following:

1. *Medicago marinae-Ammophilo australis geopermasigmetum*, psammophilous thermo-Mediterranean dry geopermaseries occurring on mobile or stabilized sandy coast dunes of western and central Mediterranean, dominated mainly by perennial grasses, chamaephytes, geophytes and also therophytes (Fig. 21.4). The permanent more characteristic associations of this series, from the most pioneers to the more mature ones, are *Salsolo tragi-Cakiletum maritimae, Eryngio maritimi-Sporoboletum arenarii, Cypero capitati-Agropyretum juncei, Medicagini marinae-Ammophiletum australis* and *Centaureo sphaerocephalae-Ononidetum ramosissimae*.
2. *Crithmo maritimi-Limonio virgati geopermasigmetum*, halophilous rupicolous thermo-Mediterranean dry geopermaseries, exclusive of marly or marly-clayey rochy coasts and strongly affected by sea aerosol (Fig. 21.5.1). It is physiognomically characterized by shrubs, often succulent, as *Crithmum maritimum. Limbarda crithmoides* subsp. *longifolia*, *Limonium virgatum*, *Darniella melitensis*, *Suaeda vera*, etc. The permanent associations of this series are *Crithmo maritimi-Limonietum virgati* and *Suaedo verae-Darnielletum melitensis,* sometimes mixed to *Silenetum melitensis*.

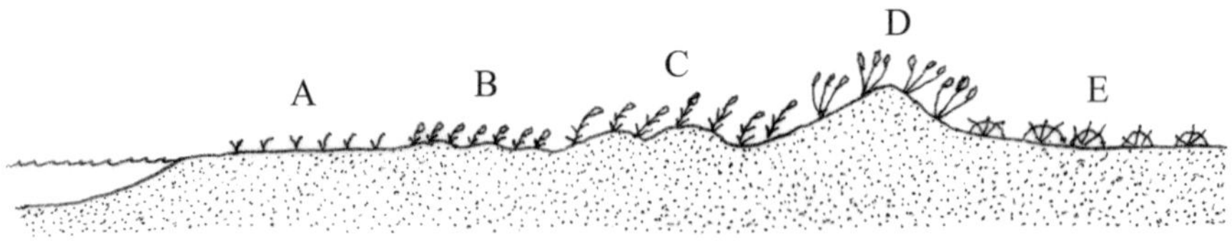

Fig. 21.4 Psammophilous geopermaseries in the Maltese Islands (*Medicago marinae-Ammophilo australis geopermasigmetum*). (A) *Salsolo tragi-Cakiletum maritimae*, (B) *Eryngio maritimi-Sporoboletum arenarii,* (C) *Cypero capitati-Agropyretum juncei,* (D) *Medicagini marinae-Ammophiletum australis,* (**e**) *Centaureo sphaerocephalae-Ononidetum ramosissimae*

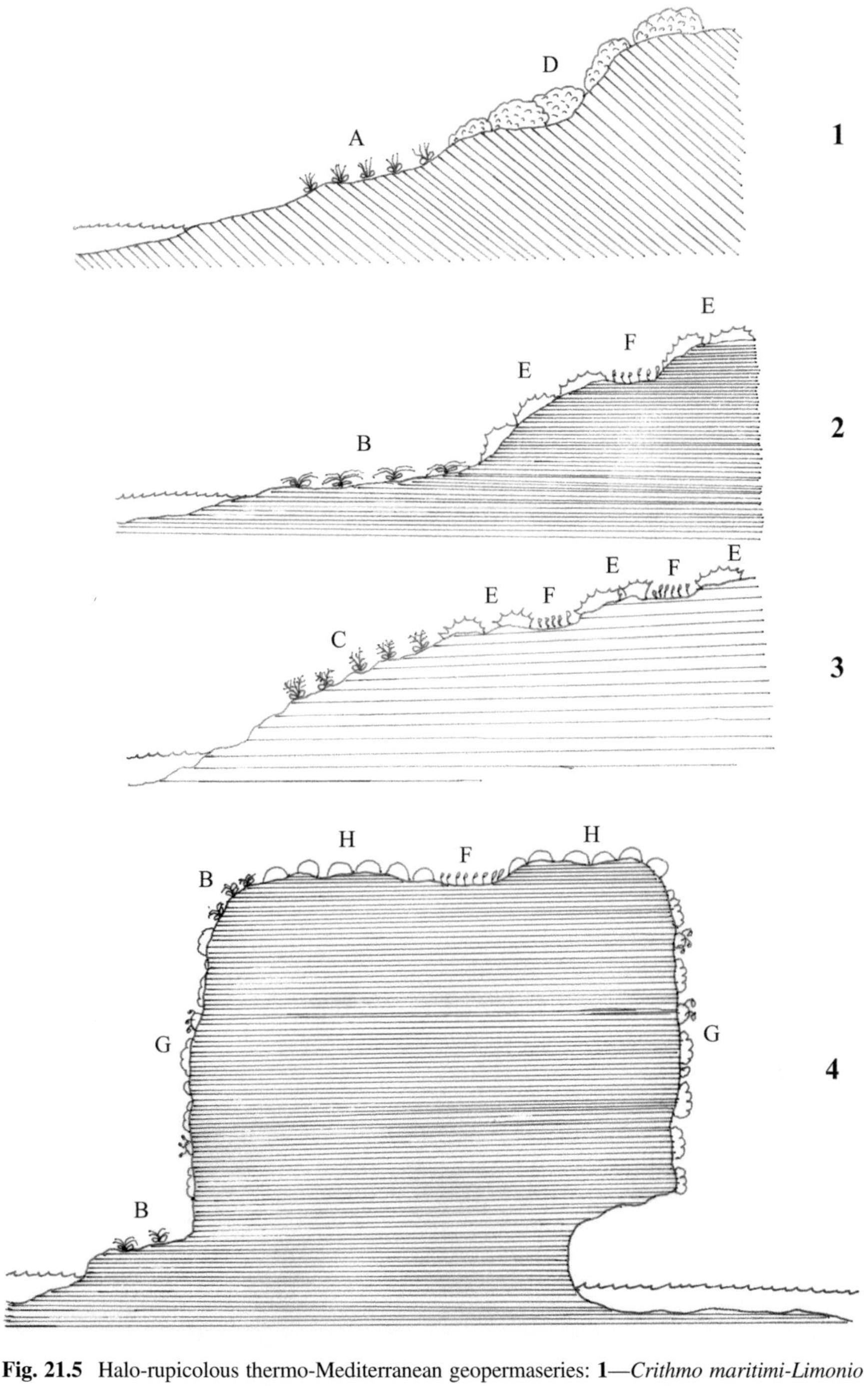

Fig. 21.5 Halo-rupicolous thermo-Mediterranean geopermaseries: **1**—*Crithmo maritimi-Limonio virgati geopermasigmetum*, **2**—*Limonio zeraphae geopermasigmetum*, **3**—*Limono melitensis geopermasigmetum*, **4**—*Crucianello rupestris-Helichryso melitensis geopermasigmetum.*

3. *Limonio zeraphae geopermasigmetum*, halophilous rupicolous thermo-Mediterranean dry geopermaseries, localized on compact coralline limestones of the rocky coast, habitat directly affected by marine agents (Fig. 21.5.2). The endemic permanent associations of this series are *Limonietum zeraphae*, *Anthyllido melitensis-Euphorbietum melitensis* and *Allietum lojaconoi*.
4. *Limono melitensis geopermasigmetum*, geopermaseries very similar to the previous one but linked to calcareos-marly substrata, represented mainly by globigerine limestones (Fig. 21.5.3). This series is differentiated by the *Limonietum melitensis*, permanent association endemic to Malta, which also shows catenal contact with *Anthyllido melitensis-Euphorbietum melitensis* and *Allietum lojaconoi.*
5. *Crucianello rupestris-Helichryso melitensis geopermasigmetum*, halo-rupicolous thermo-Mediterranean dry geopermaseries, circumscribed on high sea cliffs constitutes by compact coralline limestones, markedly affected by sea aerosol (Fig. 21.5.4). In this series only peculiar rupestrian associations are localizzed, represented by permanent endemic communities growing on rocky surfaces, washed out and without soil, such as *Limonietum zeraphae, Crucianello rupestris-Helichrysetum melitensis, Hyperico webbii-Chiliadenetum bocconei* and more rarely *Allietum lojaconoi*.
6. *Arthrocnemo macrostachyi-Junco subulati geopermasigmetum,* halo-xerophilous marshy thermo-Mediterranean dry geopermaseries, linked to coastal salt-marhes with silty-clay soils, periodically submerged by brackish waters and dominated by crassulent shrubs, helophytes and also crassulent therophytes (Fig. 21.6.1). The permanent associations belongin to this series are *Ruppietum drepanensis*, *Suaedo pruinosae-Salicornietum patulae*, *Arthrocnemo macrostachyi-Juncetum subulati* and *Agropyro scirpei-Inuletum longifolia.*
7. *Inulo longifoliae-Junco maritimi geopermasigmetum*, halo-hygrophilous marshy thermo-Mediterranean dry geopermaseries, observed on coastal salt marshes with silty-sandy soils, for long periods submerged by weakly salty waters and physiognomically characterized by helophytes (Fig. 21.6.2). The permanent associations of this series are *Enteromorpho intestinalis-Ruppientum maitimae, Juncetum maritimo-acuti, Caricetum divisae* and *Inulo longifoliae-Juncetum maritimi.*
8. *Halimiono portulacoidis-Suaedetum verae geopermasigmetum,* halo-nitrophilous marshy thermo-Mediterranean dry geopermaseries, localized in coastal salt marshes, temporarily submerged by brackish waters on soils rich in organic material (Fig. 21.6.3). The vegetation is dominated by succulent suffruticose halophytes, This series is represented by *Cressetum creticae*, *Suaedetum spicatae* and *Halimiono portulacoides-Suaedetum verae.*

Fig. 21.5 (continued) (A) *Crithmo maritimi-Limonietum virgati*, (B) *Limonietum zeraphae*, (C) *Limonietum melitensis*, (D) *Suaedo verae-Darnielletum melitensis*, (E) *Anthyllido melitensis-Euphorbietum melitensis*, (F) *Allietum lojaconoi*, (G) *Hyperico webbii-Chiliadenetum bocconei*, (H) *Crucianello rupestris-Helichrysetum melitensis*

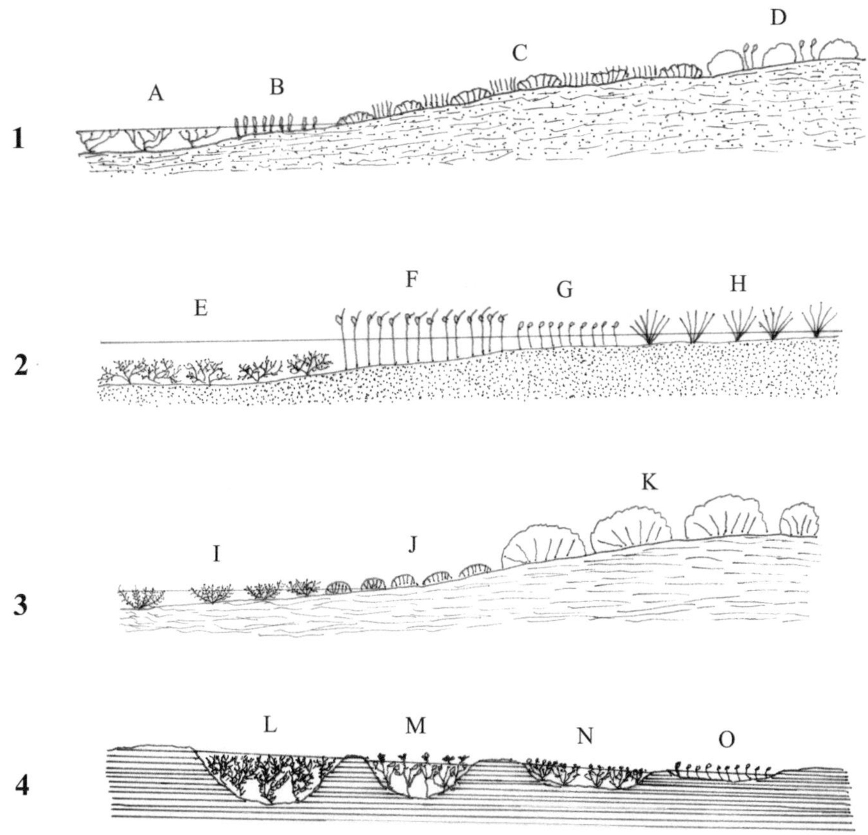

Fig. 21.6 Halophilous and hygrophilous thermo-Mediterranean geopermaseries: **1**—*Arthrocnemo macrostachyi-Junco subulati geopermasigmetum*, **2**—*Inulo longifoliae-Junco maritimi geopermasigmetum*, **3**—*Halimiono portulacoidis-Suaedetum verae geopermasigmetum*, **4**—*Crassulo buillardii-Elatino gussonei geopermasigmetum*. (A) *Ruppietum drepanensis*, (B) *Suaedo pruinosae-Salicornietum patulae*, (C) *Arthrocnemo macrostachyi-Juncetum subulati*, (D) *Agropyro scirpei-Inuletum longifoliae*, (E) *Enteromorpho intestinalis-Ruppientum maitimae*, (F) *Juncetum maritimo-acuti*, (G) *Caricetum divisae*, (H) *Inulo longifoliae-Juncetum maritimi*, (I) *Cressetum creticae* (J) *Suaedetum spicatae*, (K) *Halimiono portulacoides-Suaedetum verae*, (L) *Zanichellietum melitensis*, (M) *Damasonio bourgaei-Ranunculetum saniculifolii*, (N) *Crassulo buillardii-Elatinetum gussonei*, (O) *Triglochino laxiflori-Romuleetum melitensis*

9. Finally, another quite relevant hygrophilous geopermaseries is that established in the temporary freshwater rock pools, localized in the hollows of the inland calcareous plateaux. It is proposed as *Crassulo buillardii-Elatino gussonei geopermasigmetum* and is constituted by permanent communities, mostly endemic, in close catenal contact according to a gradient related to the submerging period (Fig. 21.6.4). In this pool, with decreasing water depth, various associations colonize the muddy bottom, initially they are characterized by totally

immerged hydrophites and subsequently substituted by hygrophites less linked to submersion. The most relevant permasigmeta beloging to this geopermaseries are *Zanichellietum melitensis, Damasonio bourgaei-Ranunculetum saniculifolii, Crassulo buillardii-Elatinetum gussonei* and *Triglochino laxiflori-Romuleetum melitensis*.

Chapter 22
Maltese Habitat of Community Interest

According to MEPA (2010) and ERA (2017), in the Maltese Islands within a management plan for the preservation of areas deemed deserving of protection, 34 terrestrial "Natura 2000" sites have been recognized, of which 27 are Special Areas of Conservation (SAC) and 7 are Species Protection Areas (SPAs). Inside of these protected areas there are species and habitats of Community Interest, as required by Habitat Directive of EU. On the base of our phytosociological researches and of the literature data, the habitats currently occurring in the Maltese territory are the following:

1120 Posidonia beds (*Posidonion oceanicae*)—This habitat is represented by *Posidonietum oceanicae.*

1150 Coastal lagoons with or without vegetation from *Ruppietea maritimae*, *Potametea*, *Zosteretea* or *Charetea*—This habitat is represented by *Enteromorpho intestinalis-Ruppietum maritimae, Ruppietum drepanensis* and *Lamprothamnietum papulosi.*

1210 Annual vegetation of drift lines, found at the foredune or embryo dunes and close to the sea with input of organic matter such as the Posidonia sea grass, *Cakiletea maritimae*—This habitat is represented by *Salsolo tragi-Cakiletum maritimae.*

1240 Vegetated sea cliffs of the Mediterranean coasts with endemic Sea Lavender species, *Crithmo-Limonietea*—This habitat is represented by *Limonietum zeraphae*, *Limonietum melitensis* and *Crithmo maritimi-Limonietm virgati.*

1310 *Salicornia* (Twiggy Glasswort) and other annuals (plants that complete their lifecycle within 1 year) colonising mud and sand, Subtype: 15.12 (*Frankenion pulverulentae*) and 15.11—Glasswort swards (*Thero-Salicornietalia*)—This habitat is represented by *Suaedo spicatae-Salicornirtum patulae*, *Parapholidetum filiformis*, *Desmazerio pignattii-Senecionetum pygmaei* and *Anthemido urvilleanae-Frankenietum pulverulentae*.

S. Brullo et al., *The Vegetation of the Maltese Islands* , Geobotany Studies,
https://doi.org/10.1007/978-3-030-34525-9_22

1410 Mediterranean salt meadows (subject to flooding by sea water) (*Juncetalia maritimi*)—This habitat is represented by *Inulo longifoliae-Juncetum maritimi*, *Juncetum maritimo-acuti* and *Caricetum divisae*.

1420 Mediterranean and thermo-Atlantic halophilous scrubs (perennial vegetation of marine saline muds) (*Sarcocornetea fruticosae*)—This habitat is represented by *Arthrocnemo macrostrachyi-Juncetum subulati*, *Agropyro scirpei-Inuletum longifoliae* and *Halimiono partulacoidis-Suaedetum verae*.

1430 Halo-nitrophilous scrubs (*Pegano-Salsoletea*)—This habitat is represented by *Suaedo verae-Darnielletum melitensis*.

2110 Embryonic shifting dunes *Agropyron junceum* e *Sporoboletum*—This habitat is represented by *Eryngio maritimi-Sporoboletum arenarii* and *Cypero capitati-Agropyretum juncei*.

2120 Shifting dunes along the shoreline with *Ammophila arenaria* (white dunes)—This habitat is represented by *Medicagini marinae-Ammophiletum australis*.

2210 *Crucianellion maritimae* fixed beach dune—This habitat is represented by *Centaureo spaerocephalae-Ononidetum ramosissimae*.

2230 *Malcolmietalia* dune grasslands—This habitat is represented by *Cutandio divaricatae-Pseudorlayetum pumilae*.

3140 Hard oligo-mesotrophic waters with benthic vegetation of *Chara* spp.—This habitat is represented by *Charetum vulgaris*.

3150 Natural eutrophic lakes with *Magnopotamion* or *Hydrocharition*—type vegetation—This habitat is represented by *Lemnetum minoris*.

3170 Mediterranean temporary ponds—This habitat is represented by *Crassulo vaillantii-Elatinetum gussonei, Triglochino laxiflori-Romuleetum melitensis* and *Damasonio bourgaei-Crypsietum aculeatae.*

3260 Water courses of plain to montane levels with the *Ranunculion fluitantis* and *Callitricho-Batrachion* vegetation—This habitat is represented by *Damasonio bourgaei-Ranunculetum saniculifolii*, *Zanichellietum melitensis* and *Ranunculetum trichophylli.*

3280 Constantly flowing Mediterranean rivers with *Paspalo-Agrostidion* species and hanging curtains of *Salix* and *Populus alba*—This habitat is represented by *Potentillo repentis-Panicetum repentis*.

5330 Thermo-Mediterranean and pre-desert scrub subtype; 32.25—Mediterranean pre-desert scrub. *Periplocion angustifoliae*; 32.22—Tree-spurge formations *Euphorbia dendroides*; 32.441p, Spiny spurge garrigues *Euphorbia melitensis* garrigues of Malta—This habitat is represented by *Periploco angustifoliae-Euphorbietum dendroidis* and *Anthyllido melitensis-Euphorbietum melitensis*.

5410 West Mediterranean clifftop phryganas (*AstragaloPlantaginetum subulatae*)—This habitat is represented by *Erico multiflorae-Anthyllidetum melitensis* and *Erico multiflorae-Coronilletum glaucae*.

5430 Endemic phryganas of the *Euphorbio-Verbascion* subtypes 33.5—*Hypericum aegyptiacum* phryganas—This habitat is represented by *Crucianello rupestris-Helichrysetum melitensis.*

6220 Pseudo-steppe with grasses and annuals of the *Thero-Brachypodietea*—This habitat is represented by *Hyparrhenietum hirto-sinaicae, Chamaemeleo-Brachypodietum retusi, Leontodonto tuberosi-Lygeetum sparti, Convolvulo oleifolii-Lygeetum sparti, Allietum lojaconoi, Silenetum melitensis, Phalaridetum paradoxae, Catapodio balearici-Linarietum pseudolaxiflorae, Thero-Sedetum caerulei, Vulpio-Trisetarietum aurei* and *Ononido sieberi-Stipelletum capensis.*

6420 Mediterranean tall humid herb grasslands of the *Molinio-Holoschoenion*—This habitat is represented by *Festuco arundinaceae-Caricetum distantis, Phalarido caerulescentis-Schedonoretum arundunacei* and *Schedonoro arundinacei- Caricetum divisae.*

6430 Hydrophilous tall herb fringe communities of plains and of the montane to alpine levels—This habitat is represented by *Calystegio sylvaticae-Arundinatum donacis.*

8210 Calcareous rocky slopes with chasmophytic vegetation; 62.14—Southern Italian communities (*Dianthion rupicolae*)—This habitat is represented by *Hyperico webbii-Chiliadenetum bocconei* and *Putorio calabricae-Micromerietum m*icrphyllae.

8310 Caves not open to the public—This habitat is represented by mosses and algal carpets at the entry of caves, sometimes mixed to ferns (e.g. *Asplenium sagittatum*).

92A0 Salix alba and Populus alba galleries—This habitat is represented by *Ulmo canescentis-Salicetum pedicellatae.*

92D0 Southern riparian galleries and thickets (*Nerio-Tamaricetea* and *Securinegion tinctoriae*)—This habitat is represented by *Tamarici gallicae-Viticetum agni-casti.*

9340 *Quercus ilex* and *Quercus rotundifolia* forests—This habitat is represented by *Pistacio lentisci-Quercetum ilicis.*

9540 Mediterranean pine forests with endemic Mesogean pines; 42.846—Sicilian Aleppo pine woods—This habitat is represented by *Thymo capitati-Pinetum halepensis.*

9570 *Tetraclinis articulata* forests—This habitat is represented by *Asparago aphylli-Tetraclinetum articulatae.*

References

Abela GF (1647) Della descritione di Malta: isola nel mare Siciliano con le sue Antichità, ed altre Notizie. Paolo Bonacota, Malta. 573 pp

Adams AL (1870) Notes of a naturalist in the Nile Valley and Malta. Edmonston and Douglas, Edinburgh. 295 pp

Alcaraz Ariza FJ (2011) Asociaciones de hábitats en el Sureste de la Península Ibérica. Vegetación y tipos de hábitats de interés en la Unión Europea, Universidad de Murcia

Alexander D (1988) A review of the physical geography of Malta and its significance for tectonic geomorphology. Quat Sci Rev 7:41–53

Amigo J, Romero MI (1997) Comportamiento fitosociológico de *Smyrnium olusatrum* L. en el noroeste ibérico. Lazaroa 18:153–164

Anderson EW, Schembri PJ (1989) Coastal zone survey of the Maltese Islands report. Planning Services Division, Works Department, Beltisebh, Malta

Angius R, Bacchetta G (2009) Boschi e boscaglie ripariali del Sulcis-Iglesiente (Sardegna sud-occidentale, Italia). Braun-Blanquetia 45:1–63

Antonelli M, Franciosi R, Pezzi G, Querci A, Ronco GP, Vezzani F (1988) Paleogeographic evolution and structural setting of the of the northern side of Sicily channel. Mem Soc Geol Ital 41:141–157

Arrigoni PV, Di Tommaso PL, Mele A (1990) Caratteri fisionomici e fitosociologici delle leccete delle montagne calcaree della Sardegna centro-orientale. Boll Soc Sarda Sci Nat 27:205–219

Augier A (1982) Inventaire et classification des biocénoses marines benthiques de la Méditerranée. Conseil Eur Collect Sauv Nat 29:1–59

Bagella S, Caria MC, Farris E, Filigheddu R (2009) Phytosociological analysis in Sardinian Mediterranean temporary wet habitats. Fitosociologia 46(1):11–26

Baldoni M, Biondi E (1993) La vegetazione del medio e basso corso del Fiume Esino (Marche – Italia centrale). Stud Bot 11:209–257

Ballou MM (1893) The story of Malta. The Riherside Press, Cambridge. 318 pp

Barbagallo C (1983) Vegetazione di alcuni boschi di sughera (*Quercus suber*) della Sicilia meridionale-orientale. Boll Acc Gioenia Sci Nat Catania 16(321):289–296

Barbagallo C, Brullo S, Fagotto F (1979) Vegetazione a *Platanus orientalis* L. e altri aspetti igrofili dei fiumi iblei. Pubbl Ist Bot Catania, pp 1–28

Barbagallo C, Brullo S, Guglielmo A (1980) Esempi di cartografia della vegetazione di alcune aree della Sicilia. Carta della vegetazione di Monte Cofano. Quaderni CNR ser. AQ/1/39:43–52, Roma

Barbagallo C, Brullo S, Furnari F (1987) Le associazioni nitrofile dell'*Onopordion illyrici* in Tunisia. Boll Acc Gioenia Sci Nat Catania 20(330):219–225

S. Brullo et al., *The Vegetation of the Maltese Islands*, Geobotany Studies,
https://doi.org/10.1007/978-3-030-34525-9

Barbagallo C, Brullo S, Furnari F (1990) La vegetazione alofila palustre della Tunisia. Boll Acc Gioenia Sci Nat Catania 23(336):581–652

Barbéro M, Quézel P, Rivas-Martinez S (1981) Contribution à l'étude des groupements forestiers et préforestiers du Maroc. Phytocoenologia 9(3):311–412

Bardat J, Bioret F, Botineau M, Boullet V, Delpech R, Géhu J-M, Haury J, Lacoste A,Rameau J-C, Royer J-M, Roux G, Touffet J (2004). Prodrome des végétations de France. Mus Nat Hist Nat Paris, pp 1–143

Bartolo G, Brullo S (1986) La classe *Parietarietea judaicae* in Sicilia. Arch Bot Biogeogr It 62(1–2):31–50

Bartolo G, Brullo S (1993) La classe *Crithmo-Limonietea* in Sicilia. Boll Acc Gioenia Sci Nat Catania 26(342):5–47

Bartolo G, Brullo S, Lo Cicero E, Marcenò C, Piccione V (1978) Osservazioni fitosociologiche sulla Pineta a "*Pinus halepensis*" di Vittoria (Sicilia meridionale). Arch Bot Biogeogr Ital 54:137–153

Bartolo G, Brullo S, Marcenò C (1982) La vegetazione costiera della Sicilia sud-orientale. Contributo alla interpretazione delle fasce di vegetazione delle coste mediterranee. Quad CNR ser. AQ/1/226, Roma

Bartolo G, Brullo S, Fagotto F, Grillo M (1983) La vegetazione segetale della Sicilia sud-orientale. Boll Acc Gioenia Sci Nat Catania 15(320):307–321

Bartolo G, Brullo S, Minissale P, Spampinato G (1985) Osservazioni fitosociologiche sulle pinete a *Pinus halepensis* Miller del bacino del fiume Tellaro (Sicilia sud-orientale). Boll Acc Gioenia Sci Nat Catania 18:255–270

Bartolo G, Brullo S, Minissale P, Spampinato G (1990) Flora e vegetazione dell'Isola di Lampedusa. Boll Acc Gioenia Sci Nat Catania 21(334):119–255

Bartolo G, Lanfranco E, Pulvirenti S, Stevens DT (2001) The Orchidaceae of the Maltese Archipelago (Central Mediterranean). J Eur Orch 33(3):743–870

Bazan G, Ilardi V, Sciandrello S, Minissale P (2006) La biodiversità vegetale di Monte Gibliscemi (Mazzarino, Sicilia). Quad Bot Amb Appl 17(2):121–140

Bazan G, Brullo S, Raimondo FM, Schicchi R (2010) La serie di vegetazione della Regione Sicilia. In: Blasi C (ed) La vegetazione d'Italia. Palombi & Partner s.r.l, Roma, pp 429–469

Béguinot A (1907) Diagnoses of new or less known Romulea species. Bot Jahrb Syst 38:322–339

Béguinot A (1908) Monographic revision of the genus Romulea (Maratti) – Biological study: part II. Systematic enumeration and illustrations of the species of the genus Romulea. Malpighia 22:377–469

Benabid A (1984) Etude phytoécologique des peuplements forestiers et préforestiers du Rif centro-occidental (Maroc). Trav Inst Scient Rabat, ser Bot 34:1–64

Bensettiti F, Bioret F, Roland J, Lacoste J-P (2004) « Cahiers d'habitats » Natura 2000. Connaissance et gestion des habitats et des espèces d'intérêt communautaire. Tome 2. Habitats côtiers. La Documentation Française, Paris, pp 1–399

Bergmeier E (2001) Seasonal pools in the vegetation of Gavdos (Greece) – in situ conservation required. Bocconea 13:511–516

Bergmeier E (2008) Diversity of aestival plant communities of irrigated garden croplands in Cretan villages. Braunsch Geobot Arb 9:65–80

Bergmeier E, Raus T (1999) Verbreitung und Einnischung von Arten der *Isoeto-Nanojuncetea* in Griechenland. Mitt Bad Landesver Naturk Natursch 17:463–479

Bertoloni A (1829) Florae Melitensis Thesaurus sive plantarum enumeratio, quae in Melitae Gaulosque insulis aut indigene aut vulgatissimae occurrunt ecc. Curante Stephano Zerapha Med. Doct. Fasc. 1°. Ann Stor Nat 2:359–364

Bianco P, Bedalov M, Medagli P, Mastropasqua L (1984) Un contributo alla conoscenza dell'associazione *Oleo-Euphorbietum dendroidis* Trinajstic nelle stazioni pugliesi e confronto con quelle dell'Adriatico Orientale jugoslavo. Not Fitosoc 19(2):23–28

Billy F (2000) Prairies et pâturages en Basse-Auvergne. Bull Soc Bot Centre-Ouest, NS 20:1–258

Biondi E (1988) Aspetti di vegetazione alo-nitrofila sulle coste del Gargano e delle isole Tremiti. Arch Bot Biogeograf Ital 64:19–33

Biondi E (1999) Diversità fitocenotica degli ambienti costieri italiani. Suppl Boll Mus Civ St Nat Venezia 49:39–105

Biondi E (2011) Phytosociology today: methodological and conceptual evolution. Plant Biosyst 145:19–29

Biondi E, Bagella S (2005) Vegetazione e paesaggio vegetale dell'arcipelago di La Maddalena (Sardegna nord-orientale). Fitosociologia 42(2) suppl. 1:3–99

Biondi E, Gèhu J-M (1987) A study of some phytocoenoses of *Euphorbia dendroides* L. found in Greece. Acta Bot Croat 46:81–84

Biondi E, Mossa L (1992) Studio fitosociologico del promontorio di Capo S.Elia e dei Colli di Cagliari (Sardegna). Doc Phytosoc 14:1–44

Biondi E, Allegrezza M, Filigheddu R (1988) Su alcune formazioni ad *Artemisia arborescens* L. della Sardegna settentrionale. Boll Soc Sarda Sci Nat 26:177–185

Biondi E, Allegrezza M, Filigheddu R (1989) *Smyrnium olusatrum* vegetation in Italy. Braun-Blanquetia 3:219–222

Biondi E, Allegrezza M, Filigheddu R (1990) Su alcune associazioni di vegetazione nitrofila della Sardegna settentrionale. Boll Soc Sard Sci Nat 27:221–236

Biondi E, Ballelli S, Taffetani F (1992) La vegetazione di alcuni territori calanchivi in Basilicata (Italia meridionale). Doc Phytosoc, n s 14:489–498

Biondi E, Ballelli S, Allegrezza M, Taffetani F, Francalancia C (1994a) La vegetazione delle fiumare del versante ionico lucano-calabro. Fitosociologia 27:51–66

Biondi E, Blasi C, Brugiapaglia E, Fogu MC, Mossa L (1994b) La vegetazione nitrofila della citta` di Cagliari (Sardegna). Allionia 32:303–323

Biondi E, Filigheddu R, Farris E (2001) Il paesaggio vegetale della Nurra (Sardegna nord occidentale). Fitosociologia 38(2, suppl. 2):3–105

Biondi E, Vagge I, Baldoni M, Taffetani F (2004) Biodiversità fitocenotica e paesaggistica dei fiumi dell'Italia centro-settentrionale: aspetti fitosociologici e sinfitosociologici. Studi Trent Sci Nat Acta Biol 80:13–21

Biondi E, Blasi C, Allegrezza M, Anzellotti I, Azzella MM, Carli E, Casavecchia S, Copiz R, Delvico E, Facioni L, Galdenzi D, Gasparri R, Lasen C, Pesaresi S, Poldini L, Sburlino G, Taffetani F, Vagge I, Zitti S, Zivkovic L (2014) Plant communities of Italy: the vegetation prodrome. Plant Biosyst 148(4):728–814

Biondi E, Pesaresi S, Galdenzi D, Gasparri R, Biscotti N, del Viscio G, Casavecchia S (2016) Post-abandonment dynamic on Mediterranean and sub-Mediterranean perennial grasslands: the edge of vegetation of the new class *Charybdido pancratii-Asphodeletea ramosi*. Plant Sociol 53(2):3–18

Biondi E, Pesaresi S, Gasparri R, Biscotti N, del Viscio G, Bonsanto D, Casavecchia S (2017) New contributions to the class *Charybdido pancratii-Asphodeletea ramosi* Biondi 2016. Plant Sociol 54(1):137–144

Biurrun I (1999) Flora y vegetaciòn de los rios y humedidales de Navarra. Guineana 5:1–338

Blasi C, Carranza ML, Frondoni R, Rosati L (2000) Ecosystem classification and mapping: a proposal for Italian landscape. Appl Veg Sci 3:233–242

Boccone P (1674) Icones et descriptiones rariorum plantarum Siciliae, Melitae, Galliae et Italiae quarum unaquaeque proprio charactere signata, ab aliis ejusdem classis facile distinguitur. Theatro Sheldonianum, Oxoniae

Boccone P (1697) Museo di piante rare della Sicilia, Malta, Corsica, Italia, Piemonte e Germania. Baptista Zuccato, Venetia

Bocquet G, Widler B, Kiefer H (1978) The Messinian model. A new outlook for the floristics and systematics of the Mediterranean area. Candollea 33:269–287

Bonamicus JF (1670) Brevis notitia plantarum quae in Melita et Gaulos insulis observantur. Manuscript

Borg J (1927) Descriptive flora of the Maltese Islands. Government Printing Office, Malta

Borg JA, Schembri PJ (1995) The state of *Posidonia oceanica* (L.) Delile meadows in the Maltese Islands (Central Mediterranean). Rapp Congr Comm Intern Expl Sci Mer Médit 34:123

Borg JA, Schembri PJ (2002) Alignment of marine habitat data of the Maltese Islands to conform to the requirements of the EU habitats directive (Council Directive 92/43/EEC). [Report commissioned by the Malta Environment & Planning Authority]. Independent Consultants, Malta, pp 1–366

Borg JA, Rowden AA, Attrill MJ, Schembri PJ, Jones MB (2009) Occurrence and distribution of different bed types of seagrass *Posidonia oceanica* around the Maltese Islands. Mediterr Mar Sci 10(2):45–61

Borum J, Greve TM (2004) The four European seagrass species. In: Borum J et al (eds) European seagrasses: an introduction to monitoring and management. The M&MS project, Hillerød, pp 1–7. http://www.seagrasses.org

Botschantzev V (1976) Conspectus specierum sectionis Coccosalsola Fenzl generis Salsola L. Nov Syst Pl Vasc (Leningrad) 13:74–102

Brandes D (1988) Kenntnis der ruderal Vegetation von Mallorca I. Die Vegetation der Mauern und Mauerfüssepar. Doc Phytosoc, n s 11:111–123

Brandes D (1991) Soziologie und Ökologie von *Oxalis pes-caprae* L. im Mitterleergebiet unter besonderer Berück-fichtigung von Malta. Pytocoenologia 19(3):285–306

Brandes D, Brandes E (1999) The flora of Maltese walls, pp 1–25. http://www.doku.biblio.etc.tu-bs.de/opus/volltexte/1999/55

Braun Blanquet J (1928) Pflanzensoziologie. Grundzüge der vegetationnskunde. Springer, Berlin, p 632

Braun Blanquet J (1964) Pflanzensoziologie, 3rd edn. Springer, New York, p 865

Braun-Blanquet J (1933) Prodrome des groupments vègètaux I: *Ammophiletalia* et *Salicornetalia* Medit. Montpellier

Braun-Blanquet J (1936) Un joyau floristique et phytosociologique: « l'*Isoëtion* » méditerranéen. Bull Soc Étude Sci Nat Nîmes 47:1–23

Braun-Blanquet J, de Bolós O (1958) Les groupements vègètaux du bassin moyen de l'Ebre et leur dynamisme. Anal Estac Exp Aula Dei 5:1–266

Braun-Blanquet J, Maire R (1924) Etudes sur la végétation et la flore marocaine. Mém Soc Sci Nat Maroc 8(1ère partie):1–244

Braun-Blanquet J, Molinier R (1935) Une exursion phytosociologique à l'île de Porquerolles. Le Chêne 40:169–181

Braun-Blanquet J, Gajewski W, Wraber M, Walas J (1936) Prodrome des groupements végétaux. Fasc. 3 (Classe des *Rudereto-Secalinetales*, groupements messicoles, culturaux et nitrophiles-rudérales du cercle de végétation méditerranéen). Imprimerie Mari-Lavic, Montpellier

Braun-Blanquet J, Roussine N, Nègre R (1952) Les groupments vègètaux de la France mèditerranèenne. Centre National de la Recherche Scientifique, Montpellier

Brenner P (1838) Botany. In: Bager GP (ed) Description of Malta and Gozo. M.Weiss, Malta, pp 61–65

Briffa M (1986) Two interesting additions to the flora of the Maltese Islands. Cen Medit Nat 1(4):85–86

Brullo S (1975) Aspetti di vegetazione tero-xerofitica del territorio ibleo (Sicilia meridionale-orientale). Boll Acc Gioenia Sci Nat Catania 12:5–16

Brullo S (1979) Taxonomic and nomenclatural notes on the genera "*Jasonia*" Cass. and "*Chiliadenus*" Cass. (Compositae). Webbia 34(1):289–308

Brullo S (1980a) La classe *Polygono-Poetea annuae* in Sicilia. Not Fitosoc 15:117–123

Brullo S (1980b) Taxonomic and nomenclatural notes on the genus *Limonium* in Siciliy. Bot Notiser 133:281–293

Brullo S (1983a) Le associazioni subnitrofile dell'*Echio-Galactition tomentosae* in Sicilia. Boll Acc Gioenia Sci Nat Catania 15(320):405–452

Brullo S (1983b) L'*Hordeion leporini* in Sicilia. Arch Bot Biogeogr Ital 58(1–2):55–88

Brullo S (1984a) Taxonomic considerations on the genus *Darniella* (Chenopodiaceae). Webbia 38:301–329

Brullo S (1984b) L'alleanza *Bromo-Oryzopsion miliaceae* in Sicilia. Boll Acc Gioenia Sci Nat 17(323):239–258
Brullo S (1985) Sur la syntaxonomie des pelouses therophytiques des territoires steppiques de l'Europe sud-occidentale. Doc Phytosoc, n s 9:1–24
Brullo S (1988a) Miscellaneous notes on the genus *Limonium* (Plumbaginaceae). Willdenowia 17:11–18
Brullo S (1988b) Le associazioni della classe *Frankenietea pulverulentae* nel Mediterraneo centrale. Acta Bot Barc 37:45–57
Brullo S, Di Martino A (1974) Vegetazione dell'Isola Grande dello Stagnone (Marsala). Boll St Inform Giard Bot Palermo 26:15–62
Brullo S, Furnari F (1970) Vegetazione psammofila presso il Capo Isola delle Correnti (Sicilia sud-orientale). Pubbl Ist Bot Univ Catania, Catania
Brullo S, Furnari F (1976) Le associazioni vegetali degli ambienti palustri costieri della Sicilia. Not Fitosoc 11:1–43
Brullo S, Furnari F (1981) Phytogeographical considerations on the coastal vegetation of Cyrenaica. An Jard Bot Madr 37(2):65–772
Brullo S, Furnari F (1990) La vegetazione costiera della Cirenaica. Boll Acc Gioenia Sci Nat Catania 21(334):37–117
Brullo S, Furnari F (1996) L,a vegetazione del Gebel el-Akhdar (Cirenaica settentrionale). Boll Acc Gioenia Sci Nat Catania 27:197–412
Brullo S, Giusso del Galdo G (2003) La classe *Saginetea maritimae* in Italia. Fitosociologia 40(2):29–41
Brullo S, Giusso del Galdo G (2006) Taxonomic remarks on the *Anthyllis hermanniae* L. (Fabaceae, Faboideae) species complex of the Mediterranean flora. Novon 16(3):304–314
Brullo S, Grillo M (1985) Le associazioni psammofile effimere dei *Malcolmietalia* rinvenute in Sicilia. Boll Acc Gioenia Sci Nat Catania 18:271–282
Brullo S, Guarino R (1998) Syntaxonomy of the *Parietarietea judaicae* class in Europe. Ann Bot (Roma) 56:110–146
Brullo S, Guarino R (2002) La classe *Parietarietea judaicae* Oberd. 1977 in Italia. Fitosociologia 39(1):5–27
Brullo S, Guarino R (2007) The Mediterranean weedy vegetation and its origin. Ann Bot (Roma) 7:101–110
Brullo S, Marcenò C (1976) Osservazioni su una micro-associazione a Cariofillacee annue rinvenuta in Sicilia. Boll Acc Gioenia Sci Nat Catania 12(9–10):105–108
Brullo S, Marcenò C (1979) *Dianthion rupicolae* nouvelle alliance sud-tyrrhenienne des *Asplenietalia glandulosi*. Doc Phytosoc, n s 4:131–146
Brullo S, Marcenò C (1980) Il *Diplotaxion erucoidis* in Sicilia, con considerazioni sulla sintassonomia e distribuzione. Notiz Fitosoc 15:27–44
Brullo S, Marcenò C (1983) Osservazioni fitosociologiche sull'isola di Marettimo (Arcipelago delle Egadi). Boll Acc Gioenia Sci Nat Catania 15(320):201–228
Brullo S, Marcenò C (1985a) Contributo alla conoscenza della classe *Quercetea ilicis* in Sicilia. Not Fitosoc 19(1):183–229
Brullo S, Marcenò C (1985b) Contributo alla conoscenza della vegetazione nitrofila della Sicilia. Coll Phytosoc 12:23–148
Brullo S, Minissale P (1998) Considerazioni sintassonomiche sulla classe *Isoeto-Nanojuncetea*. Itin Geobot 11:263–290
Brullo S, Pavone P (1985) Taxonomic considerations on the genus Desmazeria (Gramineae) with description of a new species: Desmazeria pignattii. Willdenowia 15(1):99–106
Brullo S, Pavone P (1987) Cremnophyton lanfrancoi: a new genus and species of Chenopodiaceae from Malta. Candollea 42:621–625
Brullo S, Pavone P (1988) Hyoseris frutescens, a new species from Gozo (Malta). Candollea 43:717–726

Brullo S, Piccione V (1980) Esempi di cartografia della vegetazione di alcune aree della Sicilia. Carta della vegetazione di Linosa. Quaderni C.N.R. AQ/1/40:53–66

Brullo S, Ronsisvalle GA (1975) La vegetazione dei Gorghi Tondi e del Lago Preola, presso Mazara del Vallo (Sicilia occidentale). Not Fitosoc 10:45–67

Brullo S, Scelsi F (1998) Contributo alla conoscenza della vegetazione terofitica costiera del territorio ibleo. Boll Acc Gioenia Sci Nat Catania 29:151–168

Brullo S, Sciandrello S (2006) La vegetazione del bacino lacustre "Biviere di Gela" (Sicilia meridionale). Fitosociologia 43(2):21–40

Brullo S, Siracusa G (1996) Studio fitosociologico dell' isola di Linosa. Doc Phytosoc, n s 16:123–174

Brullo S, Siracusa G (2000) Indagine fitosociologica su di un'area umida del versante sud-occidentale dell'Etna di notevole interesse naturalistico. Arch Geobot 4(1):71–90

Brullo S, Spampinato G (1990) La vegetazione dei corsi d'acqua della Sicilia. Boll Acc Gioenia Sci Nat Catania 23(336):119–252

Brullo S, Spampinato G (1997) Indagine fitosociologica sulle ripisilve della Calabria (Italia meridionale). Lazaroa 18:105–151

Brullo S, Spampinato G (2004) Indagine sintassonomica sulla vegetazione a *Quercus calliprinos* Webb del Mediterraneo. Coll Phytosoc 28:539–559

Brullo S, Grillo M, Terrasi MC (1976) Ricerche fitosociologiche sui pascoli di Monte Lauro (Sicilia meridionale). Boll Acc Gioenia Sci Nat Catania 12(9–10):84–104

Brullo S, Di Martino A, Marcenò C (1977) La vegetazione di Pantelleria. Pubbl Ist Bot Univ Catania, Catania

Brullo S, Lanfranco E, Pavone P (1982) *Allium lojaconoi* sp.nov. e sue affinità con *Allium parciflorum* Viv. Webbia 35(2):295–306

Brullo S, Guglielmo A, Pavone P (1985) La classe *Pegano-Salsoletea* in Sicilia. Boll Acc Gioenia Sci Nat Catania 18:247–254

Brullo S, De Santis C, Furnari F, Longhitano N, Ronsisvalle GA (1988a) La vegetazione dell'Oasi della Foce del Simeto (Sicilia orientale). Braun-Blanquetia 2:165–188

Brullo S, Lanfranco E, Pavone P, Ronsisvalle GA (1988b) Taxonomical notes on the endemic flora of Malta. G Bot Ital 122:45

Brullo S, Lo Giudice R, Privitera M (1989) La classe *Adiantetea* in Sicilia. Arch Bot Ital 65(1/2):81–99

Brullo S, De Marco G, Signorello P (1990) Studio fitosociologico delle praterie a *Lygeum spartum* dell'Italia meridionale. Boll Acc Gioenia Sci Nat Catania 23(336):561–579

Brullo S, Privitera M, Puglisi M (1993) *Thamnobryo alopecuri-Phyllitidetum scolopendrium* nuova associazione centro-mediterranea della classe *Adianthetea*. Arch Bot Ital 68(1/2):35–43

Brullo S, Scelsi F, Siracusa G (1994) Contributo alla conoscenza della vegetazione terofitica della Sicilia occidentale. Boll Acc Gioenia Sci Nat Catania 27:341–365

Brullo S, Minissale P, Spampinato G (1995) Considerazioni fitogeografiche sulla flora della Sicilia. Ecol Medit 21(1/2):95–117

Brullo S, Minissale P, Spampinato G (1997) La classe *Cisto-Micromerietea* nel Mediterraneo centrale e orientale. Fitosociologia 32:207–238

Brullo S, Guarino R, Ronsisvalle G (2000) La vegetazione del litorale di Manfria, presso Gela (Sicilia), area soggetta a vincolo archeologico. Arch Geobot 4(1):101–107

Brullo S, Giusso del Galdo G, Lanfranco E (2001a) A new species of *Zannichellia* L. (Zannichelliaceae) from Malta. Fl Medit 11:379–384

Brullo S, Giusso del Galdo G, Siracusa G, Spampinato G (2001b) Considerazioni fitogeografiche sulla vegetazione psammofila dei litorali italiani. Biogeographia 12:93–137

Brullo S, Scelsi F, Spampinato G (2001c) La vegetazione dell'Aspromonte. Laruffa Editore, Reggio Calabria

Brullo S, Giusso del Galdo G, Guarino R (2002a) Phytosociological notes on the *Lygeum spartum* grassland from Crete. Lazaroa 23:65–72

Brullo S, Minissale P, Spampinato G, Giusso del Galdo G, Siracusa G (2002b) Considerazioni sintassonomiche e fitogeografiche sulla vegetazione della Sicilia. Boll Acc Gioenia Sci Nat Catania 361(35):325–359

Brullo S, Giusso del Galdo G, Minissale P, Spampinato G (2003) Considerazioni tassonomiche sui generi *Catapodium* Link, *Desmazeria* Dumort. e *Castellia* Tineo (Poaceae) in Italia. Inform Bot Ital 35(1):158–170

Brullo S, Guarino R, Minissale P, Scelsi F, Spampinato G (2004a) Indagine fitosociologica sulla vegetazione forestale dell'Egeo meridionale. Coll Phytosoc 28:401–466

Brullo S, Marcenò C, Siracusa G (2004b) La classe *Asplenietea trichomanis* in Sicilia. Coll Phytosoc 28:467–538

Brullo C, Brullo S, Giusso del Galdo G, Tomaselli V (2006) Contributo alla conoscenza delle praterie a *Brachypodium retusum* del Mediterraneo centro-orientale. Quad Bot Amb Appl 17(2):49–64

Brullo S, Giusso del Galdo G, Guarino R, Minissale P, Spampinato G (2007) A survey of the weedy communities of Sicily. Ann Bot (Roma), n s 7:127–161

Brullo C, Brullo S, Giusso del Galdo G (2009a) Sulla presenza in Sicilia di Romulea melitensis (Iridaceae), specie critica della flora mediterranea. In: Peccenini S, Domina G (eds) Gruppi critici per la flora italiana. Tipolitografia Luxograph, Palermo, pp 27–28

Brullo S, Gianguzzi L, La Mantia A, Siracusa G (2009b) La classe *Quercetea ilicis* in Sicilia. Boll Acc Gioenia Sci Nat Catania 41(369):1–77

Brullo S, Giusso del Galdo G, Guarino R, Minissale P, Siracusa G, Scuderi L, Spampinato G (2009c) The *Lygeo-Stipetea* class in Sicily. XVIII International Workshop European Vegetation Science, Rome, p 86

Brullo C, Brullo S, Giusso del Galdo G, Guarino R, Minissale P, Scuderi L, Siracusa G, Sciandrello S, Spampinato G (2010) The *Lygeo-Stipetea* class in Sicily. Ann Bot (Roma) 0:57–84

Brullo S, Giusso del Galdo G, Guarino R, Minissale P, Sciandrello S, Spampinato G (2012) Syntaxonomic survey of the class *Pegano harmalae-Salsoletea vermiculatae* Br.-Bl. & O. Bolòs 1958 in Italy. Plant Biosyst 147(2):472–492

Brullo C, Brullo S, Gaskin JF, Giusso del Galdo G, Hrusa GF, Salmeri C (2015a) A new species of *Kali* (Salsoloideae, Chenopodiaceae) from Sicily, supported by molecular analysis. Phytotaxa 201(4):256–277

Brullo C, Brullo S, Ilardi V, Giusso del Galdo G (2015b) *Kali dodecanesicum* (Chenopodiaceae, Salsoloideae) a new species from Greece. Phytotaxa 218(1):61–68

Brullo S, Brullo C, Cambria S, Giusso del Galdo G, Minissale P (2017a) Phytosociological investigation on the class *Crithmo maritimi-Limonietea* in Greece. Plant Sociol 54(1):3–57

Brullo S, Brullo C, Cambria S, Lanfranco E, Lanfranco S, Minissale P, Sciandrello S, Giusso del Galdo G (2017b) A new species of *Silene* sect. Dipterosperma (Caryophyllaceae) from Malta. Phytotaxa 297(3):245–256

Brullo S, Brullo C, Cambria S, Giusso del Galdo G, Salmeri C, Bacchetta G (2018) A new species of Ferula (Apiaceae) from Malta. Phytotaxa 382(1):74–88

Buchwald R (1994) Vegetazione e odontofauna negli ambienti acquatici dell'Italia Centrale. Braun-Blanquetia 11:3–77

Bueno A (1997) Flora y vegetaciòn de los estuarios Asturianos. Quad Medio Amb, Nat Oviedo 3:1–334

Buttigieg R, Lanfranco E (2001) New records for the Maltese flora: *Centaurea acaulis* L. (family Asteraceae). Cen Medit Nat 3(3):147–148

Candolle de AG (1838) Prodromus Systematis Naturalis Regni Vegetabilis, vol 6. Treuttel et Würtz, Parisiis

Caneva G, Dinelli A, De Marco G, Vinci M (1995) Le classi *Parietarietea diffusae* (Rivas-Martínez 1964) Oberd. 1977 e *Adiantetea* Br.-Bl. 1947 nelle aree archeologiche romane. Fitosociologia 29:165–179

Caniglia G, Chiesura Lorenzoni F, Curti L, Lorenzoni GG, Marchiori S, Razzara S, Tornadore Marchiori N (1984) Contributo allo studio fitosociologico del Salento meridionale (Puglia-Italia meridionale). Arch Bot Biogeograf Ital 60(1–2):1–40

Cano-Ortiz A, Biondi E, Gomes CJP, Del Río González S, Cano E (2014) Soil and phytosociological characterisation of grasslands in the western Mediterranean. Am J Plant Sci 5(21):3213–3240

Carroll FA, Hunt CO, Schembri PJ, Bonanno A (2012) Holocene climate change, vegetation history and human impact in the Central Mediterranean: evidence from the Maltese Islands. Quat Sci Rev 52:24–40

Casha A (2009) Contributions to Maltese flora. Cen Medit Nat 5(1):35–43

Casha A (2015) Flora of the Maltese Islands. Lulu. com, Malta

Casha A, Mifsud S (2013) *Althenia orientalis* (Tzvelev) Garcia-Mur. & Talavera (Zanichelliaceae Dum.): an addition to the Maltese Flora (Central Mediterranean). Webbia 68:57–61

Cassar C (2000) A concise history of Malta. Mireva Publications, Msida. 304 pp

Cassar J (2010) The use of limestone in a historic context – the experience of Malta. In: Smith BJ, Gomez-Heras M, Viles HA, Cassar J (eds) Limestone in the built environment: present-day challenges for the preservation of the past, vol 331. Geological Society, Special Publications, London, pp 13–25

Castroviejo S, Porta J (1976) Apport à l'écologie de la végétation des zones salées des rives de la Gigüela (Ciudad Real - Espagne). Coll Phytosoc 4:115–139

Cavallini FP (1689) Pugillus Militeus seu herbarum omnium in insula Melitaejusque districtis enascentium perbrevis enumeratio. J.B. Melo, Romae

Ceschin S, Cutini M, Caneva G (2003) La vegetazione ruderale dell'area archeologica del Palatino (Roma). Fitosociologia 40(1):73–96

Chetcuti D, Buhagiar A, Schembri PJ, Ventura F (1992) The climate of the Maltese Islands. A review. Malta University Press, Msida. 108 pp

Chronoupoulos G, Theocharopoulos M, Christodoulakis D (2005) Phytosociological study of *Hirschfeldia incana* (L.) Lagraze-Fossat (Cruciferae) communities in mainland Greece. Acta Bot Croat 64(1):75–114

Chytrý M, Schaminée JHJ, Schwabe A (2011) Vegetation survey: a new focus for applied vegetation science. Appl Veg Sci 14:435–439

Ciferri R, Giacomini V (1950) Nomenclator florae Italicae: seu, Plantae vasculares in Italia sponte nascentes, advenae, aut saepius cultae, vol 1.C. Ticini, Busca

Cirujano S (1980) Las lagunas manchegas y su vegetación. I Anal Jard Bot Madrid 37(l):155–192

Conard HS (1935) The plant association of Central Long Island. A study in descriptive plant sociology. Am Midl Nat 16(4):433–516

Corbetta F (1970) Lineamenti della vegetazione macrofitica dei Laghi di Lesina e di Varano. Nuov Giorn Bot Ital, n s 104(3):165–191

Corbetta F, Lorenzoni GG (1976) La vegetazione degli stagni del golfo di Oristano (Sardegna). Ricerc Biol Selvag 7(Suppl):271–319

Corbetta F, Pirone G (1989) La vegetazione del fiume Tirino (Abruzzo). Arch Bot Ital 65(3/4):121–153

Corillion R (1957) Les Charophycées de France e de l'Europe occidentale. Bull Soc Sci Bretagne 32:1–499

Costa M, Mansanet J (1981) Los ecosistemas dunares levantinos: La Dehesa de La Albufera. Anales Jard Bot Madrid 37(2):277–299

Costa JC, Capelo JH, Aguiar C, Neto C, Lousã M, Espírito-Santo MD (1997) An overview of the *Pegano harmalae-Salsoletea vermiculatae* Br.-Bl. & O. Bolòs 1958, vegetation class in continental Portugal. Colloq Phytosoc 27:81–93

Costa JC, Lousa M, Capelo J, Espirito Santo MD, Izco Sevillano J, Arsenio P (2000) The coastal vegetation of the Potuguese divisory sector: dunes cliffs and low-schub communities. Finisterra 35:69–93

Costa JC, Capelo J, Espirito Santo MD, Lousã M (2001) Corrección nomenclatural de los sintaxones basados en *Hyparrhenia hirta* del sector Divisorio portugués. Lazaroa 22:135–136
Costa JC, Neto C, Aguiar C, Capelo J, Espírito Santo MD, Honrado J, Pinto-Gomes C, Monteiro-Henriques T, Sequeira M, Lousã M (2012) Vascular plant communities in Portugal (Continental, the Azores and Madeira). Glob Geobot 2:1–180
Cristofolini G, Lausi D, Tarabocchia M, Pignatti S (1967) Flora e vegetazione dell'Isola Piana (Isole Tremiti). Giorn Bot Ital 101:189–198
Curcò I, Masip A (2000) La vegetaciò del delta del'Ebre (IV): les comunitats nitròfiles (Classes *Asplenitea trichomanis* I *Ruderali-Secaletea*). Acta Bot Barc 46:143–178
Cutajar D (1982) Interpretation of Maltese prehistory. Hyphen 3(4):176–185
Dakskobler I, Martinčič A, Rojšek D (2014) Phytosociological analysis of communities with *Adiantum capillus veneris* in the foothills of the Julian Alps (Western Slovenia). Hacquetia 13(2):235–258
de Bolós O (1956) De vegetazione notulae, II. Collect Bot 5(1):195–268
de Bolós O (1957) De vegetationes valentina. Collect Bot 5:527–596
de Bolós O (1962) El paisaje vegetal Barcelones. Universidad de Barcelona, Barcelona
de Bolós O (1967) Comunidades vegetales de las Comarcas proximas al litoral situatas entre los rios Llobregat y Segura. Mem Real Acad Sci Barcelona 38(1):3–281
de Bolós O (1975a) Contribution à l'étude du *Bromo-Oryziopsion miliaceae*. Phytocoenologia 2:141–145
de Bolós O (1975b) De vegetazione valentina II. Anal Inst Bot Cavanilles 32(2):477–488
de Bolós O (1996) La vegetació de les illes Balears. Comunitats de plantes. Arxius Secc Ci Inst Estud Catalans 114:1–267
de Bolós A, de Bolós O (1950) Vegetacion de las comarcas barcelonesas. Inst Esp Estud Medit Barcelona
de Bolós O, Masclans F (1955) La vegetaciòn de los arrozales en la región mediterránea. Collect Bot 4(3):1–32
de Bolós O, Molinier R (1958) Recherches phytosociologiques dans l'Ile de Majorque. Collect Bot 5(3):699–865
de Bolós O, Molinier R (1984) Vegetation of the Pityusic Islands. In: Kuhbier H, Alcover JA, Guerau d'Arellano C (eds) Biogeography and ecology of the Pityusic Islands. Monogr Biol 52:185–221
de Bolós O, Vigo J (1984) Flora vascular y vegetaciò de les Illes Medes. In: Ros J, Olivella I, Gill JM (eds) Els sistemes naturals de les Illes Medes. Arx Sec Cienc Inst Estud Catalans 73:131–208
de Bolós O, Molinier R, Montserrat P (1970) Observations phytosociologiques dans l'Ille de Minorque. Acta Bot Barcionon 5:1–150
de Foucault B (1993a) Données phytosociologiques. In: de Foucault B, Cornu J-P, Itinéraire botanique de Linnéens en Créte, 29 avril-13 mai 1991 (2 partie). Bull Soc N Pic 11:85–105
de Foucault B (1993b) Remarques sur la végétation du Maroc et de l'Algérie. Bull Soc Bot Centre-Ouest, n s 24:267–285
de Foucault B (2011) Contribution au prodrome des végétations de France: les *Filipendulo ulmariae-Convolvuletea sepium* Géhu & Géhu-Franck 1987. J Bot Soc Bot France 53:73–137
de Foucault B (2012) Contribution au prodrome des végétations de France: les *Sisymbrietea officinalis* Korneck 1974. J Bot Soc Bot France 58:55–116
de Foucault B (2013a) Contribution au prodrome des végétations de France: les *Isoëtetea velatae* de Foucault 1988 et les *Juncetea bufonii* de Foucault 1988 (« *Isoëto-Nanojuncetea bufonii*») (Partie 1). J Bot Soc Bot France 62:35–70
de Foucault B (2013b) Les journées de l'association tunisienne de physiologie et de biosurveillance de l'environnement. Compte rendu phytosociologique des herborisations (12-15 mai 2012). Rev Ecol Env 9:1–15
de Foucault B (2015) Contribution au prodrome des végétations de France: les *Pegano harmalae-Salsoletea vermiculatae* Braun Blanq. & O. Bolòs 1958. J Bot Soc Bot France 72:41–60

de Foucault B, Catteau E (2012) Contribution au prodrome des végétations de France: les *Agrostietea stoloniferae* Oberd. 1983. J Bot Soc Bot France 59:5–131

Deil U (1996) Zur Kenntnis der *Adiantetea*-Gesellschaften des Mittelmeerraumes und angrenzender Gebiete-mit allgemeinen Überlegung zur ökologischen Skalierung ihrer Standorte und zur Sättigung von Pflanzengesellschaften. Phytocoenologia 26(4):481–536

Deil U (1998) The class *Adiantetea* in the Mediterranean area-a state of knowledge report. Ann Bot 56(1):73–78

Deil U (2005) A review on habitats, plant traits and vegetation of ephemeral wetlands-A global perspective. Phytocoenologia 35:533–705

Del Guacchio E, Brullo S, Iamonico D (2018) (2586) Proposal to reject the name *Statice reticulata* (*Limonium reticulatum*) (Plantaginaceae). Taxon 67(1):213–214

Del Prete C, Mazzola P, Miceli P (1984) Anacamptis Urvilleana Sommíer et Car. Gatto (orchídaceae). Gior Bot ltal 118(suppl. 2):70

Delassus L, Magnanon S (2014) Classification physionomique et phytosociologique des végétations de Basse-Normandie, Bretagne et Pays de la Loire. Chaiers Scient Tecn 1:1–260

Den Hartog C (1976) Structure of seagrass communities and its impact on the phytosociological classification systems. Coll Phytosoc 4:249–256

Den Hartog C (2003) Phytosociological classification of seagrass communities. Phytocoenologia 33(2–3):203–229

Den Hartog C (2016) Seagrass communities: structure, distribution and classification. Phytocoenologia 46(4):397–414

Devillers P, Devillers-Terschuren J (1994) Essai d'analyse systématique du genre Ophrys. Natural Belges 75(Orchid.7, suppl.):303–304

Devillers P, Devillers-Terschuren J (2001) Application and development of the Palaearctic habitat classification in the course of the setting up of the Emerald Project-Malta-Convention on the conservation of European wildlive and natural habitats. Strasbourg

Devoto S, Biolchi S, Bruschi VM, Furlani S, Mantovani M, Piacentini D, Pasuto A, Soldati M (2012) Geomorphological map of the NW Coast of the Island of Malta (Mediterranean Sea). J Maps 8(1):33–40

Díaz Gonzalez TE, Fernández-Prieto JA (1994) La vegetación de Asturias. Itinera Geobot 8:243–520

Diaz Gonzalez TE, Penas Merino A, Luis Herrero Cembranos L, Carmen Perez Morales C, Llamas Garcia F, Terron Alfonso A (1988) Estudio de los herbazales nitrofilos vivaces y cominidades ruderal viarias de la Provincia de Leon (NW de Espana). Acta Bot Barc 37:113–131

Díaz-Gonzales TE (1975) La vegetación del litoral occidental asturiano. Rev Fac Ci Oviedo 15(16):369–545

Díez-Garretas B, Asensi A (1999) Syntaxonomical analysis of the *Andropogon* rich grasslands (*Hyparrhenietalia hirtae*) in the western mediterranean region. Fol Geobot 34:307–320

Díez-Garretas B, Asensi A, Gavilán R (2003) Sabulicolous therophytic plant communities in the Mediterranean Region: a proposal of phytosociological synthesis. Phytocoenologia 33(2–3):495–526

Ditè D, Peric R, Jun PE, Melečkova Z (2015) Distribution and community of *Suaeda pannonica* in Serbia. Bull Nat Hist Mus 7:101–117

Djamalì M, Gambin B, Marriner N, Andrieu-Ponel V, Gambin T, Gandouin E, Lanfranco S, Médail F, Pavon D, Ponel P, Morhange C (2013) Vegetation dynamics during the early to mid-Holocene transition in NW Malta, human impact versus climatic forcing. Veg Hist Archaeobotany 22:367–380

Dostal J (1976) New nomenclatural combinations and taxa of the compositae subtribe centaureinae in Europe. In: Heywood VH (ed) Flora Europaea. Notulae Systematicae ad Floram Europaeam spectantes. Bot J Linn Soc 71:191–210

Dumont D'Urville J (1822) Enumeratio Plantarumquas in insulis Archipelagi aut littoribus Ponti Euxini annis 1919 et 1820 collegit atque detexit. Mém Soc Linn Paris 1:1–135

Duthie JF (1872) Notes on the Flora of Malta and Gozo. J Bot 10:206–210

Duthie JF (1874–1875) On the botany of the Maltese Islands in 1874: Part One, Two. J Bot 12:321–326, 13:36–42
Duvigneaud J (1967) Flore et végétation halophiles de la Lorraine orientale. Mém Soc Roy Bot Belg 3:1–122
Eig K (2015) Malta a country shaped by limestone (and a bit of very old poo). Geoexpro 12(3):86–88
ERA (Environment & Resources Authority) (2017) Natura 2000 management planning for terrestrial sites in Malta & Gozo. Natura 2000 Management Plan, Malta. https://era.org.mt/en/Pages/Natura-2000-Management-Planning.aspx
Espírito-Santo MD, Costa JC, Jardim R, Segueira M (2003) A vegetação da Madeira VI: Comunidades nitrófilas dos campos agrícolas, dos pousios e das suas margens. Silva Lusitana 11(2):241–251
Farris E, Secchi Z, Filigheddu R (2007) Caratterizzazione fitosociologica dell'habitat prioritario 6220∗–"Percorsi substeppici di graminacee e piante annue dei *Thero-Brachypodietea*": caso di studio della Sardegna settentrionale. Fitosociologia 44(2) suppl. 1:271–278
Feldmann J (1937) Recherches sur la végétation marine de la Méditerranée: la côte des Albères. Rev Algol 10(1–4):1–339
Felix R (1973) Oligo-Miocene stratigraphy of Malta and Gozo, vol 73, issue 20. Medeel Landb-Geschool, Wageningen, pp 1–104
Felzines J-C (2012) Contribution au prodrome des végétations de France: les *Lemnetea minoris* Tüxen ex O. Bolòs & Masclans 1955. J Bot Soc Bot France 59:189–240
Fenech K (2009) Human-Induced Changes in the Environment and Landscape of the Maltese Islands from the Neolithic to the 15th Century AD as Inferred from a Scientific Study of Sediments from Marsa, Malta. BAR international series 1682. Archaeopress, Oxford
Fennane M (1986) Contribution a l'etude phytosociologique des tetraclinaies marocaines. Bull Inst Scient Rabat 10:57–78
Ferrez Y, Bailly G, Beaufils T, Collaud R, Caillet M, Fernez T, Gillet F, Guyonneau J, Hennequin C, Royer J-M, Schmitt A, Vergon-Trivaudey MJ, Vadam J-C, Vuillemenot M (2011) Synopsis des groupements végétaux de Franche-Comté. Besançon: Société botanique de Franche-Comté, Conservatoire botanique national de Franche-Comté, col. Les Nouvelles Archives de la Flore jurassienne et du nord-est de la France, n° spécial 1
Ferro G (1980) La vegetazione di Butera (Sicilia meridionale). Atti Ist Bot Lab Critt Univ Pavia 13:51–118
Ferro G (1990) Revisione della vegetazione segetale mediterranea ed europaea dell'ordine *Secaletalia*. Braun-Blanquetia 6:1–122
Ferro G (2004) Nuovi dati sulla flora e sulla vegetazione dei coltivi e degli incolti di Lipari (Isole Eolie). Quad Bot Amb Appl 15:21–39
Fici S, Gianguzzi L (2011) Aspetti vegetazionali e paesaggistici di Settefrati (Cefalù, Sicilia Settentrionale) biotopo meritevole di conservazione. Nat Sicil, s 4 35(2):233–255
Filigheddu R, Farris E, Biondi E (2000) The vegetation of S'Ena Arrubia lagoon (centre-western Sardinia). Fitosociologia 37(1):39–59
Flahaut C, Schröter C (1910) Rapport sur la nomenclature phytogéographique, vol 1. Third International Botanical Congress (IBC), Brussels, pp 131–164
Forskåll P (1775) Flora Aegyptiaco-Arabica, sive descriptiones plantarum quas per Aegyptum inferiorem et Arabiam felicem detexit, illustravit. Mölleri, Hauniae
Frei M (1937) Studi fitosociologici su alcune associazioni littorali in Sicilia. N Giorn Bot Ital 44:273–295
Galán de Mera A, Sánchez García I, Vicente Orellana JA (1997) Coastal plant communities of the South-Western Iberian Peninsula, Spain and Portugal. Phytocoenologia 27:313–352
Galdies C (2011) The climate of Malta: statistics, trends and analysis 1951–2010. National Statistics Office, Malta. 45 pp
Galdies C (2012) Temperature trends in Malta (central Mediterranean) from 1951 to 2010. Meteorog Atmos Phys 117:135–143

Galdies C, Saida A, Camilleri L, Caruanaa M (2016) Climate change trends in Malta and related beliefs, concerns and attitudes toward adaptation among Gozitan farmers. Eur J Agron 74:18–28

Gambi MC, Barbieri F, Bianchi CN (2009) New record of the alien seagrass *Halophila stipulacea* (Hydrocharitaceae) in the western Mediterranean: a further clue to changing Mediterranean Sea biogeography. Mar Biodivers Rec 2:1–7

Gambin B, Andrieu-Ponel V, Médail F, Marriner N, Peyron O, Montade V, Gambin T, Morhange C, Belkacem D, Djamali M (1998) 7300 years of vegetation history and climate for NW Malta: a Holocene perspective. Clim Past 12:273–297

Gaudillat V, Louvel-Glaser J (2015) Note méthodologique sur les correspondances entre les associations des *Pegano harmalae-Salsoletea vermiculatae* Braun-Blanq. & Bolos 1958 et les classifications d'habitats européennes. J Bot Soc Bot Fr 72:61–68

Gèhu J-M (1979) Pour une approce nouvelle de paysages végétaux: la symphytosociologie. Bull Soc Bot France Lettres Bot 126(2):213–223

Gèhu J-M (1991) L'analyse phytosociolgique et gèosymphytosocioloque de l'espace. Théorie et méthodogie. Coll Phytosoc 17:11–46

Géhu J-M, Biondi E (1986) Sur les formations méditerranéennes à *Artemisia arborescens*. VI Jornades de fitosociologia. Vegetaciò nitrofila, Barcelona, pp 77–78

Géhu J-M, Biondi E (1988) Données sur la végétation des ceintures d'atterissement des lacs Alimini (Salento Italie). Doc Phytosoc 11:353–380

Géhu J-M, Biondi E (1994) Végétation du littoral de la Corse: essai de synthèse phytosociologique. Braun-Blanquetia 13:1–149

Géhu J-M, Géhu-Franck J (1961) Recherche sur la végétation et le sol de la réserve de l'île des Landes et de quelques îlots de la côte nord Bretagne. Bull Lab Marit Dinard 47:19–57

Géhu J-M, Géhu-Franck J (1969) Les association végétales des dunes mobiles et des bordures de plages de la côte atlantique française. Vegetatio 18(1–6):122–166

Géhu JM, Rivas-Martinez S (1981) Notions fondamentales de phytosociologie. Ber. Int. Simp. Int. Ver Veg Syntax, Rinteln, pp 5–33

Géhu J-M, Costa M, Scoppola A, Biondi E, Marchiori S, Peris GB, Franck J, Caniglia G, Veri L (1984a) Essai synsystematique et syncorologique sur les végétations littorales italiennes dans un but conservatoire. 1. Dunes et vases salées. Doc Phytosoc, n s 8:393–474

Géhu J-M, Costa M, Biondi E, Peris JB, Arnold N (1984b) Données sur la végétation maritime des côtes meridionales de l'Ile de Chypre (Plages, Dunes, Lacs salés et falaises). Doc Phytosoc, n s 8:343–363

Géhu J-M, Biondi E, Géhu-Frank J, Arnold-Apostolides N (1986a) Donnés synsystématiques et synchorologiques sur la végétation du littoral sedimentaire de la Grèce continentale. Doc Phytosoc 10:43–91

Géhu J-M, Biondi E, Géhu-Franck J (1986b) Les végétations nitrohalophiles des falaises de Bonifacio (Corse). VI Jornades de fitosociologia. Vegetaciò nitròfila, Barcelona, pp 79–82

Géhu J-M, Biondi E, Géhu-Franck J (1988) Les végétations nitrohalophiles des falaises de Bonifacio (Corse). Acta Bot Barc 37:237–243

Giaccone G (1968) Contributo allo studio fitosociologico dei popolamenti algali del Mediterraneo orientale. Giorn Bot Ital 102:485–506

Giaccone G, Pignatti S (1967) Studi sulla produttività primaria del fitobenthos nel Golfo di Trieste. II. La vegetazione del Golfo di Trieste. Nova Thalassia 3:1–28

Giaccone G, Alongi G, Pizzuto F, Cossu A, Vito L (1994) La vegetazione marina bentonica fotofila del Mediterraneo: 2.: Infralitorale e circalitorale: proposte di aggiornamento. Boll Acc Gioenia Sci Nat Catania 27(346):111–157

Giacinto C (1811) Saggio di agricoltura per le isole di Malta se del Gozo. Giovanni del Nobolo, Malta e Messina

Gianguzzi L (1999) Vegetazione e Bioclimatologia dell'Isola di Pantelleria (Canale di Sicilia). Braun-Blanquetia 22:1–70

Giusso del Galdo G, Marcenò C, Musarella CM, Sciandrello S (2008) La vegetazione costiera della R.N.O. "Torre Salsa" (Siculiana - AG). Inform Bot Ital 40(1):73–89

Government of Malta (2012) National climate change adaptation policy. Government of Malta. http://mra.org.mt/climate-change/climate-change-introduction/
Grech Delicata JC (1853) Flora Melitensis, sistens stirpes phanerogames in Melitae insulisque adiacentibus hucusque delectas, secundum systemae Candolleanum digestas. Ex typis F. W. Franz, Melitae
Greuter W (2003) The Euro+Med treatment of Gnaphalieae and Inuleae (Compositae)-generic concepts and required new names. Willdenowia 33:239–244
Guarino R, Cusimano D, Ilardi V, Pasta S (2017) Syntaxonomic list of the vegetation units. In: Guarino R, Pasta S (eds) Botanical excursions in Central and Western Sicily, field guide for the 60th IAVS symposium Palermo, 20–24 June 2017. Officine Grafiche, Palermo, pp 281–370
Guyonneau J, Mady M, Fernez T (2008) Typologie, cartographie et évaluation de l'état de conservation des habitats naturels et semi-naturels et des populations d'espèces végétales remarquables du site ENS 25MA04 « Lac et marais de l'Entonnoir » (communes de Bouverans et de Bonnevaux, Doubs). Conservatoire Botanique National de Franche-Comté, Besancon
Harmsen GW (1936) Systematische Beobachtungen der Nord West-Europaeischen Seegrasformen. Nederl Kruidk Arch 46:852–877
Haslam SM (1969) Malta's plant life. Progress Press Co, Malta. 129 pp
Haslam SM, Borg J (1998) The river valleys of the Maltese Islands. The Islands and Small States Institute of the Foundation for International Studies, Malta and CIHEAM, Bari. 330 pp
Haslam SM, Sell PD, Wolseley PA (1977) A flora of the Maltese Islands. Malta University Press, Msida
Hećimović S (1984) Vegetation der Inseln Bobara und Mrkan. Acta Bot Croat 43(1):109–118
Henwood J (2014) A first record of *Chara vulgaris* var. *papillata* Wallroth. ex A. Braun. (Charales) in the Maltese Islands (Central Mediterranean). J Malta Chamb Scient. www.xjenza.org; http://dx.medra.org/10.7423/XJENZA.2014.1.07
Hertig E, Jacobeit J (2008) Downscaling future climate change: temperature scenarios for the Mediterranean area. Glob Planet Chang 63:127–131
Horvatić S (1934) Flora y vegetacija otoka Paga. Jugosl Akad Znan Umjetn Zagr 19:116–372
Horvatić S (1958) Tipolosko resclanjenje primoske vegetacije gariga y borovih suma. Acta Bot Croat 17:7–98
Horvatić M (1963) Vegetacjeska karta otoka Paga s opcim pregledom vegetacijeskih jedenica hrvatskog primorja. Acta Biol 4:7–187
House MR, Dunham KC, Wigglesworth JC (1961) Geology and structure of the Maltese Islands. In: Bowen-Jones H, Dewdney JC, Fisher WB (eds) Malta; background for development. Beaufont Ltd, London, pp 25–33
Hrivnák R, Oťaheľová H, Kochjarová J, Blanár D, Husák S (2005) Plant communities of the class *Charetea fragilis* Fukarek ex Krausch 1964 in Slovakia: new information on their distribution and ecology. Thaiszia – J Bot Košice 15:117–128
Hruska K (1987) Synatoxonomical study of Italian walls vegetation. Vegetatio 73:13–20
Humboldt de A, Bonpland A (1807) Essai sur la Géographie des Plantes. Schoelle, Paris, p 155
Hunt CO (1997) Quaternary deposits in the Maltese Islands: a microcosm of environmental change in Mediterranean lands. GeoJournal 41(2):101–109
Hunt CO (2000) Pollen. In: Bonanno A, Frendo AJ (eds) Excavations in Tas-Silg, vol 13. Mediterranean Archaeology, Malta, pp 111–114
Hunt CO, Schembri PJ (1999) Quaternary environments and biogeography of the Maltese Islands. In: Mifsud A, Savona Ventura C (eds) Facets of Maltese prehistory. Prehistoric Society of Malta, Valletta
Hunt C, Vella NC (2004–2005) A view from the countryside: pollen from a field at Mistra Valley, Malta. Malta Archaeol Rev 7:57–65
Hyde HPT (1955) Geology of the Maltese Islands. Lux Press, Malta. 135 pp
IUCN (2017) The IUCN red list of threatened species. Version 2017–3. http://www.iucnredlist.org
Izco J, Géhu J-M (1978) Un exemple de végétation macrohémithérophytique phytonitrophile: le *Smyrnietum olusatri* du littoral de la Manche occidentale. Coll Phytosoc 6:263–267

Jasprica N, Caric M, Batistic M (2003) The Marshland Vegetation (*Phragmito-Magnocaricetea*, *Isoeto-Nanojuncetea*) and Hydrology in the Hutovo Blato Natural Park (Neretva River Delta, Bosnia and Herzegovina). Phyton (Horn) 43(2):281–294

Jasprica N, Milović M, Romić M (2015) Phytosociology and ecology of *Cressa cretica* L. (Convolvulaceae) on the eastern Adriatic coast. Hacquetia 14(2):265–276

Jasprica N, Bogdanović S, Dolina K, Ruščić M, Pandža M, Kovačić S (2016) Syntaxonomy of Arundo stands along the eastern Adriatic coast. Plant Biosyst 150(5):887–903

Julve P (1993) Synopsis phytosociologique de la France (communautés de plantes vasculaires). Lejeunia, n s 140:1–160

Junikka L, Uotila P, Lahti T (2006) A phytogeographical comparison of the major Mediterranean islands on the basis of Atlas Florae Europaeae. Willdenowia 36(Special Issue):379–388

Kadereit G, Mavrodiev EV, Zacharias EH, Sukhorukov AP (2010) Molecular phylogeny of Atripliceae (Chenopodioideae, Chenopodiaceae): implications for systematics, biogeography, flower and fruit evolution, and the origin of C4 photosynthesis. Am J Bot 97(10):1664–1687

Kramer KU, Westra LYT, Kliphuis E, Gadella TWJ (1972) Floristic and cytotaxonomic notes on the flora of the Maltese Islands. Acta Bot Neerl 21(1):54–66

Krausch H-D (1964) Die Pflanzengesellschaften des Stechlinseegebietes. In: Die Gesellschaften des offenen Wassers. Limnologica 2:145–203

Krause W, Lang G (1977) Klasse: *Charetea fragilis* (Fukarek 1961 n.n.) Krausch 1964. Gesellschaften aus Armleuchteralgaen. In: Oberdorfer E (ed) Süddeutsche Pflanzengesellschaften. Teil I. G. Fischer Verlag, Stuttgart, pp 78–88

Kubiena WL (1953) The soil of Europe. Thomas Murby and Compagny, London. 318 pp

La Mantia T, Pasta S, Rühl J (2009) Quadro conoscitivo e proposte gestionali relative agli aspetti floristici, vegetazionali e agro-forestali. Piano di Gestione "Isole Pelagie" SIC ITA040002 "Isole di Lampedusa e Lampione" e ZPS ITA040013 "Arcipelago delle Pelagie. Area marina e terrestre"

Ladero M, Socorro O, Molero Mesa J, Lòpez Guadalupe M, Zafra L, Marìn G, Hurtado J, Raya FP (1981) Algunas consideraciones sobre las comunidades nitrófilas de Granada (España). Anal Jard Bot Madrid 37(2):737–763

Lalov SV, Seginkova A, Sultana J (2008) Rediscovery of *Asplenium sagittatum* and *A. marinum* (Pteridophyta: Aspleniaceae) in the Maltese Islands (Central Mediterranean). Cen Medit Nat 4(4):235–241

Landucci F, Gigante D, Venanzoni R, Chytry M (2013) Wetland vegetation of the class *Phragmito-Magno-Caricetea* in central Italy. Phytocoenologia 43(1–2):67–100

Lanfranco E (1969a) A revised check-list of Maltese algae. National Press, Pietà

Lanfranco GG (1969b) Guide to the flora of Malta with 300 illustrations. Progress Press Co, Malta

Lanfranco E (1970) The occurrence of *Halophila stipulacea* (Forskal) Ascherson (Fam. Hydrocaritaceae) in Malta water. Malt Nat 1(1):16–17

Lanfranco E (1971) Some new flowering plants from Malta. Malt Nat 1(2):3–7

Lanfranco E (1972) Additions and corrections to the Maltese flora. Malt Nat 1(3):14–20

Lanfranco E (1973) Additions and corrections to the Maltese flora. Malt Nat 1(4):8–10

Lanfranco E (1975) Some additions to the Maltese flora. Malt Nat 2(2):47–48

Lanfranco E (1984) Guida alla escursioni a Malta-Aprile 1984. Società Botanica Italiana sezione Siciliana, Malta

Lanfranco E (1989) The flora. In: Schembri PJ, Sultana J (eds) Red data book for the Maltese Islands. Environment Division, Malta, pp 5–70

Lanfranco E (1995) The vegetation of the Maltese Islands. In: Giusti F, Manganelli G, Schembri PG (eds) The non-marine molluscs of the Maltese Islands, vol 15. Museo Regionale di Scienze Naturali, Monografie, pp 27–29

Lanfranco E (2002) A contribution to the freshwater macroalgal flora of the Maltes Islands. Cen Medit Nat 3(4):203–206

Lanfranco E, Bonnett G (2015) Wild flowers of the Maltese Islands nature guide series. Print Iti Paola, Malta

Lanfranco E, Schembri PJ (1986) Maltese wetlands and wetland biota. Potamon 15:122–125

Lanfranco E, Spiteri HJ (2001) New records for the Maltese flora: pancratium foetidum Pomel (Family Amaryllidaceae). Cen Medit Nat 3(3):147

Lanfranco S, Lanfranco E, Westermeier R, Zammit MA, Mifsud MA (2013). The vascular flora of the Maltese islands. In: Cardona Pons E, Estaun Clariso I, Comas Casademont M, Fraga i Arguimbau P (eds) Islands and plants: preservation and understanding of flora on Mediterranean Islands. Menorca, pp 149–156

Lang DM (1960) Soils of Malta and Gozo. Colon Res Stud 29:1–112

Lang D (1961) Soils of Malta and Gozo. In: Bowen-Jones H, Dewdney JC, Fisher WB (eds) Malta; background for development. Beaufont Ltd, London, pp 83–98

Lastrucci L, Paci F, Raffaelli M (2010) The wetland vegetation of the natural reserves and neighbouring stretches of the Arno river in the Arezzo province (Tuscany, Central Italy). Fitosociologia 47(1):29–59

Licandro G, Marino P, Raimondo FM (2011) Flora e vegetazione della Riserva Naturale Orientata "Laghetti di Marinello" (Sicilia Nord-Orientale). Inform Bot Ital 43(2):333–351

Linnaeus C (1753) Species plantarum. Laurentii Salvii, Holmiae

Loidi J (2017) The vegetation of the Iberian Peninsula, vol 1–2. Springer, Cham

Loidi J, Herrera M, Biurrun I, Garcia-Mijangos I (1999) Relationship between sintaxonomy of *Thero-Salicornietea* and taxonomy of the genera *Salicornia* and *Suaeda* in the Iberian Peninsula. Fol Geobot 34:97–114

Lopez G (1978) Contribuciòn al conocimiento fitosociològico de la Serrania de Cuenca II. Anal Inst Bot Cavanilles 34:597–702

Lososová Z (2009) XBF Spergulo arvensis-Erodion cicutariae J. Tüxen in Passarge 1964. In: Chytrý M (ed) Vegetace České republiky, vol 2. Ruderální, plevelová, skalní a suťová vegetace. Academia, Praha, pp 126–127

Lumbreras A, Navarro G, Pardo C, Molina JA (2011) Aquatic Ranunculus communities in the northern hemisphere: A global review. Plant Biosyst 145:118–122

Luqa Weather average (2015) Luqa, Malta. https://www.holiday-weather.com/luqa/averages/

Magri O (2006) A geological and geomorphological review of the Maltese islands with special reference to the coastal zone. Territoris Universitat de les Illes Balears 6:7–26

Maiorca G, Spampinato G, Crisafulli A, Cameriere P (2007) Flora vascolare e vegetazione della Riserva Naturale Regionale "Foce del Fiume Crati" (Calabria, Italia meridionale). Webbia 62(2):121–174

Maire R (1924) Études sur la végétation et la flore du Grand Atlas et du Moyen Atlas marocains. Missions du Ministère de l'Instruction Publique en 1921. Mém Soc Sci Nat Maroc 7:1–220

Malta Weather (2013) Bringing you accurate Maltese weather forecasts. http://www.maltaweather.com/

Marcenò C, Colombo P (1982) Su alcuni esempi di vegetazione ad *Erica multiflora* L. (*Erico-Polygaletum preslii* dei *Cisto-Ericetalia*) sui monti di Palermo (Sicilia). Rev Biol Ecol Medit 9:85–94

Marchioni Ortu A (1982) Indagine tassonomica ed ecologica in *Ruppia drepanensis* Tineo (Ruppiaceae). Atti Soc Tosc Sci Nat Mem, s B 89:153–163

Marriner N, Gambin T, Djamali M, Morhange C, Spiteri M (2012) Geoarchaeology of the Burmarrad ria and early Holocene human impacts in western Malta. Palaeogeogr Palaeoclimatol Palaeoecol 339–341:52–65

Martinez Parras JM, Peinado Lorca M (1993) Vegetacion de los arrozales de las marismas de Guadalquivir. Lagascalia 17(1):21–35

Maugeri G (1975) Una nuova associazione dell'Echio-Galactition nei prati di Sulla della piana di Catania. Arch Bot Biogeogr Ital 20(3):83–113

Maugeri G (1980) La vegetazione infestante gli agrumeti dell'Etna. Not Fitosoc 15:45–54

Maugeri G, Ronsisvalle GA, Leonardi S, Cafarella N (1979) La presenza dell'*Amarantho-Cyperetum rotundi* nella colture di alcune isole del distretto siculo (Malta, Favignana, Linosa). Boll Acc Gioenia Sci Nat Catania, s 4(13):127–136

Mehrvarz SS, Naqinezhad A, Ravanbakhsh M, Maghsoudi M (2016) The survey of macrophytes diversity in wetland zone of Boujagh National Park, Guilan. Iran Biodiv J 7(3):353–358

Meinesz A, Boudouresque CF, Falconeiti C, Astrier JM, Bay D, Blanc JJ, Bourcier M, Cinelu F, Cirik S, Cristiani G, Di Geronimo I, Giaccone G, Liarmelin JG, Laubier L, Lovric AZ, Mounier R, Sover J, Vamvakas C (1983) Normalisation des symboles pour la représentation et la cartographie des biocénoses benthiques littorales de Méditerranée. Ann Inst Océanogr 59(2):155–172

Melendo M, Cano E, Valle F (2003) Synopsis of aquatic plant-communities of the class *Potametea* in the southern Iberian Peninsula. Acta Bot Gall 150(4):429–444

MEPA (Malta Environment & Planning Authority) (2010) The environment report 2008. MEPA, Floriana, Malta. https://era.org.mt/en/Pages/The-Environment-Report-2008.aspx

Mereu L, Lastrucci L, Viciani D (2010) Contributo alla conoscenza della vegetazione del fiume Pesa (Toscana, Italia centrale). Stud Bot 29:105–143

Merxmüller H (1968) Melitella (Cichoriaceae)–über ein Vorkommen in Australien und die taxonomische Einreihung. Mitt Bot Staatssamml München 7:271–275

Mifsud S (2007) Updates in the flora of the Maltese Islands (Central Mediterranean). Centr Medit Natur 4:171–179

Mifsud S (2008) Four new records for the flora of the Maltese Islands-Dactyloctenium aegyptium (L.) P. Beauv. (Poaceae), Amaranthus muricatus (Gillies ex Moq.) Hieron (Amaranthaceae), Fumaria reuteri Boiss. (Fumariaceae) and Spirodela oligorrhiza (Kurtz) Hegelm. (Lemnaceae). Cen Medit Nat 4(4):263–273

Mifsud S (2009) Rediscovery and present distribution of *Valantia hispida* L. (fam. Rubiaceae) in the Maltese Islands. Cen Medit Nat 5(1):17–23

Mifsud S (2010) First occurrence of *Lemna minuta* in the Maltese Islands. Cen Medit Nat 5(2):1–4

Mifsud S (2011) A floristic survey on the gozitan islets of Tac-Cawl and Tal-Halfa in the Maltese slands. MaltaWildPlants.com Online Publications; Ref: MWPOP-001. URL http://www.MaltaWildPlants.com/publ/index.php#W01

Mifsud S (2012) *Fumana laevis* (Cav.) Pau (Cistaceae): new to the flora of Malta, and considerations on *Fumana* (Dunal) Spach from Maltese Flora. Webbia 67:101–107

Mifsud S (2013) Distribution of some rare or endemic chasmophytic and rupestral species growing along the coastal cliffs of the Maltese Islands. Webbia 68(1):35–50

Mifsud S (2015) A review of *Romulea* Maratti (Iridaceae) in the Maltese Islands. Webbia 70(2):247–287

Mifsud S, Stephenson R, Thiede J (2015) *Sedum album* subsp. *rupi-melitense* (Crassulaceae), a new vegetatively reproducing subspecies from Malta (Maltese Islands, Central Mediterranean). Phytotaxa 227(2):135–146

Mifsud S, Lanfranco E, Fiorentino J, Mifsud SD (2016a) An updated flora of Selmunett (St. Paul's Island) including Mosses and Lichens. Xjenza Online 4:142–159

Mifsud S, Napier M, Fenech S, Cassar LF (2016b) Current status of *Asplenium sagittatum* (Aspleniaceae) in the Maltese Islands. Fl Medit 26:69–80

Mile O, Walter J (2003) A *Suaeda* Foskal ex Scop. (Chenopodiaceae) nemzetsèg Magyaroszagon. Fl Pann 1(1):29–43

Minissale P, Sciandrello S (2013) A relic wood of *Juniperus turbinata* Guss. (Cupressaceae) in Sicily: structural and ecological features, conservation perspectives. Plant Biosyst 147(1):145–157

Minissale P, Sciandrello S (2015) The sabulicolous therophytic associations in Sicily: new insights through the statistical approach, stressing the continuum versus discrete model of plant communities. Acta Bot Gall Bot Lett 162(1):55–78

Minissale P, Sciandrello S (2016) Ecological features affect patterns of plant communities in Mediterranean temporary rock pools. Plant Biosyst 150(1):171–179

Minissale P, Sciandrello S (2017) Flora and habitats of Vendicari Islet (Isola di Vendicari) in South East Sicily. Nat Croat 26(1):1–16

Minissale P, Spampinato G (1990) Osservazioni fitosociologiche sul "Pantano Gurna" presso Mascali (Sicilia orientale). Boll Acc Gioenia Sci Nat 23(336):317–336

Mitchell PK, Dewdney JC (1961) The Maltese climate and weather. In: Bowen-Jones H, Dewdney JC, Fisher WB (eds) Malta; background for development. Beaufont Ltd, London, pp 48–82

Molinier R (1960) Observations sur les Phanérogames marines méditerranéennes. Rapp PV Réun Commiss internation Explor Sci Médit 15(2):165–170

Molinier R, Tallon G (1970) Prodrome des unités phytosociologiques observées en Camargue. Bull Mus Hist Nat Marseille 30:5–110

Molnar AV, Popiela A, Andràs Lukàcs B (2014) *Elatine gussonei* (Sommier) Brullo et al. (Elatinaceae) in Sicily. Plant Biosyst 148:27–30

Mucina L et al (2016) Vegetation of Europe: hierarchical floristic classification system of vascular plant, bryophyte, lichen, and algal communities. Appl Veg Sci 19(Suppl. 1):3–264

Murray J (1890) The Maltese Islands, with special reference to their geological structure. Scott Geogr Mag 6(9):449–488

Nicolson M (2013) Community concepts in plant ecology: from Humboldtian plant geography to the superorganism and beyond. Web Ecol 13(95–102):2013

Ninot JM, Font X, Masalles RM, Vigo J (2010) Syntaxonomic conspectus of the vegetation of Catalonia and Andorra. II. Ruderal communities. Acta Bot Barcin 53:113–189

Nyman CF (1844) Observationes in floram Siculam quas itinere anno 1844 adnotavit. Linnaea 18:625–663

Oberdorfer E (1954) Nordaegaeische kraut und zwergstrauchfluren im vergleich mit den entsprechenden vegetationseinheiten des westlichen mittelmeergebietes. Vegetatio 5–6:88–96

Oberdorfer E (1977) Süddeutsche Pflanzengesellschafte. Gustav Fischer, Stuttgart, p 1

Ortiz S (1993) Cakile. In: Castroviejo S, Aedo C, Gomez Campo C, Lainz M, Montserrat P, Morales R, Muñoz Garmendia F, Nieto Feliner G, Rico E, Talavera S, Villar L (eds) Flora Iberica, vol 4. Real Jardín Botánico, C.S.I.C, Madrid, pp 423–426

Pandža M, Franjić J, Škvorc Ž (2005) Weed and ruderal vegetation (Stellarietea mediae R. Tx. et al. ex von Rochow 1951) in central part of the East Adriatic Coast. Period Biol 107(3):361–372

Parlatore F (1869) Flora Italiana, 4 (2). Le Monnier, Firenze

Pasta S, Perez-Graber A, Fazan L, de Montmollin B (eds) (2017) The top 50 Mediterranean Island Plants UPDATE 2017. IUCN/SSC/Mediterranean Plant Specialist Group, Neuchâtel. E-book and online. 141 pp. top50.iucn-mpsg.org

Pedley HM (1993) Geological map of the Islands of Malta, 1:25,000 scale. British Geological Survey, Keyworth

Pedley HM, House MR, Waugh B (1976) The geology of Malta and Gozo. Proc Geol Assoc 87(3):325–341

Pedley HM, House MR, Waugh B (1978) The geology of the Pelagian block: the Maltese Islands. In: Nairn AEM, Kanes WH, Stehli FG (eds) The ocean basins and margins: the Western Mediterranean, vol 4B. Plenum Press, New York, pp 417–433

Pedrol J, Castroviejo S (1990) Suaeda. In: Castroviejo S, Lainz M, Lopez Gonzalez G, Montserrat P, Muñoz Garmendia F, Paiva J, Villar L (eds) Flora Iberica, vol 2. Real Jardín Botánico, C.S.I.C, Madrid, pp 536–541

Pedrotti F (2008) La vegetazione delle marcite di Norcia (Italia centrale). Braun-Blanquetia 44:1–31

Pedrotti F (2013) Plant and vegetation mapping. Geobotany series. Spinger Verlag, Berlin

Peinado M, Martinez Parras JM, Bortolomé C, Alcaraz F (1989) Sintesis sintaxonomica de la clase *Pegano-Salsoletea* en España. Doc Phytosoc, n s 11:283–301

Perez Latorre AV, Caballero G, Casimiro-Soriguer Solanas F, Gavira O, Cabezudo B (2008) Vegetaciòn del sector Malacitano-Axarquiense (Comarca de la Axaquia, Montes de Malaga y Corredor de Colmenar). Malaga (España). Acta Bot Malac 33:215–270

Peroni A, Peroni G, Mifsud S (2013) *Polypodium vulgare* subsp. *melitense* new subspecies from Gozo, Maltese Islands (Pteridophyta: *Polypodiaceae*). Bot Complut 37:41–46

Pignatti S (1953) Introduzione allo studio fitosociologico della pianura veneta orientale con particolare riguardo alla vegetazione litoranea. Continuazione. Arch Bot 29(2):65–98

Pignatti S (1966) La vegetazione alofila della Laguna Veneta. Mem Ist Ven Sc Lett Arti 33:1–174
Pignatti E, Pignatti S, Nimis P, Avanzini A (1980) La vegetazione ad arbusti spinosi emisferici: contributo alla interpretazione delle fasce di vegetazione delle alte montagne dell'Italia mediterranea. Quad CNR ser. AQ/1/79, Roma, pp 1–130
Pirone G (1995) La vegetazione alofila della costa abruzzese (Adriatico centrale). Fitosociologia 30:233–256
Pirone G, Ciaschetti G, Frattaroli AR, Corbetta F (2003) La vegetazione della Riserva Naturale Regionale «Lago di Serranella» (Abruzzo-Italia). Fitosociologia 40(2):55–71
Pisanu S, Farris E, Caria MC, Filigheddu R, Urbani M, Bagella S (2014) Vegetation and plant landscape of Asinara National Park (Italy). Plant Sociol 51(1):31–57
Pott R (2011) Phytosociology: a modern geobotanical method. Plant Biosyst 145:9–18
Pottier AG (1954) L'île de Zembra: excursion phytosociologique. Mém Soc Sci Nat Tunisie 2:35–44
Prampolinia M, Foglini F, Biolchi S, Devoto S, Angelini S, Soldati M (2017) Geomorphological mapping of terrestrial and marine areas, northern Malta and Comino (Central Mediterranean Sea). J Maps 13(2):457–469
Raffaelli M, Ricceri C (1988) Su alcune specie mediterranee del genere *Euphorbia* L.: *E. bivonae* Steud., *E. papillaris* (Boiss.) Raffaelli et Ricceri, stat. nov., *E. melitensis* Parl. Webbia 42(1):1–13
Renfrew C (1972) Malta and the calibrated radiocarbon chronology. Antiquity 46(182):150–154
Rivas Goday S (1964) Vegetacion y florula de la cuenca extremeña del Guadiana (vegetacíon y florula de la provincia dc Badajoz). Publ Exem Diput Prov Badajoz, Madrid
Rivas-Martinez S (1960) Roca, clima y comunidades rupícolas. Anal Real Acad Farm 2:153–168
Rivas-Martinez S (1969) Vegetatio Hispaniae. Notula I. Publ Inst Biol Apl 46:5–34
Rivas-Martinez S (1975) La Vegetación de la clase *Quercetea ilicis* en España y Portugal. Anal Inst Bot Cavanilles 31(2):205–259
Rivas-Martınez S (1976) Sinfitosociologıa, una nueva metodologıa para el estudio del paisaje vegetal. Anal Inst Bot Cavanilles 33:179–188
Rivas-Martinez S (1978a) Sur la syntaxonomie des pelouses thérophytiques de l'Europe occidentale. Coll Phytosoc 6:55–69
Rivas-Martinez S (1978b) Sobre la vegetacion nitrofila del *Chenopodion muralis*. Acta Bot Malac 4:71–78
Rivas-Martinez S (1978c) La vegetacion del *Hordeion leporini* en Espana. Doc Phytosoc, n s 2:377–392
Rivas-Martinez S (1983) Series de vegetaciòn de la region eurosiberiana de la Peninsula iberica. Lazaroa 4:155–166
Rivas-Martinez S (1985) Biogeografia y vegetaciò. Discurso de ingreso come Académico de Numero. Real Academia de Ciencias Exacta, Fisicas y Natuales, Madrid
Rivas-Martinez S (2005) Notions on dynamic-catenal phytosociology as a basis of landscape science. Plant Biosyst 139(2):135–144
Rivas-Martínez S, coauthores (2011) Mapa de series, geoseries y permaseries de vegetaciòn de España. Itin Geobot 18:5–800
Rivas-Martínez S, Costa M (1976) Datos sobre la vegetación halófila de la Mancha (España). Coll Phytosoc 4:81–97
Rivas-Martinez S, Costa M (1984) Sinopsis sintaxonomica de la clase *Arthrocnemetea* Br.-Bl. & R. Tx. 1943 en la Peninsula Iberica. Doc Phytosoc 8:15–26
Rivas-Martinez S, Izco J (1977) Sobre la vegetacion terofitica subnitrofila mediterranea (Brometalia rubentitectori). Anal Inst Bot Cavanilles 34:355–381
Rivas-Martínez S, Costa M, Castroviejo S, Valdés E (1980) La vegetación de Doñana (Huelva, España). Lazaroa 2:5–190
Rivas-Martínez S, Wildpret W, del Arco M, Rodriguez O, Pérez de Paz PL, Garcıà Gallo A, Acebes JR, Dıaz TE, Fernandez-Gonzalez F (1993) Las comunidades vegetales de la Isla de Tenerife (Islas Canarias). Itin Geobot 7:169–374

Rivas-Martínez S, Fernández González F, Loidi J (1999) Checklist of plant communities of Iberian Peninsula, Balearic and Canary Islands to suballiance level. Itin Geobot 13:353–451

Rivas-Martínez S, Fernández-González F, Loidi J, Lousã M, Penas A (2001) Syntaxonomical checklist of vascular plant communities of Spain and Portugal to association level. Itin Geobot 14:5–341

Rivas-Martínez S, Diaz TE, Fernández-González F, Izco J, Loidi J, Lousã M, Penas A (2002) Vascular plant communities of Spain and Portugal. Addenda to the syntaxonomical checklist of 2001. Itin Geobot 15(1):5–432

Rivas-Martínez S, Penas A, Díaz TE (2004) Bioclimatic & biogeographic maps of Europe. University of León, E-24071, Spain. http://www.globalbioclimatics.org/form/maps.htm

Rivas-Martínez S, Rivas-Sáenz S, Penas Merino A (2011) Worldwide bioclimatic classification system. Glob Geobotany 1:1–634. + 4 Maps

Robson NKB (1993) Studies in *Hypericum*: validation of new names. Bull Nat Hist Mus Lond Bot 23:67–70

Rodwell JS (2000) British plant communities: volume 5. Maritime communities and vegetation of open habitats. Cambridge University Press, Cambridge

Rotevan A, Thorshein E, Bastesen E, Fossmark HSS, Torabi A, Saelen G (2016) Sequential growth of deformation band in carbonate grainstones in the hangingwall of an active fault: implications for deformation mechanisms in different tectonic regimes. J Struct Geol 90:27–47

Ruiz-Téllez T (1991) Contribución al conocimiento de las comunidades de "*Oryzopsi-Anthirrinetum granitici*" corr. Rivas-Martínez 1969, del Centro y Sur del Portugal. Acta Bot Malac 16(2):391–403

Salkowski HE (1992) Ophrys sphegodes Mill. subsp. melitensis Salkowski, subsp. nova auf Malta und Gozo. Mitt Bl Arbeitskr Heim Orch Baden-Württ 24(4):633–643

Sammut S (2002–2004) MALSIS, a soil information system for the Maltese Islands (LIFE00 TCY/M/036). Malta

Savić NM (1926) Luga Kol'skogo poluostrova. Izvest Geograf Instit Leningrad 6:56–72

Savona-Ventura C (2001) The Maltese Islands, the natural environmental. Malta. http://www.shadowservices.com/nature/Maltese/index.htm

Sburlino G, Tomasella M, Oriolo G, Poldini L (2004) La vegetazione acquatica e palustre dell'Italia nord-orientale. 1- La classe *Lemnetea* R.Tx. ex O.Bolòs & Masclans 1955. Fitosociologia 41(1):27–42

Schembri PJ (1993) Physical geography and ecology of the Maltese Islands: a brief overview. In: Busuttil S, Lerin F, Mizzi L (eds) Malta: food, agriculture, fisheries and the environment. Options Méditerranéennes: Série B. Etudes et Recherches, vol 7. CIHEAM, Montpellier, pp 27–39

Schembri PJ (1994) Malta's natural heritage. In: Frendo H, Friggieri O (eds) Malta culture and identity. Ministry of Youth and the Arts, Malta, pp 105–124

Schembri PJ (1995) Treatned habitats as a critereion for selecting coastal protected areas in the Maltese Islands. Rapport du Congrés de la Commision International pour l'Exploration Scientifique de la Mer Méditerranée 34:128

Schembri PJ (1997) The Maltese islands: climate, vegetation and landscape. GeoJournal 41(2):115–125

Sciandrello S (2007) La vegetazione alofila di Piana del Signore presso Gela (Sicilia meridionale): proposte di conservazione e gestione del biotopo. Inform Botan Ital 39(1):129–141

Sciberras J, Sciberras A (2010) A contribution to the knownledge of alien flora in the Maltese Islands. Cen Medit Nat 5(2):44–48

Sciberras J, Sciberras A, Pisani L (2012) Updated checklist of flora of the satellite islets surrounding the Maltese Archipelago. Biodivers J 3(4):385–396

Scoppola A (1982) Considérations nouvelles sur les végétations des *Lemnetea minoris* (R.Tx. 1955) em. A. Schwabe et R.Tx. 1981 et contribution à l'etude de cette classe en Italie centrale. Doc Phytosoc, n s 6:1–130

Scossiroli RE, Ronsisvalle GA (1991) Origine e dispersione dei popolamenti animali e vegetali nel Mediterraneo centro-occidentale. Nat Mont 38(1–2):45–52

Segal S (1969) Ecological notes on wall vegetation. Junk, The Hague

Sghaier YR, Zakhama-Sraieb R, Benamer I, Charfi-Cheikhrouha F (2011) Occurrence of the seagrass Halophila stipulacea (Hydocharitaceae) in the southern Mediterranean Sea. Bot Mar 54(6):575–582

Shackleton JC, van Andel HT, Runnels CN (1984) Coastal paleogeography of the Central and Western Mediterranean during the Last 125,000 years and its archaeological implications. J Field Archaeol 11(3):307–314

Silverwood JH (1971) Maltese flora-recent additions and notes. Maltese Nat 1(2):8–10

Sivarajasingham S (1971) The soils of Malta. [UNOP/SF Project MAT/5, Water disposal and water supply]. Food and Agriculture Organization of the United Nations, Rome. 82 pp

Sommier S (1907) Un gioiello della flora Maltese. Nuovo genere e nuova specie di Composite. Nuovo Giorn Bot Ital (ns) 14:496–505

Sommier S, Caruana Gatto A (1915) Flora Melitensis nova. Stabilimento Pellas, Firenze

Soò von Bere R (1927) Geobotanische Monographie von Kolozsvàr (Klausenburg). I. Veröff. Tisza Ges 15/16

Soò von Bere R (1933) Über Probleme, Richtungen und Literatus der modernen Geobotanik. Die Pflanzensoziologie in Ungarn. Mag Bot Kut Int 1(S):1–51

Spratt TAB (1843) On the geology of the Maltese Islands. Proc Geol Soc Lond 4(2):225–230

Spratt TAB (1852) On the geology of Malta and Gozo with maps. Malta Mail Office, Malta. 24 pp

Spratt TAB (1854) Geology of Malta and Gozo, 2nd edn. Bradbury and Evans Printers, Whitefriars. 16 pp

Stancić Z, Brigić A, Liber Z, Rusak G, Franjić J, Skvorc Z (2008) Adriatic coastal plant taxa and communities of Croatia and their threat status. Acta Bot Gall 155(2):179–199

Stevens DT (2003) Orchids with special reference to orchids in the Maltese Islands. Archipelago 1(3–4):94–95, 99–104

Stevens DT, Lanfranco E (1995) Flowering plants in Malta: 136 species extinct and 291 seriously threatened. Il-Balluta 31:1–10

Šumberová K (2011) MCA04 *Phragmitetum australis* Savič 1926. In: Chytrý M (ed) Vegetace České republiky. 3: 405–409. Vodní a mokřadní vegetace [Vegetation of the Czech Republic 3. Aquatic and wetland vegetation]. Academia, Praha

Susanna A, Garnatje T, Garcia-Jacas N (1999) Molecular phylogeny of *Cheirolophus* (Asteraceae: Cardueae-Centaureinae) based on ITS sequences of nuclear ribosomal DNA. Pl Syst Evol 214:147–160

Tabone TJ (2007) A list of records of some rare vascular plant species occuring in the Maltese Islands. (Central Mediterranean). Cen Medit Nat 4(3):181–189

Tabone TJ (2008) A list of records of some rare vascular plant species occuring in the Maltese Islands. (Central Mediterranean). Part II. Cen Medit Nat 4(4):311–337

Tansley AG (1965) The British Islands and their vegetation. Cambridge Univ Press, Cambridge

Theurillat J-P, Moravec J (1996) Index of new names of syntaxa published in 1993. Fol Geobot Phytotax 31:473–516

Thornthwaite CW, Mather JR (1955) The water balance. Drexel Inst Techn Pubbl Climate 8(1):1–104

Tomaselli V, Sciandrello S (2016) Contribution to the knowledge of the coastal vegetation of the SIC IT9110005 "Zone umide della Capitanata" (Apulia, Italy). Plant Biosyst (in press)

Tomaselli V, Urbano M, Sciandrello S, Wagensommer RP, Costanzo E, Albano A, Medagli P, Mele C, Di Pietro R (2010) Cartografia tematica ed analisi del paesaggio vegetale ed agricolo del Parco Naturale Regionale "Saline di Punta della Contessa" (Brindisi–Puglia). Quad Bot Amb Appl 21:53–76

Tomaselli V, Di Pietro R, Sciandrello S (2011) Plant communities structure and composition in three coastal wetlands in southern Apulia (Italy). Biologia 66:1027–1043

Trechmann CT (1938) Quaternary conditions in Malta. Geol Mag 75:1–26

Trinajstić I (1974) O zoni sveze *Oleo-Ceratonion* u istocnojadranskom dijelu Balkanskog poluotoka. Ekologija 8(2):283–294

Trinajstić I (1975) Novi prilog poznavanju rasprostranjenosti asocijacie *Oleo-Euphorbietum dendroidis* (*Oleo-Ceratonion*) u Jadranskom primorju. Jugoslavije. Acta Bot Croat 34:121–125

Trinajstić I, Sugar I (1977) Contribution a la connaisance de la vegetacion de l'alliance *Oleo-Ceratonion* de presqu'ile de Salerno au Sud de Naples (Italie). Acta Bot Croat 36:135–141

Tüxen R (1974) Die Pflanzengesellschaften Nordwestdeutschlands. 2(1). Cramer, Vaduz

Tüxen R (1977) Zur homogenita¨t von sigmassoziationen, ihner syntaxonomischen ordnung und ihrer verwendung in der vegetationskartierung. Doc Phytosoc, n s 1:321–328

Tüxen R (1979) Sigmeten und geosigmeten, ihre ordnung und ihre bedeutug fu¨ r wissenschaft, naturschutz und planung. Biogeographie 16:79–92

Tüxen R, Oberdorfer E (1958) Eurosibirische Phanerogamen-Gesellschaften Spaniens mit Ausblicken auf die Mediterran-Region dieses Landes. Veröff Geobot Inst Rübel Zürich 32:1–328

Ubriszy G (1948) A rizs hazai gyomnövényzete. Acta Agrobot Hungar 1(4):1–43

Valsecchi F (1983) Compendio delle ricerche fitosociologiche in Sardegna. Boll Soc Sarda Sci Nat 22:231–245

Valsecchi F, Diana Corrias S (1973) La vegetazione degli stagni della zona di Olbia (Sardegna Nord-orientale). Giorn Bot Ital 107(5):223–241

Vázquez JA, Buján MIR (1997) Comportamiento fitosociológico de *Smyrnium olusatrum* L. en el noroeste ibérico. Lazaroa 18:153–164

Vella S (2001) Soil information in the Maltese Islands. In: Zdruli P, Stedu to P, Lacirignola C & Montanarella L (eds) Soil resources of Southern and Eastern Mediterranean countries. Options Médit, Sér B, Etud Rech 34:171–191

Vella SJ (2005) Soil survey and soil maping in the Maltese Islands: the 2003 position. In: Jones RJA, Houšková B, Bullock B, Montanarella L (eds) Soil resources of Europe. European soil bureau research report no. 9, 2nd edn. The European Soil Bureau, Joint Research Centre, Ispra, pp 235–244

Venanzoni R, Gigante D (2000) Contributo alla conoscenza della vegetazione degli ambienti umidi dell'Umbria (Italia). Fitosociologia 37(2):13–63

Venanzoni R, Apruzzese A, Gigante D, Suanno G, Vale F (2003) Contributo alla conoscenza della vegetazione acquatica e igrofitica dei Laghi di Monticchio. Inform Bot Ital 35(1):69–80

Vigo J, Terradas J (1969) Sobre la vegetaciòn de la zona de acantilados triasicos del Baix Llobregat. Acta Geobot Barcin 4:1–31

Walter H, Lieth H (1967) Klimadiagramm Weltatlas. VEB Gustaf Fischer Verlag, Jena

Weber HC, Kenzior B (2006) Flora of the Maltese Islands. Margraf Pubblishers, Weikersheim

Weber HE, Moravec J, Theurillat J-P (2000) International code of phytosociological nomenclature. 3rd ed. J Veg Sci 11(5):739–768

Wendelberger G (1943) Die Salzpflanzengesellschaften des Neusiedler Sees. Wien Bot Zeitschr 92(3):124–144

Westhoff V, Dijk JW, Passchier H, Sissingh G (1946) Overzicht der plantengemeenschappen in Nederland. B.W. Breuchel, Amsterdam

World Weather Information Service (2013) Luqa, Malta. http://worldweather.wmo.int/en/city.html?cityId=42

Zammit-Maempel G (1997) An outline of Maltese geology. Progress Press, Malta. 44 pp

Zerapha S (1827–1831) Flora Melitensis Thesaurus, sive plantarum enumeratio quae in Melitae Gaulosquae insulis aut indigenae aut vulgatissimae occurrunt. Fasc. 1,2. Regia Typographia Melitae

Index

A

Acanthus mollis L., 42, 47, 114, 137, 201, 204
Adiantum capillus-veneris L., 66
Adonis microcarpa DC., 220, 221, 223
Aegilops geniculata Roth., 143, 144, 148, 152, 159, 239, 240
Agave americana L., 216
Agrostis stolonifera L., 105, 108, 110, 113–116, 118
Aira cupaniana Guss., 165, 221
Ajuga iva (L.) Schreber, 174, 194, 195
Alisma lanceolatum With., 107–111, 114
Allium chamaemoly L.
subsp. *chamaemoly*, 175, 176
Allium commutatum Guss., 54, 68, 80, 81, 162
Allium lojaconoi Brullo, Lanfranco, Pavone, 23, 28, 30, 77, 145, 146, 150
Allium melitense (Sommier & Caruana ex Borg) Cif. & Giacom., 22, 26, 48, 50–52, 73–75, 159, 162, 165, 174
Allium nigrum L., 220, 221
Allium roseum L., 166
Allium subhirsutum L., 45, 48, 51, 165, 168, 205, 213
Allium trifoliatum Cyr., 220, 223
Amaranthus albus L., 231
Amaranthus deflexus L., 211, 212
Amaranthus graecizans L.
subsp. *sylvestris* (Vill.) Brenan, 212, 227, 229–232
Amaranthus retroflexus L., 229, 231, 232
Ammoides pusilla (Brot.) Breistr., 144
Ammophila australis Mabill., 83, 86, 87
Anacamptis coriophora (L.) R.M. Bateman, Pridgeon & M.W. Chase, 144
Anacamptis pyramidalis (L.) Rich., 51, 77, 165, 168
Anacamptis urvilleana Sommier & Caruana, 23, 149, 165, 168
Anacyclus clavatus (Desf.) Pers., 200
Anagallis arvensis L., 118, 132, 144, 148, 152, 155, 176, 209, 221, 223, 225, 226, 228, 229, 232, 238
Anchusa italica Retz, 220, 221, 223
Andrachne telephioides L., 227, 229
Andropogon distachyos L., 47, 157, 163, 164, 173
Anisantha diandra (Roth) Tzvelev, 170, 198, 200, 202–204, 209, 212, 225, 233, 237
Anisantha fasciculata (C. Presl) Nevski, 146, 150, 151, 159, 187
Anisantha madritensis (L.) Nevski, 143, 159, 182, 198, 200, 202, 203, 227, 237, 239
Anisantha rigida (Roth) Hyl, 85, 88, 155
Anisantha rubens (L.) Nevski, 144, 149, 165, 200
Anisantha sterilis (L.) Nevski, 174
Anthemis urvilleana (DC.) Sommier & Caruana, 22, 23, 30, 53, 69, 72, 77, 130, 141, 145, 146, 150, 151, 186, 188, 189, 201
Anthyllis hermanniae L.
subsp. *melitensis* Brullo & Giusso, 24, 25, 30, 43, 45, 47–49, 52, 74, 75, 79
Anthyllis vulneraria L.
subsp. *maura* (Back) Lindb., 49, 76, 158, 165, 167, 170, 172
Antirrhinum siculum Miller, 45, 58, 60, 64, 65, 180–184
Antirrhinum tortuosum Bosc., 59, 62, 65, 169, 171, 174, 179–184

S. Brullo et al., *The Vegetation of the Maltese Islands*, Geobotany Studies,
https://doi.org/10.1007/978-3-030-34525-9

Apium nodiflorum (L.) Lag., 104–106, 108–110
Arenaria leptoclados (Rchb.) Guss., 140, 143, 147, 152, 191, 192
Arisarum vulgare Targ.-Tozz., 45, 47, 50, 165, 174–176, 204, 221, 228, 233
Arthrocnemum macrostachyum (Moric.) K. Koch, 30, 59, 61, 73, 92–94, 99
Arum italicum Miller, 45, 105, 108, 110, 112, 114, 118, 137, 201, 204, 214, 233
Arundo donax L., 30, 136, 214, 215
Asparagus acutifolius L., 42, 65, 69, 165, 170, 173
Asparagus aphyllus L., 42, 44, 45, 47, 49, 51, 53, 76, 81, 137, 159, 162, 167, 170
Asperula aristata L. fil., 48, 49, 51, 59, 61, 64, 65, 77, 166, 167
Asphodelus ramosus L., 45, 47, 49, 51, 54, 59, 61, 76, 157, 158, 161, 163, 164, 167–169, 171, 172, 176
Asteriscus aquaticus (L.) Less., 141, 145, 147, 159, 187
Astragalus hamosus L., 200, 202, 239
Astragalus sesameum L., 147
Atriplex halimus L., 52, 53, 99, 136
Atriplex littoralis L., 89, 98, 99, 132
Atriplex prostrata DC.
var. *salina* (Wallr.) O. Bolòs & Vigo, 98, 99
Avena barbata Pott. ex Link, 54, 115, 149, 159, 162, 170, 198, 200, 202, 203, 221, 223, 228, 235, 237, 239
Avena sterilis L., 200, 209, 233, 237

B
Ballota nigra L.
subsp. *uncinata* (Fiori & Beg.) Patzke, 214
Bellardia trixago (L.), 143, 144
Bellis annua L., 129–131, 141, 143, 145, 146, 150–152, 176, 185, 186, 189
Bellis sylvestris Cyr., 167, 168, 176
Beta maritima L., 53, 89, 107, 152, 159, 162, 180, 182, 188, 196, 199, 202, 206, 208, 209, 211, 227, 229, 231, 238–240
Bifora testiculata (L.) Roth, 220, 223
Bituminaria bituminosa (L.) C.H. Stirton, 42, 47, 49, 59, 61, 65, 149, 153, 161, 163, 164, 167–169, 171, 172
Blackstonia serotina (W.D.J.Koch ex Rchb.) Beck, 144
Bolboschoenus maritimus (L.) Palla
subsp. *compactus* (Hoffm.) Hejný in Dostàl, 104, 106, 107
Borago officinalis L., 174, 198, 200, 204, 206, 209, 225, 233, 239
Brachypodium retusum (Pers.) Beauv., 30, 49, 65, 157, 163, 167–169, 173
Brachypodium sylvaticum (Hudson) Beauv., 42, 137
Brassica rapa L.
subsp. *sylvestris* (L.) Janchen, 225, 228, 232, 233
Bromus racemosus L., 117, 205
Bromus scoparius L., 192, 193, 198, 202
Buglossoides arvensis (L.) Johnston, 220, 223
Bupleurum lancifolium Hornem, 220, 221
Bupleurum semicompositum L., 147, 186

C
Cakile maritima Scop., 85, 88, 89
Calendula arvensis L., 224–226, 228
Calendula tripterocarpa Rupr., 221–223
Callitriche truncata Guss., 126–129
Colocasia antiquorum Schott, 110
Calystegia sepium (L.) R. Br., 136
Calystegia soldanella (L.) Roem. & Schult., 84, 86–88
Calystegia sylvatica (Kit.) Griseb., 215
Campanula erinus L., 213
Capparis spinosa L., 42, 45, 47, 50, 53, 58, 60, 65, 69, 72, 74, 77, 81, 179–183
Capsella rubella Reuter, 192, 198, 202, 203, 208
Cardamine hirsuta L., 213
Carduus australis L. fil.
subsp. *marmoratus* (Boiss. & Heldr.) Kazmi., 197, 200
Carduus pycnocephalus L.
subsp. *pycnocephalus*, 196, 199, 200, 202
Carex cuprina (Heufff.) A. Kern, 109–114, 118, 139
Carex cuspidata Host, 47, 49
Carex distans L., 111–114
Carex divisa Hudson, 96, 97, 110, 111, 113–116
Carex extensa Good, 95–97
Carex hispida Willd., 112, 113, 137
Carex spicata Hudson, 114
Carlina corymbosa L., 47, 49, 81, 158, 161, 163, 164, 167–169, 172, 176, 180
Carlina lanata L., 170
Carlina sicula Ten., 51, 76
Carthamus lanatus L., 196, 202, 209
Catapodium balearicum (Willk.) Scholz, 53, 69, 73, 74, 140, 142, 145, 146, 150–154, 176, 185, 186, 192–195
Catapodium pauciflorum (Merino) Brullo, Giusso, Miniss. & Spamp., 153, 154

Catapodium rigidum ((L.) C.E. Hubb., 141, 143, 148, 151, 176, 192–194, 198, 200, 221, 224, 225, 227, 237, 239
Catapodium zwierleinii (Lojac.) Brullo, 146, 154
Centaurea calcitrapa L., 196
Centaurea sicula L., 78, 158, 161, 163, 165, 167
Centaurium erythraea Rafin., 51
Centaurium pulchellum (Swartz) Druce, 130, 144, 149, 151, 165
Cephalozia bicuspidata (L.) Dum., 66
Ceratonia siliqua L., 30, 42, 44, 45, 47, 65, 136
Ceterach officinarum L., 58, 60, 65
Chamaeleon gummifer (L.) Cass., 48, 49, 51, 76, 158, 161, 163, 165, 167, 168, 173
Chamaesyce peplis (L.) Prokh, 87, 89, 153, 229
Chara vulgaris L., 119, 125, 126
Charybdis pancration (Steinh.) Speta, 30, 45, 47, 50, 51, 54, 81, 136, 149, 157, 158, 161, 163, 164, 167–169, 172, 176
Cheatomorpha linum Ktzg., 101, 102
Cheiranthus cheiri L., 180, 183
Cheirolophus crassifolius (Bertol.) Susanna, 23, 28, 30, 51, 58, 60, 63
Chenopodium album L., 208, 211, 212, 226, 229, 231, 232
Chenopodium murale L., 198, 206, 208, 209, 211, 212, 229, 233
Chenopodium opulifolium Schreber, 229, 231
Chenopodium vulvaria L., 211, 212, 229
Chiliadenus bocconei Brullo, 22, 24, 45, 48, 50, 58, 60, 63, 65, 66, 242, 243
Chrozophora tictoria (L.) Juss., 229
Cichorium intybus L., 196
Cichorium pumilum Jacq., 130, 131
Cichorium spinosum L., 30, 67, 68, 70–72, 75, 79, 80, 149
Cistus creticus L.
  subsp. *creticus*, 46, 49
  subsp. *eriocephalus* (Viv.) Greuter & Burdet, 46
Cistus monspeliensis L., 46, 49
Citrus limon (L.) Burn. fil, 232, 233
Citrus sinensis (L.) Osbeck, 232, 233
Clematis cirrhosa L., 42, 43
Clinopodium nepeta (L.) Kuntze, 65, 170, 173
Colchicum cupanii Guss., 149, 153, 175, 176
Colocasia antiquorum Schott, 105, 108
Conium maculatum L., 214
Convolvulus althaeoides L., 164, 167, 172, 238
Convolvulus arvensis L., 118, 208, 209, 211, 224–226, 228, 229, 231–233
Convolvulus cantabrica L., 50
Convolvulus elegantissimus Miller, 49, 144, 164, 167, 169, 171, 173
Convolvulus lineatus L., 145, 147, 161, 163
Convolvulus oleifolius Desr., 26, 28, 30, 45, 49, 51, 52, 59, 61, 74, 75, 161, 163, 246
Conyza bonariensis (L.) Cronq., 192, 206, 208, 211, 212, 229, 232
Conyza canadensis (L.) Cronq., 215, 229
Coronilla scorpioides (L.) Koch, 220, 223
Coronilla valentina L.
  subsp. *glauca* (L.) Batt., 43, 50–52, 59, 62
Coronopus didymus (L.) Sm., 191, 192
Crassula vaillantii (Willd.) Roth. (= *Bulliarda vaillantii* (Willd.) DC.), 129, 131, 132
Crataegus azalorus L., 42
Cremnophyton lanfrancoi Brullo & Pavone, 23, 24, 28, 30, 58, 60, 63
Crepis bursifolia L., 193
Crepis pusilla (Somm.) Merxm., 28, 145, 146
Cressa cretica L., 92, 93, 98–100, 132, 189
Crithmum maritimum L., 30, 59, 61, 67, 68, 70, 72, 74, 79–81, 246
Crucianella rupestris Guss., 26, 28, 30, 54, 58, 61, 67, 68, 70–72, 75, 79, 80
Crupina crupinastrum (Moris) Vis., 147
Crypsis aculeata (L.) Aiton, 132
Cuscuta epithymum (L.) L., 49, 51, 77
Cutandia maritima (L.) Benth., 84, 87, 88, 154, 155
Cynara cardunculus L., 159, 162, 166, 196, 239, 240
Cynodon dactylon (L.) Pers., 87, 105, 208, 212, 229, 231, 232
Cynoglossum creticum Miller, 204
Cyperus distachyos All., 106
Cyperus longus L.
  subsp. *longus*, 104, 105, 108–111
Cyperus rotundus L., 230–232

D

Dactylis hispanica Roth., 49, 53, 69, 74, 76, 148, 157, 158, 161, 163, 164, 167–169, 171, 172
Damasonium bourgaei Cosson, 30, 128, 129, 132
Darniella melitensis (Botsch.) Brullo, 24, 30, 43, 52, 53, 58, 60, 63, 77, 80, 162, 210, 246
Daucus carota L., 48, 117, 144, 149, 162, 165, 168, 170, 173, 180, 182, 183, 206, 209, 221, 224, 233

Daucus gingidium L., 53, 67, 68, 72, 74, 75, 79, 162
Daucus rupestris Guss., 25, 27, 30, 51, 52, 54, 58, 60, 64, 68, 71, 72, 75, 79, 80, 159
Desmazeria pignattii Brullo & Pavone, 22, 27, 30, 72, 149, 185, 186, 188
Digitaria sanguinalis (L.) Scop., 230–232
Dioscorea communis (L.) Caddick & Wilkin, 47, 137
Diplotaxis erucoides (L.) DC., 198, 220–222, 224–226, 228, 229, 231, 237, 240
Diplotaxis tenuifolia (L.) DC., 162, 169, 171, 181, 229
Dittrichia graveolens (L.) Greuter, 230
Dittrichia viscosa (L.) Greuter, 95, 97, 104, 105, 107, 108, 110–112, 116, 136, 158, 164, 168, 169, 172, 181, 182, 184
Dysphania ambrosioides (L.) Mosyakin & Clemants, 211, 212, 229

E

Ecballium elaterium (L.) Richard, 199, 202, 205, 208, 209, 211, 212, 229
Echinochloa colonum (L.) Link, 230, 231
Echinochloa crus-galli (L.) Beauv., 230–232
Echinophora spinosa L., 84, 86–88
Echium arenarium Guss., 84, 148
Echium parviflorum Moench, 146, 153, 154, 170, 176, 187, 201, 238
Echium plantagineum L., 198, 235, 237
Echium sabulicola Pomel, 151, 154, 155
Elatine gussonei (Sommier) Brullo et al., 25, 27, 30, 126, 128–130, 132
Eleocharis palustris (L.) R. & S., 108–111
Elytrigia juncea (L.) Nevski, 30, 84, 86–89
Elytrigia scirpea (Presl) Holub, 92–94
Emex spinosa (L.) Campd., 203
Enteromorpha intestinalis (L.) Link, 101, 102, 248
Epilobium parviflorum Schreber, 111–113
Epilobium tetragonum L., 107, 109–111, 113, 114, 118
Equisetum ramosissimum Desf., 118
Erodium chium (L.) Willd., 153
Erodium ciconium (L.) L'Hér., 153, 198, 200
Erodium cicutarium (L.) L'Hér., 192–194, 209
Erica multiflora L., 30, 43, 45–47, 49–51, 59, 61, 65, 77, 136
Erodium laciniatum (Cav.) Willd., 85, 88, 154, 155
Erodium malacoides (L.) L'Hér., 174, 198, 200, 202, 208, 209, 225, 226, 237
Eryngium maritimum L., 30, 83, 84, 86–88
Eucladium verticillatum (Brid.) B. S. & C., 66
Euphorbia aleppica L., 227, 229
Euphorbia dendroides L., 30, 41, 43, 46, 47, 50, 59, 61, 65, 81, 136, 252
Euphorbia exigua L., 140, 143, 144, 147, 176, 220, 239
Euphorbia helioscopia L., 195, 206, 224, 226, 228, 237
Euphorbia melitensis Parl., 25, 26, 30, 45, 48, 59, 61, 74, 162, 166
Euphorbia paralias L, 84
Euphorbia peplus L., 226, 228, 233
Euphorbia pinea L., 42, 49, 69, 73, 74, 76, 81, 88, 95, 157, 158, 161, 163, 165, 167, 169, 172
Euphorbia sommieriana C. Brullo & Brullo, 144, 146, 151, 187
Euphorbia terracina L., 84, 87, 88, 155
Evax pygmaea (L.) Pers, 140, 143, 144, 147, 176, 187, 194

F

Fagonia cretica L., 26, 158, 160
Fedia graciliflora Fisch. & C.A. Mey, 167, 221
Ferula melitensis Brullo et al., 24, 25, 30, 153, 158, 162–164, 167, 171, 172
Ficus carica L., 42, 45, 59, 61, 65, 136, 137, 179–184
Filago cf. *cosyrensis* Lojac., 149
Filago congesta Guss., 194, 195
Filago pyramidata L., 148
Foeniculum piperitum (Ucria) C.Presl, 50, 54, 78, 105, 112, 136, 158, 161, 163, 164, 169, 171, 172, 199, 201, 202, 205, 239
Foeniculum vulgare Mill., 165, 167, 216
Frankenia hirsuta L., 54, 67, 68, 72, 75, 79–81, 149, 187
Frankenia pulverulenta L., 185, 186, 188, 189
Fumana arabica (L.) Spach, 46, 48, 49, 77
Fumana thymifolia (L.) Spach ex Webb, 46, 49, 77
Fumaria agraria Lag., 224, 226, 228
Fumaria bicolor Sommier ex Nicotra, 224, 226
Fumaria capreolata L., 204, 233
Fumaria densiflora DC., 198, 221, 222, 224–226, 228
Fumaria gaillardotii Boiss., 222, 225, 226, 228
Fumaria parviflora Lam., 222, 224–226, 228
Fumaria wirtgenii Koch, 201, 221, 222, 224–226, 228

G

Galactites elegans (All.) Soldano, 54, 144, 159, 166, 168, 206, 209, 235
Galium aparine L., 108, 110, 200–202, 204, 214, 215, 233
Galium murale (L.) All., 148, 192–194, 213
Galium tricornutum Dandy, 219–221, 223, 226, 233
Galium verrucosum Hudson, 226, 228
Gastridium ventricosum (Gouan) Schinz & Thell, 144
Geranium dissectum L., 110, 114–116
Geranium molle L., 192, 198, 203, 204
Geranium purpureum Vill., 213
Gladiolus dubius Guss., 220, 223
Gladiolus italicus Miller, 165, 173, 220, 223
Glaucium flavum Crantz, 211
Glebionis coronaria (L.) Spach, 54, 85, 174, 197–200, 202–204, 206, 208, 209, 211, 216, 221, 223, 225, 226, 228, 229, 235, 237, 239
Glyceria plicata Fries, 107, 109

H

Hainardia cylindrica (Will.) Greuter, 145, 147
Halimione portulacoides (L.) Aellen, 73, 93, 94
Hedera helix L., 42, 137
Hedypnois cretica (L.) Dum.-Cours., 87, 88, 141, 143, 148, 151, 153, 155, 170, 198, 200, 221, 235, 237, 240
Hedypnois rhagadioloides (L.) F.W. Schmidt, 144, 153
Helichrysum melitense (Pignatti) Brullo et al., 24, 25, 28, 30, 58, 60, 64, 80
Heliotropium europaeum L., 227, 229, 231, 232
Helminthotheca echioides (L.) Holub, 95, 111, 117, 118, 206, 229
Hippocrepis biflora Spreng., 153
Hippocrepis multisiliquosa L., 144, 153, 174, 221, 235, 237
Hirschfeldia incana (L.) Lagr.-Foss., 198, 200, 203, 235, 237
Hordeum leporinum Link, 149, 192, 197–199, 201–203, 206, 208, 209, 237
Hordeum marinum Hudson, 94, 95, 97–99, 117, 189, 238–240
Hymenolobus revelierei (Jord.) Brullo subsp. *sommierianum* (Pamp.) Brullo, 25, 185, 186
Hyoscyamus albus L., 180, 182–184, 208, 209, 211
Hyoseris frutescens Brullo & Pavone, 23, 25, 30, 60, 63, 66, 73, 80
Hyoseris radiata L., 42, 50, 59, 61, 65, 164, 167, 170, 179–183
Hyoseris scabra L., 140, 143, 147, 154
Hyparrhenia hirta (L.) Stapf., 30, 31, 45, 47, 49, 59, 61, 65, 76, 157, 158, 161, 163, 164, 167, 171, 172
Hyparrhenia sinaica (Delile) Llauradó ex G. López, 164, 173
Hypecoum procumbens L., 222, 223
Hypericum aegyptiacum L.
subsp. *webbii* (Spack) Robson, 28, 51, 52, 58, 60, 64, 66–68, 70–72, 79, 80
Hypericum pubescens Boiss., 128
Hypericum triquetrifolium Turra, 170, 221, 223, 229
Hypochoeris achyrophorus L., 140, 143, 144, 147, 151–154, 176, 199, 237

I

Ipomoea indica (Burm.) Merr., 215, 216

J

Jacobaea maritima subsp. *bicolor* (Willd.) Tod., 47, 49, 51, 54, 59, 61, 65, 68, 72, 74, 75, 159, 166, 170, 173, 180–183
Juncus acutus L., 95–98, 110, 112, 117
Juncus bufonius L., 93, 96, 97, 108, 114, 128, 129, 131, 132
Juncus capitatus Weigel, 129, 131
Juncus hybridus Brot., 186, 188, 189
Juncus maritimus Lam., 92–97, 116
Juncus subulatus Forsk., 91, 92

K

Kickxia spuria (L.) Dum.
subsp. *integrifolia* (Brot.) Fern., 227, 229
Kundmannia sicula (L.) DC., 165, 167, 172

L

Lactuca serriola L., 196
Lagurus ovatus L.
subsp. *nanus* Messeri, 144–146, 151
subsp. *ovatus*, 77, 85, 88, 141, 145, 155, 165, 176
Lamium amplexicaulis L., 224, 226
Lathyrus annuus L., 137
Lathyrus aphaca L., 221
Lathyrus articulatus L., 161, 163, 164, 167, 169, 172
Lathyrus cicera L., 220
Lathyrus ochrus (L.) DC., 220, 227, 238
Laurus nobilis L., 29, 137
Lavatera arborea L., 54, 207–209, 211

Lavatera cretica L., 206–209
Leontodon tuberosus L., 48, 50, 149, 157, 158, 165, 167, 175, 176, 246
Leopoldia comosa (L.) Parl., 165, 173, 220, 223
Lepidium graminifolium L., 169, 171
Leucaena leucocephala (Lam.) de Wit, 216
Limbarda crithmoides (L.) Dumort.
  subsp. *longifolia* (Arcang.) Greuter, 53, 58, 61, 66, 69, 72, 74, 92–96, 99, 112, 136, 161, 163, 180–182, 196, 211, 246
Limonium melitense Brullo, 22, 51, 52, 54, 59, 61, 67, 68, 71, 72, 76, 80, 159, 162
Limonium virgatum (Willd.) Fourr., 53, 67, 68, 70–72, 74, 92, 93, 246
Limonium zeraphae Brullo, 22, 24, 59, 61, 67, 68, 70–72, 75, 187
Linaria chalepensis (L.) Miller, 220
Linaria pseudolaxiflora (Lojac.) Gillett., 25, 26, 28, 30, 152–154
Linum bienne Miller, 238
Linum strictum L., 77, 104, 143, 144, 147, 151–153, 176
Linum trigynum L., 145, 146
Lobularia maritima (L.) Desv., 78, 155, 157, 158, 164, 169, 171, 172, 175, 180, 182, 183, 192, 193, 201
Lolium rigidum Gaudin
  subsp. *lepturoides* (Boiss.) Sennen & Mauricio, 130
  subsp. *rigidum*, 89, 130, 151, 198, 200, 202, 203, 220, 223, 225, 227, 228, 235, 237, 239, 240
Loncomelos narbonensis (Torn. in L.) Raf., 50, 144, 165, 168, 224
Lonicera implexa Aiton, 42, 44–47, 49, 51, 77, 166
Lotus cytisoides L., 47, 50, 59, 61, 67, 68, 70, 72, 74, 75, 79, 80, 84, 87, 88, 154, 159, 162, 168, 174, 187
Lotus edulis L., 143, 144, 148, 152, 153, 159, 168, 198, 200, 221, 224, 237
Lotus ornithopodioides L., 144, 166, 167, 205, 221, 223, 235, 237, 239
Lycopersicon esculentum Miller, 231
Lygeum spartum L., 30, 31, 73, 76, 94, 150, 157, 158, 160, 161, 210, 246
Lythrum hyssopifolia L., 128–132
Lythrum junceum Banks & Sol., 107, 108, 111, 116, 118

M

Malva nicaeensis All., 198, 206, 207
Malva parviflora L., 194, 195, 197, 198, 200, 206–209, 211, 231
Malva sylvestris L., 199, 202, 203, 209, 212, 226, 228, 231, 237, 239
Matricaria chamomilla L., 192–194
Matthiola incana (L.) R. Br.
  subsp. *incana*, 180, 181, 184
  subsp. *melitensis* Brullo et al., 25, 26, 58, 60, 64, 73, 81
Matthiola tricuspidata (L.) R. Br., 54, 85
Medicago arabica (L.) Hudson, 115, 231
Medicago arborea L., 216
Medicago ciliaris (L.) All., 239, 240
Medicago littoralis Rohde ex Loisel., 84, 88, 141, 143–146, 150–152, 155, 176, 187, 188, 194
Medicago lupulina L., 153
Medicago marina L., 84, 86–88, 246
Medicago minima L., 140, 143, 144, 147, 152–154
Medicago orbicularis (L.) Bartal., 166, 221
Medicago polymorpha L., 143, 149, 198, 200, 202, 221, 226, 228, 231, 235, 237, 239
Medicago rugosa Desr., 221
Medicago truncatula Gaertner, 141, 148, 202, 237
Medicago turbinata (L.) All., 144
Melica arrecta Kuntze, 47, 51, 77
Melica minuta L., 58, 60, 65
Melilotus infesta Guss., 220, 221
Melilotus messanensis (L.) All., 106, 209, 238–240
Melilotus sulcata Desf., 143, 159, 196, 199, 221, 237, 239, 240
Mentha pulegium L., 97, 108, 111, 113, 115, 116, 118, 128–132
Merculialis annua L., 203
Mesembryanthemum crystallinum L., 210, 211
Mesembryanthemum nodiflorum L., 53, 73, 151, 159, 187, 210, 211
Micromeria graeca (L.) Bentham, 49
Micromeria microphylla (D'Urv.) Bentham, 50, 51, 64, 65, 149, 165
Moraea sisyrinchium (L.) Ker-Gawl., 53, 76, 130, 140, 142–146, 150, 151, 159, 166, 168, 176, 187, 188
Misopates orontium (L.) Rafin., 220, 223, 230

N

Narcissus serotinus L., 175, 176
Narcissus tazetta L., 97, 111, 165, 172
Nasturtium officinale R. Br., 105, 106, 108, 109
Neslia paniculata (L.) Desv., 220, 227
Nicotiana glauca Graham, 180–183, 215, 216
Nigella damascena L., 220, 223
Notobasis syriaca (L.) Cass., 196, 214

O

Ochlopoa annua (L.) H. Scholz, 191–195, 199
Ochlopoa infirma (Kunth) H. Scholz, 129, 131
Oenanthe globulosa L., 114, 116
Olea europaea L.
var. *sylvestris* Brot., 42, 45, 47
Ononis mitissima L., 115, 236, 238, 239
Ononis ornithopodiodes L., 147
Ononis ramosissima Desf., 84, 86–88, 155
Ononis reclinata L., 144, 147, 152
Ononis sieberi Besser, 142, 144, 146, 176
Ononis variegata L, 85, 88
Onopordon argolicum Boiss., 196
Orobanche ramosa L., 221, 223
Oryzopsis miliacea (L.) Asch. & Scw., 54, 157, 164, 168, 169, 172, 206, 232
Oxalis pes-caprae L., 53, 136, 153, 170, 173, 180–183, 193, 196, 198, 200, 202, 204, 208, 209, 213–216, 221–223, 225, 228, 233, 237, 240

P

Pallenis spinosa (L.) Cass., 48, 49, 77, 157, 158, 161, 164, 167, 171, 173
Pancratium maritimum L., 84, 86–88
Panicum repens L., 105, 110, 112, 118, 119
Papaver dubium L., 220, 226
Papaver hybridum L., 202, 220, 221, 223, 225, 226, 228
Papaver rhoeas L., 199, 201, 219, 220, 223, 230
Parapholis filiformis (Roth) C.E. Hubbard, 92, 93, 95, 99, 189
Parapholis incurva (L.) C.E. Hubbard, 53, 69, 73, 77, 89, 96, 99, 132, 140, 142, 143, 145, 146, 150–152, 159, 185, 186, 188, 189, 211, 240
Parietaria judaica L., 42, 65, 136, 170, 173, 180–184, 201, 202, 204, 206, 212, 214, 215, 231–233
Parietaria lusitanica L., 59, 61, 213
Pennisetum setaceum (Forssk.) Chiov., 216
Periploca angustifolia Labill., 30, 44–47, 50, 51
Phagnalon graecum Boiss. & Heldr., 45, 47, 49, 51, 52, 54, 74, 75, 159, 161, 163, 164, 168, 170, 173, 182, 183
Phagnalon rupestre (L.) DC., 49, 59, 61, 65, 167
Phalaris canariensis L., 220, 223, 226, 239
Phalaris coerulescens Desf., 113, 115
Phalaris paradoxa L., 115, 130, 151, 152, 224, 240
Phlomis fruticosa L., 47, 166
Phragmites australis (Cav.) Trin. ex Steud., 92, 95, 97, 99, 103–105, 107–116, 136, 137, 205
Pinus halepensis Mill., 29, 30, 41, 44, 48, 50
Pistacia lentiscus L., 29, 30, 42, 45–47, 49, 51, 77
Pisum sativum L., 228
Plantago afra L.
subsp. *afra*, 25, 141, 143, 154, 202
subsp. *zwierleinii* (Nicotra) Brullo, 25, 153, 154
Plantago coronopus L., 53, 69, 74, 129, 132, 140, 142, 145, 146, 150–152, 176, 185, 186, 188, 189, 191–194, 211
Plantago crypsoides Boiss., 26, 28, 186, 188
Plantago lagopus L., 130, 142–144, 148, 188, 192–195, 198–200, 202, 209
Plantago lanceolata L., 112
Plantago macrorhiza Poir., 67, 70, 72, 79–81
Plantago major L., 105, 113, 115, 116, 118
Plantago serraria L., 54, 148, 162, 167
Plumbago europaea L., 169
Poa bulbosa L., 141, 143, 148, 153, 166
Poa trivialis L., 113, 115, 116, 118
Podospermum laciniatum (L.) DC., 239, 240
Polycarpon diphyllum Cav., 85, 155
Polycarpon tetraphyllum (L.) L., 191–195, 208
Polygonum arenastrum Boreau, 191–195
Polygonum aviculare L., 208, 229
Polygonum maritimum L., 85, 89
Polygonum salicifolium Brouss. ex Willd., 104, 105, 108, 110
Polypogon monspeliensis (L.) Desf., 92, 95–97, 107, 108, 116, 132, 189
Polypogon subspathaceus Req., 128–132, 186, 192
Populus alba L., 110, 137, 252, 253
Portulaca oleracea L., 227, 229–232
Potentilla reptans L., 105, 108, 110, 113–116, 118, 205
Prasium majus L., 42, 44, 46, 47, 49, 51, 59, 61, 77
Prospero autumnale (L.) Speta, 175, 176, 187
Pseudorlaya pumila (L.) Grande, 84, 88, 154, 155
Pteranthus dichotomus Forssk., 25, 28, 161, 163
Pulicaria dysenterica (L.) Bernh., 112–114
Punica granatum L., 136
Putoria calabrica (L.) Pers., 64, 65

Q

Quercus ilex L., 9, 30, 31, 41–43

R

Ranunculus bullatus L.
subsp. *bullatus*, 175, 176
Ranunculus macrophyllus Desf., 107, 109–111, 113–116, 205
Ranunculus muricatus L., 115, 116
Ranunculus paludosus Poiret, 143, 147
Ranunculus saniculifolius Viv., 30, 126–129
Ranunculus sardous Crantz., 108, 114, 205
Ranunculus trichophyllus Chaix, 119
Raphanus raphanistrum L., 202, 219, 220, 223, 225, 226
Rapistrum rugosum (L.) Arcang., 220, 221
Reichardia picroides (L.) Roth
var. *maritima* (Boiss.) Fiori, 68, 72, 74, 75, 79, 81
var. *picroides*, 48, 54, 59, 61, 65, 68, 72, 74, 75, 84, 148, 155, 157, 158, 161, 163, 164, 167–169, 171, 172, 180, 199, 209, 212, 229, 237
Reseda alba L., 173, 182, 197, 198, 200, 209, 229
Reseda lutea L., 220, 228
Rhagadiolus stellatus (L.) Gaertner, 219, 220, 223
Rhamnus alaternus L., 42, 43, 45
Rhamnus oleoides L.
subsp. *oleoides*, 42, 44–47, 49, 59, 61, 65, 78
Rhodalsine geniculata (Poiret) Williams, 201
Ricinus communis L., 215, 216
Ridolfia segetum (Guss.) Moris, 220, 221, 223
Roccella cf. *tinctoria* DC., 59, 61, 65
Romulea columnae Sebast. & Mauri, 151, 152
Romulea melitensis Bèg., 22, 27, 30, 77, 81, 131, 140, 143, 145, 146, 150, 153, 154, 175, 176, 186, 188
Romulea ramiflora Ten., 130, 131
Rosa gallica L., 137
Rosmarinus officinalis L., 46, 48, 49, 52
Rostraria cristata (L.) Tzvelev, 141, 143, 148, 192, 193, 198, 200, 202, 203
Rubia peregrina L., 42, 45, 47, 50, 137
Rubus ulmifolius Schott, 42, 47, 135–137, 216
Rumex bucephalophorus L.
subsp. *bucephalophorus*, 88, 149, 155, 189, 201, 224, 226, 228
Rumex conglomeratus Murray, 104, 105, 107–111, 113–116, 118, 131, 205
Rumex pulcher L., 118, 231, 232
Ruppia drepanensis Tineo, 31, 101, 102
Ruppia maritima L., 101
Ruta chalepensis L., 166, 174

S

Sagina apetala Ard., 191–195
Sagina maritima G. Don., 149, 185, 186, 188, 189, 192
Salicornia patula Duv.-Jouv., 92, 93, 96–100
Salix pedicellata Desf., 31, 137, 243
Salsola soda L., 98
Salsola tragus L., 84, 87–89, 96
Salvia verbenaca L., 143, 153, 170, 173
Samolus valerandi L., 66, 108
Sanguisorba minor Scop., 77, 165, 167, 173
Sarcocornia fruticosa (L.) A.J. Scott, 69
Scandix pecten-veneris L., 174, 220, 223, 225, 226, 228
Schedonorus arundinaceus (Schreb.) Dumort.
subsp. *arundinaceus*, 30, 95, 104, 105, 107–116, 118
Schenkia spicata (L.) G. Mans, 95, 97, 111, 240
Scirpoides holoschoenus (L.) Soják
subsp. *australis* (Murray) Soják, 105, 107, 108, 110–112, 115, 116, 118
Scolymus grandiflorus Desf., 196
Scolymus hispanicus L., 84, 88, 196
Scolymus maculatus L., 196
Scorpiurus muricatus L., 143, 148, 152, 159, 162, 166, 167, 196, 238
Scrophularia auriculata L., 105, 108
Scrophularia peregrina L., 201, 204
Sedum caespitosum (Cav.) DC., 139, 140, 142
Sedum coeruleum L., 139, 140
Sedum dasyphyllum L., 58, 60
Sedum litoreum Guss., 139, 140, 142
Sedum rubens L., 139, 140, 147, 176
Sedum sediforme (Jacq.) Pau, 45, 47, 51, 59, 61, 65, 165–167, 173
Senecio leucanthemifolius Poiret, 73, 81, 148, 151, 152, 185, 186
Senecio pygmaeus DC., 25, 27, 30, 185, 186
Senecio vulgaris L., 154, 198, 206, 222, 225, 226, 228, 239
Setaria pumila (Poir.) Roem. & Schult., 230–232
Setaria verticillata (L.) Beauv., 212, 227, 229
Setaria verticilliformis Dumort., 230, 232
Setaria viridis (L.) Beauv., 230–232
Sherardia arvensis L., 221, 223, 225, 226, 233
Sideritis romana L., 144, 147, 153, 154
Silene colorata Poir., 25, 84, 88, 143, 148, 153, 155, 227
Silene fruticosa L.
subsp. *fruticosa*, 58, 60, 64
Silene melitensis Brullo et al., 25, 26, 30, 145, 151

Silene nocturna L., 147
Silene sedoides Poiret, 67, 68, 72, 74, 76, 187
Silene vulgaris (Moench) Garcke
subsp. *angustifolia* (Mill.) Hayek, 81
Silybum marianum (L.) Gaertner, 196, 199, 202
Sinapis arvensis L., 220, 223, 227, 228
Sisymbrium irio L., 200, 206, 208
Sisymbrium officinale (L.) Scop., 198, 203
Sisymbrium orientale L., 199, 203
Sixalix maritima (L.) Greuter & Burdet, 51, 65, 77, 84, 87, 88, 162, 165
Smilax aspera L., 42, 45, 47, 50, 51, 136
Smyrnium olusatrum L., 173, 201, 203, 204, 214
Soda inermis Fourr., 89, 92–95, 98–100, 189, 211
Solanum melongena L., 231
Sonchus asper (L.) Hill., 225, 226, 228, 229, 237
Sonchus glaucescens Jordan, 68, 76
Solanum nigrum L., 198, 201, 205, 206, 212, 229, 232
Sonchus oleraceus L., 53, 84, 92, 99, 117, 151, 159, 162, 167, 170, 173, 180, 198, 200, 202–204, 206, 208, 209, 211–213, 221–223, 225, 226, 228, 229, 231–233, 237, 239, 240
Sonchus tenerrimus L., 65, 179–183
Sorghum halepense (L.) Pers., 231, 232
Spergularia marina (L.) Griseb., 92, 99, 130, 132, 149, 151, 185, 186, 188, 189, 207, 208, 211
Spergularia rubra (L.) J.& C. Presl, 191, 192, 194, 195, 206
Sporobolus arenarius Duv.-Jouv., 30, 83, 84, 86, 88, 96, 98
Stellaria media (L.) Vill., 200, 204, 206, 208, 209, 213, 215, 222, 225
Stipella capensis (Thunb.) Röser & Hamasha, 140, 142–144, 147, 151–154, 166, 176, 198, 200, 235, 237
Suaeda spicata (Willd.) Moq., 95, 96, 98–100, 132, 189
Suaeda vera Forssk. ex J.F.Gmel., 30, 52, 53, 92–94, 246
Sulla capitata (Desf.) B.H. Choi & H. Ohashi, 77, 147, 240
Sulla coronaria (L.) Medik., 158, 162, 174, 198, 200, 209, 236, 239, 240
Symphyotrichum squamatum (Spreng.) G.L. Nesom, 66, 92, 96, 99, 104, 118, 130, 132, 170, 189, 208, 212, 215, 229

T

Tamarix africana Poir., 54, 135
Tamarix gallica L., 135, 136
Taraxacum minimum (Brig. ex Guss.) Terrac., 207, 208
Tetraclinis articulata (Vahl) Masters, 29, 30, 41, 43–45, 242, 253
Tetragonolobus purpureus Moench, 166
Teucrium flavum L.
subsp. *flavum*, 42, 43, 45–47, 49, 51, 58, 60, 77, 165, 167
Teucrium fruticans L., 42, 44–47, 49, 51, 59, 61, 65, 76, 167, 173
Theligonum cynocrambe L., 205, 213, 221
Thesium humile Vahl, 166
Thymbra capitata (L.) Cav., 30, 46, 47, 49, 51, 59, 61, 76, 79, 166, 167, 173
Tordylium apulum L., 154
Torilis nodosa (L.) Gaertner, 220
Trachynia distachya (L.) Link, 65, 140, 143, 147, 151, 152, 159, 166, 238–240
Tribulus terrestris L., 231
Trifolium cherleri L., 153
Trifolium fragiferum L., 116
Trifolium nigrescens Viv.
subsp. *nigrescens*, 141–143, 149, 174, 192–194, 235, 237
Trifolium pratense L.
subsp. *pratense*, 113, 114
Trifolium resupinatum L., 114, 115, 118, 203
Trifolium scabrum L., 140, 143, 144, 147, 151–154, 176, 187
Trifolium stellatum L., 143, 144, 148, 153, 200
Trifolium suffocatum L., 194
Trifolium tomentosum L., 128, 192, 193, 199, 201
Triglochin barrelieri Loisel, 92, 93
Triglochin laxiflorum Guss., 129, 131
Trigonella monspeliaca L., 140, 143, 147, 153, 154, 176
Tripodium tetraphyllum (L.) Fourr., 238
Trisetaria aurea (Ten.) Pignatti, 141–143, 148, 192–195, 199, 203
Triticum durum Desf., 220, 223
Typha domingensis L., 103, 104
Typha latifolia L., 104, 111

U

Ulmus canescens Melville, 137, 243
Umbilicus horizontalis (Guss.) DC., 213
Umbilicus rupestris (Salisb.) Dandy, 58, 60, 65

Urospermum picroides (L.) Scop., 203, 235, 237, 239
Urtica membranacea Poiret, 198, 200–202, 204, 206, 208, 212, 214, 225, 233
Urtica pilulifera L., 205
Urtica urens L., 226, 233

V
Valantia muralis L., 77, 140, 143, 147, 153, 154, 176, 182, 187, 213
Verbascum sinuatum L., 170, 171, 173, 196
Verbena officinalis L., 108, 118
Veronica anagallis-aquatica L., 104, 105, 107–110
Veronica cymbalaria Bodard, 213, 224, 226
Veronica hederifolia L., 222, 225, 228
Vicia faba L., 225, 226, 228
Vicia sativa L., 202, 221, 224, 227, 230, 238
Vitex agnus-castus L., 135, 136
Vitis vinifera L., 228
Vulpia ciliata Dumort., 142–144, 174, 200, 213, 237
Vulpia fasciculata (Forssk.) Fritsch, 85, 154, 155

Z
Zea mais L., 231
Zanichellia melitensis Brullo, Giusso & Lanfranco, 26, 30, 119, 126–128

FSC
www.fsc.org
MIX
Papier aus verantwortungsvollen Quellen
Paper from responsible sources
FSC® C105338

If you have any concerns about our products,
you can contact us on
ProductSafety@springernature.com

In case Publisher is established outside the EU,
the EU authorized representative is:
Springer Nature Customer Service Center GmbH
Europaplatz 3, 69115 Heidelberg, Germany

Printed by Libri Plureos GmbH
in Hamburg, Germany